# NATURE'S TRUTH

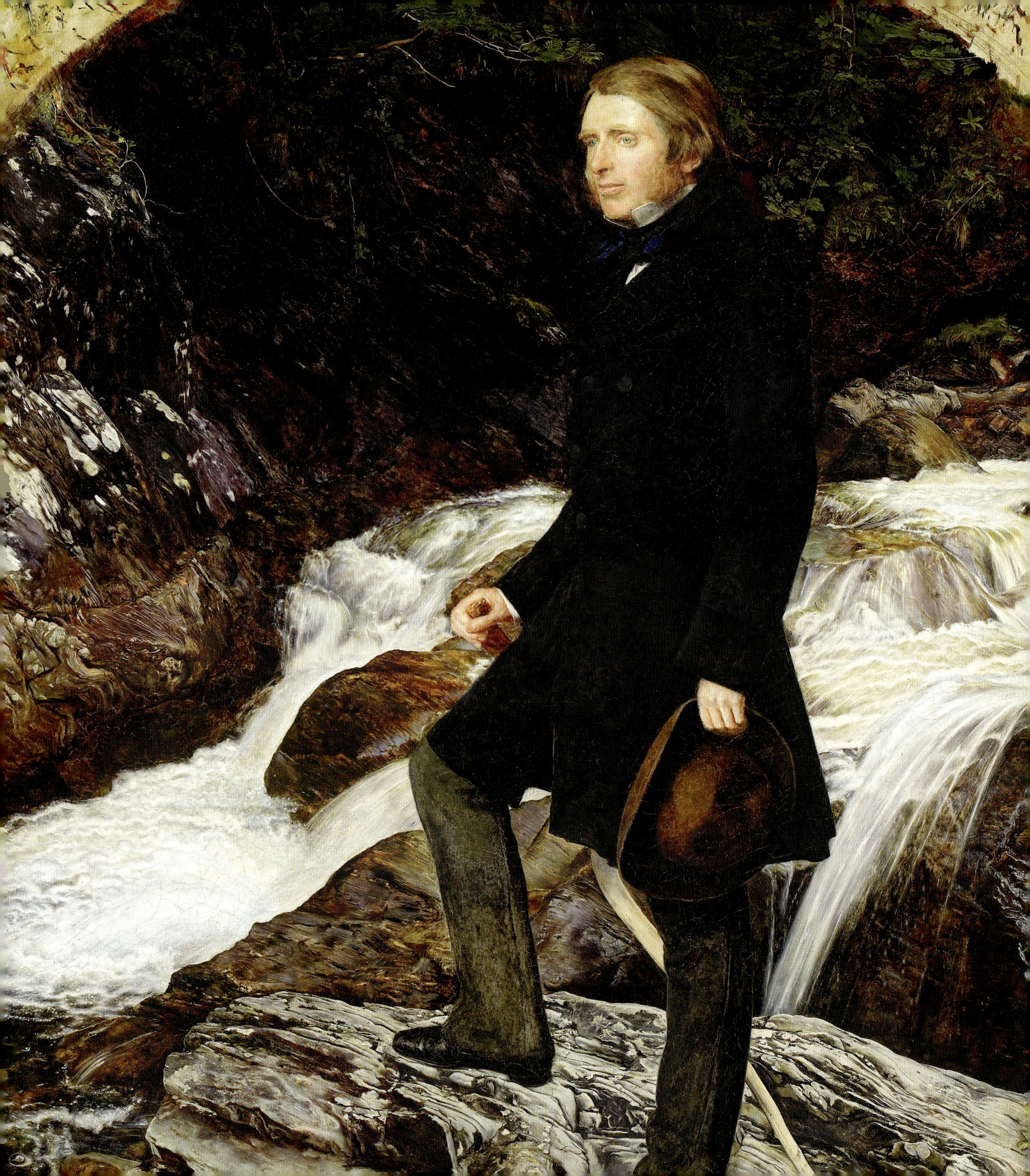

# NATURE'S TRUTH

## Photography, Painting, and Science in Victorian Britain

ANNE HELMREICH

THE PENNSYLVANIA STATE UNIVERSITY PRESS
UNIVERSITY PARK, PENNSYLVANIA

Supported by a Publications Grant from the Paul Mellon Centre for Studies in British Art.

PAUL MELLON CENTRE
*for Studies in British Art*

Publication of this book has been aided by a grant from the Millard Meiss Publication Fund of the College Art Association.

MM



Library of Congress Cataloging-in-Publication Data
Names: Helmreich, Anne, author.
Title: Nature's truth : photography, painting, and science in Victorian Britain / Anne Helmreich.
Description: University Park, Pennsylvania : The Pennsylvania State University Press, [2016] | Includes bibliographical references and index.
Summary: "Investigates why nineteenth-century British painters and photographers as diverse as the Pre-Raphaelites, P. H. Emerson, and Augustus John pursued truth to nature, and how contemporary science and philosophy informed their artistic practice and the critical reception of their work"—Provided by publisher.
Identifiers: LCCN 2015044025 | ISBN 9780271071145 (cloth : alk. paper)
Subjects: LCSH: Art and science—Great Britain—History—19th century. | Nature (Aesthetics) | Photography—Great Britain—History—19th century. | Painting, British—19th century. | Pre-Raphaelitism—Great Britain.
Classification: LCC N72.S3 H45 2016 | DDC 701/.050941—dc23
LC record available at http://lccn.loc.gov/2015044025


Printed in China by
Oceanic Graphic International
Published by The Pennsylvania State University Press,
University Park, PA
16802–1003

The Pennsylvania State University Press is a member of the Association of American University Presses.

It is the policy of The Pennsylvania State University Press to use acid-free paper. Publications on uncoated stock satisfy the minimum requirements of American National Standard for Information Sciences—Permanence of Paper for Printed Library Material, ANSI Z39.48–1992.

Frontispiece: John Everett Millais, *John Ruskin*, 1853–54. © Ashmolean Museum, University of Oxford (fig. 8). Pages iv–v: "Plate XIV, Lastrea Filix-Mas." In John Lindley and Thomas Moore, eds., *The Ferns of Great Britain and Ireland, Nature Printed by Henry Bradbury* (London: Bradbury and Evans, 1855) (fig. 18).

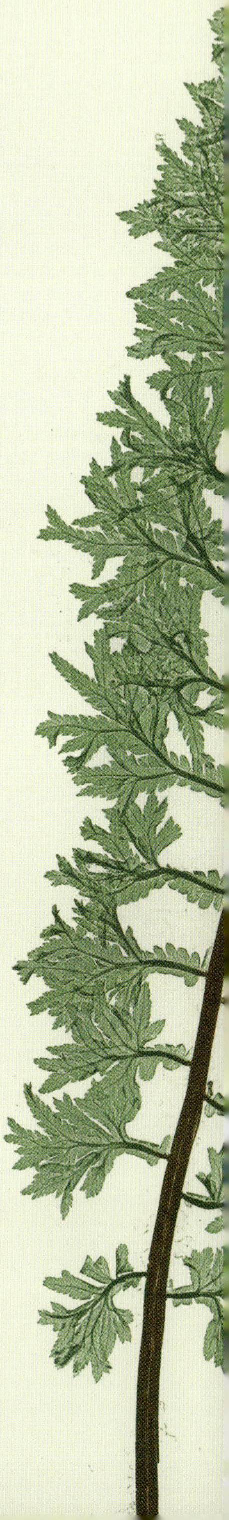

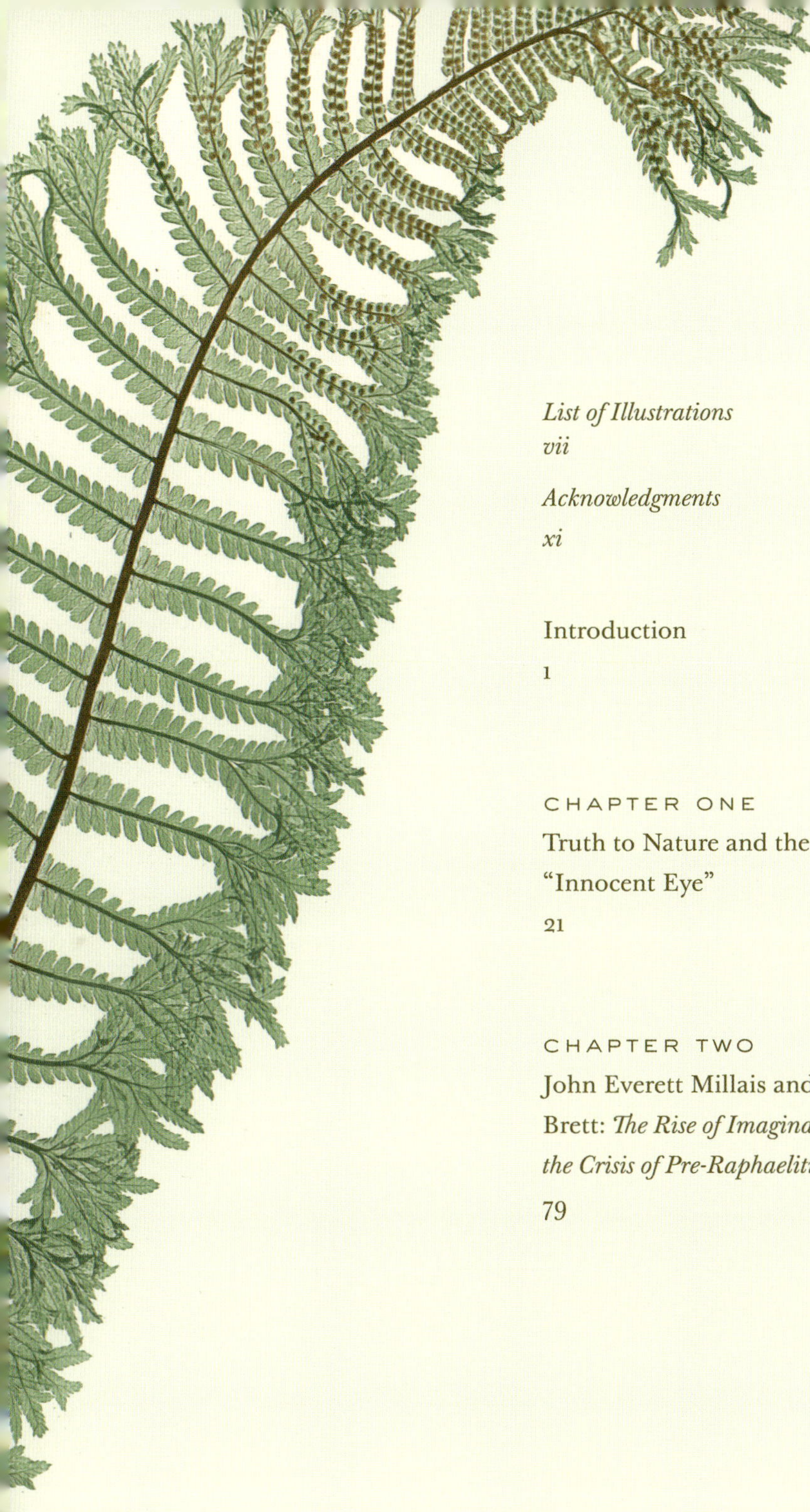

# { contents }

# { illustrations }

# { *acknowledgments* }

This project has unfolded over a long period of time but has benefited greatly from this gestation. I began my research focused on the fate of landscape painting in Great Britain in the wake of J. M. W. Turner. While landscape painting, by the 1850s, was claimed as the foremost genre in British art, by the 1880s critics proclaimed landscape painting to be in crisis. They pointed to the incursion of modernist styles of painting associated with foreign countries, namely, France, as well as the quickening pace of industrialization, which altered the British landscape and its representation. My working hypothesis was that modernity, in its many forms, forced artists to reinvent the British landscape genre. My research, at this early stage, was generously supported by a travel grant from the Paul Mellon Centre for Studies in British Art (2001) and a National Endowment for the Humanities Summer Stipend (2003), which permitted me to visit a wide range of archives and art collections in Great Britain and, most critically, to see Victorian paintings typically kept in storage. A visiting fellowship at the Yale Center for British Art (2002) allowed me to work with relevant objects from the Center's collections and superb library. I am most appreciative of the support of Linda Hughes, K. Dian Kriz, John Rohrbach, and Malcolm Warner at this stage of the project. I was ably assisted at this time by Letha Clair Robinson, then a student at Texas Christian University, and I thank colleagues there for their support. A Mellon-Meyerson Fellowship at the Harry Ransom Center at the University of Texas at Austin (2004) gave me unprecedented access to photographic materials, particularly the work of P. H. Emerson, and I am deeply appreciative of consultations with Linda Briscoe-Myers, David Coleman, Roy Flukinger, and Debbie Gresham.

Over time, I revised my hypothesis, particularly in response to Ysanne Holt's important study *British Artists and the Modernist Landscape* (2003), which addressed many of the questions that concerned me. Co-teaching in Scotland with geologist Nowell Donovan in the summer of 2003 sparked a number of new ideas for me as we explored how geologists and artists responded to the landscape. My dear friend Sally Cline mentored me through a series of writing exercises, undertaken during a semester sabbatical leave from Case Western Reserve University (CWRU). These helped me to identify the discourse of scientific naturalism in my primary sources. Historian of science Alan Rocke (CWRU) offered important guidance at this stage.

Situating my project within the history of science revealed patterns and points of significance that had previously been obscured, and allowed me to make better sense of new research materials I had uncovered in the UK. These patterns demanded a longer chronological arc than I had originally envisaged. My tenure as director of the Baker-Nord Center for the Humanities at CWRU, in particular our investigations related to the freshman writing seminar Cultures of Green: Nature and the Environment (2008–9), broadened my interdisciplinary perspective and deepened my commitment to demonstrating that historically the arts and sciences were not distinct disciplines (with humanists and scientists suspiciously eyeing one another over the parapets) but instead were deeply interconnected and interdependent. (Thanks to the National Endowment for the Humanities for an Enduring Questions Grant that provided funding for two iterations of that seminar.) While at CWRU, I also benefited from the university library and the collections of the Cleveland Museum of Art. The opportunity to co-teach a graduate seminar on nineteenth-century photography with curator Tom Hinson helped me see afresh many of the photographs I was investigating.

In the final stages of the project, I was able to take advantage of the holdings of the Getty Research Institute and the J. Paul Getty Museum, and I want to thank my colleagues there and the Getty Foundation, my new institutional setting, for their kind assistance and support. I also benefited from the resources of the Huntington Library, Art Collections and Botanical Gardens and spent many happy hours in their reading rooms. I also want to thank Adam Kujawksi for his assistance in compiling the final bibliography. I also want to express my gratitude to Charles Brett for sharing his research files and insights concerning John Brett.

The opportunity to present my research findings at the College Art Association, the Midwest Art History Society, the North American Victorian Studies Association, and the University of Pittsburgh helped me to refine my arguments. Friends and colleagues helped spur my thinking. Space limitations permit me to mention only a few names; I want to thank Julie Codell, Pamela Fletcher, Ysanne Holt, and Andrew Stephenson for their many conversations about British art. Laura Hengehold, my generous colleague in the Baker-Nord Center for the Humanities and a scholar of philosophy, helped me untangle the complexities of Bergson, as did our informal humanities reading group at CWRU. The "NASVA 2013" team of Kate Flint, Erika Rappaport, and Amy Woodson-Boulton, along with Charlotte Becker and Christopher McGeorge, kept my commitment to Victorian studies kindled. My father, once again, heroically performed the role of sounding board, bringing to bear his honed historical instinct and incisive editorial skills. My mother, siblings, nieces, and nephews endured many family holidays accompanied by this manuscript, and my partner, Christian Wulffen, may have despaired of ever seeing me untethered from the laptop. But he cheered me on to the end—thank you.

It has been a pleasure to work with Penn State Press. The manuscript has benefited greatly from the kind attention of executive editor Ellie Goodman, manuscript editor John Morris, contract editor Suzanne Wolk, managing editor Laura Reed-Morrisson, former editorial assistant Charlee Redman, and the anonymous peer reviewers. The publication of this work has been supported in part by Historians of British Art, an affiliated society of the College Art Association. I am also pleased to acknowledge the Paul Mellon Centre for Studies in British Art for their publication support through an author's grant and publisher's grant. A grant from the Millard Meiss Publication Fund of the College Art Association has also aided publication.

# Introduction

Three telling voices from the nineteenth century offer enticing evidence that modern forms of image making were inseparable from modern science. In 1839, William Henry Fox Talbot published his invention of photogenic drawing, the precursor to modern photography, declaring that, "of whatever value it may turn out in its application to the arts, it will at least be accepted as a new proof of the inductive methods of modern science."[1] In 1850, the art critic F. G. Stephens, writing in the Pre-Raphaelite journal the *Germ,* called for a new form of art making commensurate with recent scientific developments.

> *The sciences have become almost exact within the present century. Geology and chemistry are almost re-instituted. The first has been nearly created; the second expanded so widely that it now searches and measures the creation. And how has this been done but by bringing greater knowledge to bear upon a wider range of experiment; by being precise in the search after truth? If this adherence to fact, to experiment and not theory,—to begin at the beginning and not fly to the end,—has added so much to the knowledge of man in science; why may it not greatly assist the moral purposes of the Arts?*[2]

At the end of the century, the art critic James Stanley Little pointed even more emphatically to science as the decisive influence on preceding decades.

> *The nineteenth century has made its mark in history, chiefly by reason of its achievements in Science; and it is not surprising that men should look to Science with a kind of reverential awe, as the last pronouncement of the Deity, as the Do-all of the future. Science is based upon facts, its proudest boast is that it takes nothing on trust. . . . The careful and laborious method of the scientist in arriving at the truth, in the influence which the process has had on the people at large, has been in the main beneficial. . . . It has made us all searchers after the truth, made us as a nation of investigators and pioneers, each desirous to find his own answer to the terribly perplexing question propounded by Pilate, each glad if he can add something to the sum of ascertained facts.*[3]

Why was science accorded such significance within British artistic circles over the course of the long nineteenth century? Why and how did artists and art critics turn to science for inspiration and even methodology?

What was the impact of science upon art making in Britain in the period we have come to associate with the formation of modernism?

Partial answers to these questions can be found in studies devoted to specific individuals, such as the pioneering photographer Talbot, or decisive art movements, such as Pre-Raphaelitism. But these have been treated as isolated episodes. My aim is to take a longer and fuller historical view. My hypothesis is that art was profoundly shaped by radical changes in science and its discourses over the course of the long nineteenth century. As demonstrated by the writings of Talbot, Stephens, and Little, artists and critics turned to science to inform and to justify their practice.

In turn, leading scientists drew attention to points of connection between art and science, in rhetoric well suited to their cultural ambitions. The scientist Thomas Huxley told the members of the Royal Academy of Arts in 1871 that "our purpose is the same as yours—namely, to seize the idea which lies hidden under the shifting phenomena of nature, and to bind it in such fetters that it may increase the pleasure and the profit of endless generations of men. We both seek truth and we both seek beauty."[4] If we fail to understand art's dynamic relationship to science, we are left with an incomplete picture of British art making and will fail to grasp the particular conditions and expressions of modernism in Britain.

Even intellectuals outside the spheres of art and science recognized the bonds between the two disciplines. In 1863, Cardinal Wiseman, the first cardinal and archbishop of Westminster upon the reestablishment of the Catholic Church in England, chose as his topic "points of contact between science and art" for his address to the Royal Institution, a society founded in 1799 to facilitate sharing scientific and technical information with lay audiences. As befitting a man of the church, Wiseman stayed away from such controversial topics as evolutionary biology and focused upon such seemingly unremarkable themes as mathematical principles of perspective. Yet Wiseman's definitions of science and art reflected current understandings. Science, according to Wiseman, was "whatever knowledge has come to man as the result of investigation, by thought, calculation, and experiment—whether the word be referred to the more abstruse and abstract, or to the more practical, exercise of the power of observation."[5] Wiseman touched here upon a potentially controversial subject—the scientific method. At the time of his lecture, the generation of scholars later known as the scientific naturalists was coming to the fore and championing visual observation and experiment in order to ground science in empiricism and challenge natural theology.[6]

The fine arts, according to Wiseman, "approach our intellect and our feelings through the eye." More specifically, Wiseman pointed to landscape painting as the arena in which British artists had already succeeded in "educating the public eye and the public mind" by encouraging "a perception of the beautiful in nature." He attributed the contemporary appreciation of depictions of nature to the modern age—the age of railways and seaside tourism—which thrust humanity together in cities and turned nature into rural scenery detached from Britons' daily lives. In sum, landscape painting was the art form of the modern age. But, for Wiseman, the desire to gaze upon nature—mountains, lakes, and wild coasts—was not just a rejection of industrialization; it was a means by which to admire the "greatness and power of its Author and Creator."[7] Here again Wiseman touched upon a point of controversy—should the study and representation of nature be informed by natural theology, or was it the perfect subject for a secular modern age searching for utilitarian advances?

Addressing such questions requires taking the long view, investigating change over time. The artistic and scientific developments traced out in this study exceeded a single generation and do not form a straightforward linear narrative of progress. Close analysis of the evidence reveals quandaries, contestations, and reversals of opinion as well as declarations and challenges. I present here a series of case studies, focusing on artists who were understood at the time to be modern, responding to issues of the day and seeking new paths for artistic practice. These individual actors and objects are assembled into a narrative that elucidates how science was seen as the means of reforming art over the course of the long nineteenth century.

Art and science converged in large part because they both engaged directly with visuality.[8] The historian Barbara Gates explains that concern with sight was paramount in the Victorian age. Amateurs and professionals alike were "retraining their eyes to look as never before to witness what was around them in their everyday worlds. . . . Their kind of seeing became a hallmark of Victorian culture, a culture obsessed with sight."[9] However, this emphasis on visual knowledge came into conflict with developing understandings of the subjectivity of vision.[10] Debates concerning sight, and what the cultural historian Kate Flint has identified as an "increasing awareness of the instability of the visual," created a natural meeting point for the work of artists and scientists.[11]

The boundaries between art and science in the nineteenth century were porous, although both disciplines were attempting to establish their distinct professional identities. Indeed, their spatially located histories help to establish this sensibility of proximity. For example, the leading societies associated with the two disciplines—the Royal Society, dedicated to science, and the Royal Academy of Arts, devoted to the visual arts—were both located at Somerset House by 1780 and eventually relocated to Burlington House in the mid-nineteenth century. Art and science were also lexically linked. As the opening quotations suggest, key terms—including *experiment*, *fact*, *observation*, and *truth*—marked nineteenth-century debates about the visual and knowledge and were shared across the disciplines. These terms characterized not only the rise of new art forms, most notably photography, and artistic developments spearheaded by painters affiliated with Pre-Raphaelitism, naturalism, impressionism, and postimpressionism, but also what has been referred to as the "second scientific revolution," or "modern science."

The modern, in science, denotes the demise of natural theology, the birth of scientific naturalism, and the rise of the specialized disciplines that gained public visibility beginning in the 1830s. In the fine arts, the notion of the modern first came to be associated with naturalism or realism, terms coined to signify artists' direct engagement with the world around them and an implicit rejection of idealism. This shift in artistic practice reflected artists' desires to develop modes of representation in concert with the conditions of modernity. Realism—a concern with capturing external reality in all its rudeness and beauty—crossed borders and embraced such diverse artists as the German painter Adolph Menzel, the French painter Édouard Manet, and the British painter William Holman Hunt.

But modernism took on many local forms in response to differing conditions, and this has driven me to look anew at the problem of realism and its aftermath as it played out in a particular context—that of nineteenth-century Britain and the metropolis of London, home to many artistic and scientific societies formed to advance and disseminate knowledge. Scholars such as George Levine, Gillian Beer, and Peter Dale have amply described how Victorian novelists like Charles Dickens, Thomas Hardy, and George Eliot participated in scientific naturalism and its debates.[12] Here, I take up similar questions for the visual arts, but I also set my frame beyond the midcentury moment associated with Charles Darwin. In this study, I trace the formation, consolidation, and recalibration of the tightly bound relationship between art and science over the decades of the 1830s to the 1910s, when modernism took shape.

This study contributes to the ongoing revision of the history of modern British art. As the art historian Lisa Tickner has observed, the "received account" of British modernism "is suffering from fatigue," defined by shop-worn paradigms of an "embattled avant-garde . . . that casts the British permanently into the shade of the French; and . . . the plot of an inexorable progress from Victorian narrative painting to an art of pure abstraction, or, at least, of purely formal values."[13] My aim is to suggest a plot that was not inexorable but plagued with crisis, a plot, moreover, driven not by the internal exigencies of painting but by changes in science. It is the story of a particularly local form of modernism—one built on the British tradition of landscape painting as well as on British debates about scientific knowledge and practice. But the narrative is also inflected by internationalism, as neither art nor science was contiguous with national borders. In short, this is a story that relocates the study of British modernism within the cultural field of science. Both artists and scientists were active interpreters of the natural world in the nineteenth century, and each turned to the other on occasion to seek verification or to stake opposition, acts that contributed collectively to how art and science were understood to function at the time.[14]

In reconceptualizing the story of British modernism, I join several scholars who have offered new insights into their chosen subjects by investigating hitherto overlooked relationships between art and science. Pamela Smith, in *The Body of the Artisan*, argues persuasively that the artisan and the body of knowledge and experience associated with artisanal practice contributed significantly to new philosophies of nature that emerged over the course of the fourteenth to sixteenth centuries in northern Europe. This was manifested materially, she explains, in the mode of depiction understood as naturalism, which allowed these artisans "to make claims about their status as active knowers, about their knowledge of nature, and about their mode of working."[15] Craig Hanson, in his *English Virtuoso*, sets aside the focus on the rise of civic humanism, common in so many studies of early modern British art, in order to investigate the impact of empiricism. The latter was fostered by the Royal Society, home of the English virtuoso, who was "interested in everything from human anatomy, to ancient burial sites, to the technical aspects of glass production," and who often supported art, casting "pictorial representation as a form of knowledge."[16] Amy Meyers, in *Knowing Nature: Art and Science in Philadelphia, 1740–1840*, argues convincingly for reversing the traditional hierarchy that places scientists' investigations of nature above those of the artist or craftsperson, given the critical role that visual culture played in producing knowledge. Such a hierarchical view, she adds, skews our understanding of how nature was studied and analyzed during a period of momentous change in the emerging metropole.[17]

These accounts establish that what we think of today as art was crucially involved in the development of modern thought and in analyzing and explicating the external universe. The formation and validation of knowledge were highly contested, and no one discipline could claim an exclusive purview. Yet in the collection *The Organisation of Knowledge in Victorian Britain*, assembled by Martin Daunton, art and visual culture do not appear, inadvertently regarded as outside the pale. Here, I aim to prove that artists and art critics were just as committed as scientists, mathematicians, classicists, and other scholars were to the pursuit of new paths of inquiry and new knowledge.

In organizing my study, I have taken a cue from Cardinal Wiseman and focus largely on landscape representation, which was characterized by intense investigations in paint, photography, and print over the course of the long nineteenth century. The stakes were high for this genre, which had come to be associated with the leading edge of British art, largely through the efforts of J. M. W. Turner and John Constable. While the shadows of these artists loomed over the next generations, their model of individual genius, imaginative flourishes, and dazzling visual productions was increasingly seen as out of step with the drive for an empirically grounded practice in which the artist was humbled before nature and individualism subsumed to the rigorous task of reproducing the details of the scene. "Truth to nature" became the leitmotif of the Victorian age, and the major question facing artists and scientists alike was how to achieve it.

Nineteenth-century artists and scientists were still wrestling with the Cartesian divide of mind and body as they pursued their study of the natural world and sought to understand its internal mechanisms. Was the external world best apprehended with the bodily eye? Could the eye (and hand) faithfully record the world? Could mechanical instruments improve upon human sight? Or was the external world truly apprehended only by the mind, irrespective of the mechanics of vision? Did the mind interfere with purity of vision, or did it aid in apprehension and recognition? What was the role of feeling, memory, or imagination? Related to issues of perception were those of assessment—how would "truth to nature" be characterized, recognized, and judged? Was "truth" measured by faithfulness to detail or by ability to capture significant broad effects? Such questions drove lively debates in the pages of both artistic and scientific journals.

In this study, I have deliberately focused on the artistic side of this dialogue between art and science, asking how artistic practice was shaped by the leading scientific paradigms of the nineteenth century. I do not wish to imply that art did not have an equally compelling influence upon science. But the story of how artists sought to reinvigorate their practice through lessons derived from science is full and complex on its own terms and has yet to be set forth. As James Stanley Little's essay shows, science was recognized as a powerful, authoritative voice in nineteenth-century British artistic circles, and understanding how and why drives this study.

## Science and Art: Defining Terms in the Public Sphere

This argument rests upon the historical record produced by artists, scientists, critics, and popular writers who made artworks, published notices and articles in the contemporary press, and authored numerous treatises and books, creating a vast archive. My narrative begins in the 1830s, the decade in which modern science began to gain visibility through the efforts of both scientists and popular writers.[18] It is admittedly challenging to pinpoint the historic origins of modern science; indeed, one could cite the formation of the Royal Society in the 1660s as a critical starting point.[19] But at the beginning of the nineteenth century, several key developments mark a distinct shift. The profession and field of practice known as science emerged, as signified by the birth of the term *scientist*. William Whewell, a professor at Cambridge University, coined the term in 1834 in his review of the popularizing book *The Connexion of the Physical Sciences* by Mary Somerville. In his essay, Whewell recalled how he and his colleagues at the British Association for the Advancement of Science arrived at the term *scientist* in order to create an identity for "students of the knowledge of the material world collectively." They had rejected the terms *philosophers* and *savans*, as well as English translations of *Natur-forscher* (nature-poker or nature-peeper), and instead proposed that "by analogy with *artist*, they might form *scientist*."[20] That is, just as the artist was a practitioner of the arts, the scientist was a practitioner of the sciences—a form of logic that implies an intellectual equivalency between the two disciplines.

Whewell elaborated on this logic in his later text, *The Philosophy of the Inductive Sciences* (1840). In the course of his discussion of aphorisms concerning the language of science, he again explained, "We need very much a name to describe a cultivator of science in general. I should incline to call him a *Scientist*. Thus we might say, that as an Artist is a Musician, Painter, or Poet, a Scientist is a Mathematician, Physicist, or Naturalist."[21] The subcategories of science that Whewell cited are telling; they are those fields that he regarded as holding the most significance and promise.

While the term *scientist* was adopted only slowly, its etymology nonetheless correlates with an important shift in the conception of science, as the historians Jack Morrell and Arnold Thackray explain: "The word 'science' took on new and narrower meanings in the 1830s and 1840s. It ceased to be a synonym for all knowledge and became the party label of a particular mode of understanding, possessed—so it was said—of superior power."[22] This change in understanding was tied closely to the birth of distinctive subdisciplines within the broader field of science, and to a rigorous debate about scientific method. These activities dramatically shaped the contours of modern science in Britain and also permeated the cultural sphere, leading to an unprecedented interest in science and also affecting artistic practice.

In many ways, Whewell's need to define the intellectual work of science was the logical conclusion to developments that can be traced back to at least the seventeenth century and the debate between Robert Boyle and Thomas Hobbes. Steven Shapin and Simon Schaffer provide a compelling analysis of this episode in *Leviathan and the Air-Pump: Hobbes, Boyle, and the Experimental Life*. Using Boyle's development of his air-pump device and subsequent experiments on the properties of air and vacuums, and Hobbes's refutation of Boyle's mode of experimentation as a reliable source of knowledge, the authors shed light on how the concept of experimentation entered into scientific practice and became a convention. As in the later Victorian age, "the experimental production of matters of fact involved an immense amount of labour, . . . rested upon the acceptance of certain social and discursive conventions, and . . . depended upon the production and protection of a special form of social organization" that could participate in the collective confirmation of experimental findings.[23]

Shapin and Schaffer touch upon how experimental philosophers were challenged by the church; Peter Harrison, in *The Fall of Man and the Foundations of Science*, more fully elucidates the stakes involved in the dynamic interchange between science and religion in the early modern age. Harrison explains that the postlapsarian fall from grace propelled natural philosophers such as Francis Bacon "to determine whether the human mind might by any means be restored to its perfect and original condition." Critically, Adam's lost knowledge not only drove the search for knowledge but also informed methodologies by which this knowledge could be obtained. To avoid the sin of error, scholars like Robert Hooke, curator of experiments for the Royal Society, argued in favor of "mechanical and experimental philosophy." Proponents of this view, among them Bacon and Boyle, advocated intensive labor, repeated trials, and probable rather than certain conclusions.[24]

Natural philosophy went on to flourish in the Enlightenment, particularly through the efforts of scholars who upheld empiricism as Bacon described it in his *Novum Organum* (1620), in which he argued on behalf of observation and experimentation as well as the inductive method. As Roy Porter explains in his overview of medical and human science in the eighteenth century, "observation and experiment became the watchwords of many," pointing to Giorgio Baglieri and Bernard Mandeville among others.[25] Isaac Newton was an important figure for the eighteenth century as well; scholars were drawn to his certainty in the methods of mathematical reasoning and experimentalism.[26]

But while we can see continuity between the seventeenth and nineteenth centuries in the concern for scientific method, as well as a shared commitment to natural theology, it is important to recognize sharp differences as well. Whereas John Locke, in his *Thoughts Concerning Education* (1692), argued that "Works of Nature are contrived by a Wisdom, and operate by ways too far surpassing our Faculties to discover, or Capacities to conceive, for us ever to be able to reduce them into a Science," Whewell, upholding the utopian belief in progress characteristic of his age, argued that understanding the infinity of the universe and its Creator simply required improving man's faculties: "Difficulties arising from space, number, gradation, are such as we can conceive *ourselves* capable of overcoming, merely by an extension of our present faculties. Is it not then easy to imagine that such difficulties must vanish before him who made us and our faculties?"[27]

Whewell's comments build upon an Enlightenment belief in the capacities of human reason to harness observations and experiments into a larger body of knowledge, organized by laws.[28] Ephraim Chambers, in his *Cyclopaedia* (1728), described science thus:

> *To* SCIENCE, *then, belong such Things as Men may discover by the use of Reasoning, and Sense: Whatever the Mind descries in virtue of that Faculty whereby we perceive Things, and their Relations, is matter of Science: Such are the Laws of Nature, the Affections of Bodies, the Rules and Canons of Right and Wrong, Truth and Error, the Properties of Lines and Numbers* &c. Science, *in effect, is the Result of mere Reason and Sense in their general or natural State, as imparted to all Men.*[29]

One critical way in which Whewell and his colleagues believed that they could generate "a more attentive examination" was through collaborative enterprise and rigorous debate. This belief fueled the birth of new scientific societies, including the Geological Society (founded in 1807 and awarded a royal charter in 1825), the Astronomical Society (founded in 1820 and awarded a royal charter in 1831), the Zoological Society (founded in 1826 and awarded a royal charter in 1829), the Geographical Society (founded in 1830 and awarded a royal charter in 1859), the Statistical Society (founded in 1834 and awarded a royal charter in 1887), the Meteorological Society of London (founded in 1823, renewed in 1836, reconfigured in 1848, and merged with the newly formed British Meteorological Society in 1850), the Microscopical Society (founded in 1839 and awarded a royal charter in 1866), and the Chemical Society (founded in 1841 and awarded a royal charter in 1848). Both social and scholarly, such organizations held periodic meetings that were increasingly reported on in the press, organized public events, and issued their own publications, contributing to the growth of print culture.

The Geological Society and the Astronomical Society emerged as key players in the Victorian age, largely because they attracted members who were major contributors to scientific activities. William Buckland, the theologian and geologist who undertook pioneering work concerning fossils; Adam Sedgwick, who challenged Buckland's theories regarding Noah's flood and also helped to map out geological change over time; and Whewell, master of Trinity College, theologian, and major advocate for rigorous scientific method, were all members of the Geological Society's council in its early decades. Associated with leading roles in the Astronomical Society were Francis Baily, who contributed to both the reform of nautical measurements and astronomical observations; mathematician Charles Babbage, who helped to found modern computing; and John Herschel, who undertook inventive experiments in the study of light and optics contributing to the birth of photography, and whose study of the natural world led him, as it did Whewell, to address the scientific method.

The formation of these specialized societies reflected dissatisfaction with the Royal Society and a desire for more focused discussion of topics among colleagues. Early nineteenth-century reformers of the Royal Society argued that the learned body's range of interests was too diffuse and not sufficiently focused upon science. This critique was complicated by competing concepts of what constituted science: Was it a philosophy that embraced abstract or ideal concepts such as mathematics and geometry? Or was it dedicated to chronicling the contents of the natural world and speculating about how they came to be? How could this body of knowledge best be configured? For example, should the natural or physical sciences be distinguished from the human sciences as well as from the social sciences? Was science a method of investigation, and, if so, what was the approved method—strict observation, as the term *natural history* often implied, or experimentation? Critics were also dissatisfied with the Royal Society's standard of scholarship, as measured by papers delivered before the society and published in its journal, *Philosophical Transactions*, and with its poor care of expensive scientific instruments, the acquisition of which had been a major impetus for the formation of the society. The new specialized societies also argued that they could better serve the needs of British government and society. The Astronomical Society, for example, began to offer the British Admiralty and Parliament advice on astronomical matters, to the dismay of Royal Society fellows such as the astronomer Thomas Young.[30]

Criticism of the Royal Society, leveled in particular by investigators associated with the physical sciences, commingled with concerns that Great Britain was falling behind in what Herschel referred to as the "race" for scientific achievements. Fear of decline and questions of how science should best be configured and supported in Britain motivated Babbage to write *Reflections on the Decline of Science in England, and on Some of Its Causes* (1830), an endeavor in which he was supported publicly by such reformers as Herschel and David Brewster, who shared Herschel's interest in optics and astronomy.[31] Babbage argued that contemporary scientific debate was anemic, in part because the government refused to encourage it and the Royal Society's leadership was weak and poorly qualified.[32]

In 1831, a group of scientists and their supporters, including Brewster and Reverend William Harcourt, formed the British Association for the Advancement of Science, explicitly focused upon science and dedicated to creating a public platform for disseminating the results and benefits of the discipline. As Morrell and Thackray explain, the group linked its identity to "an appeal to nature." Often known as the "Gentlemen of Science" because of their generally shared Anglican background, close affiliation with the liberal Broad Church wing, centrist political attitudes, and close ties to university life, particularly Trinity College, Cambridge, they "articulated a particular ideology of science," Morrell and Thackray argue, that had "an important formative influence on the modern world." These gentlemen of science aimed to demarcate science as a distinct domain of knowledge, separate from religious or political knowledge; to privilege the physical sciences over the social or biological sciences; and to promote a rhetoric of progress linked to discovery. They also claimed a position of seeming disinterest, not relying on scientific activity for income (having personal means to support themselves) but pursuing it out of dedication to the search for truth.[33]

These scientists were aided in their ambitions by the rise of print culture. As the historian of science Richard Yeo argues, Whewell, Herschel, Brewster, and their colleagues took advantage of the establishment of such journals as the *Quarterly Review* (founded in 1809), *Blackwood's Magazine* (1817), *Literary Gazette* (1817), and the *Athenaeum* (1828) to air their debates.[34] Brewster even helped to found two journals, the *Edinburgh Philosophical Journal*, in 1819, and the *Edinburgh Journal of Science*, in 1824, in addition to contributing many papers. Members of this generation voiced their opinions with certitude, as Yeo explains, believing in "the existence of ultimate truths and the ability of the mind to find them."[35] Journals were also critically important for disseminating information about scientific activities, including meetings of scientific societies. Science also entered the public sphere through *conversazione*, public lectures, and museum and exhibition displays, which were becoming increasingly spectacular and were eagerly reported on in the press (fig. 10).[36] The editor of the *Athenaeum*, William Cooke Taylor, for example, was so interested in the British Association for the Advancement of Science's meeting in Dublin in 1835 that he reported on it in five issues, totaling "twenty-seven triple-column pages."[37] At midcentury, T. H. Huxley persuaded the editors of the *Saturday Review* (founded in 1855) to publish a column on science every two weeks and to include scientific topics in their reviews.[38]

Book publishers eventually followed the lead of periodical publishers, even launching series by authors who became household names, such as John George Wood, who wrote *The Common Objects of the Sea-Shore* (1857) (fig. 15) and *The Common Objects of the Country* (1858). Indeed, by the 1840s and 1850s, as the historian Bernard Lightman observes, "there were four times the number of titles on the sciences published annually than at the start of the century."[39]

Art publishing likewise expanded greatly over the course of the nineteenth century, as measured by the growth in specialist periodicals. The *Art Journal,* for example, was founded in 1839, the *Portfolio* in 1870, and the *Magazine of Art* in 1878. High-quality illustrations distinguished these journals from their peers; the *Art Journal,* for example, included a full-page reproductive art print in each issue.[40] These journals also offered fulsome accounts of historical and contemporary art, as well as tidbits of news and gossip. In an age in which criticism had not yet been professionalized to the degree that it would by the close of the century, writers on art came from a variety of backgrounds and experiences. Some, like William Hazlitt and F. G. Stephens, were working artists or had originally trained as artists, while others, such as William Makepeace Thackeray and Tom Taylor, were also literary or theater writers.[41] Many critics traveled in the same social circles as artists (indeed, were often close friends), and these close-knit networks stimulated the rapid exchange of information and opinion, while also fueling spirited antagonism and debate.

But readers could stay abreast of developments in the art world without recourse to these specialized journals, as most daily, weekly, and monthly periodicals reported on major events in the art world such as the annual exhibition of the Royal Academy of Arts. Art, displayed in public venues and reproduced via circulating prints, became a topic of intense interest and was even regarded as an index of community values and concerns.[42] As scholar Helene Roberts summarized the situation, "the attention that Victorian periodicals paid to art was formidable."[43]

These journals and books were part of the rapid expansion of the public sphere in the nineteenth century. The growing middle classes and even the upper tier of the working classes (the respectable or skilled working class), with both literary skills and leisure time, galvanized the proliferation of periodical literature and the publishing industry. Publishers took advantage of new technologies in paper production and printing, which brought down the cost of newspapers, journals, and books. Libraries and subscription reading rooms helped to circulate these materials beyond the small circles who could afford to buy them on a regular basis. By the middle of the century, circulation numbers had reached new heights. As Richard Altick has documented, "at least a half-dozen weekly newspapers and cheap miscellanies had attained a circulation of more than 100,000" by 1855. With the repeal of the duty on paper in 1861, these numbers quickly rose; in 1864, the publisher of the *Athenaeum* reported that in London at least nineteen monthly periodicals on "useful, entertaining, educational" topics printed 350,000 copies per month, fifty-four monthly "magazines and series of a higher class" produced 250,000 copies per month, and "serials issued by the great publishing firms, embellished and illustrated," contributed another 350,000 copies per month.[44]

The growing public interest in art and literature also fueled the rise of exhibition culture. By the early nineteenth century, the practice of annual exhibitions organized by artists' societies had become highly regularized. In addition to the exhibition of the Royal Academy, displays were organized by the British Institution (founded in 1805) and the Society of British Artists (founded in 1823, charter granted in 1846, royal prefix given in 1887). Specialist bodies also arose, including those defined by medium, among them the Old Watercolour Society (founded in 1804 and awarded a royal charter in 1881, thereafter known as the Royal Watercolour Society), the Institute for Painters in Watercolour (founded in 1832 and awarded a royal charter in 1885), the Society of Engravers (founded in 1802), the Etching Club of London (founded in 1838), and the Photographic Society of London (founded in 1853, renamed the Photographic Society of Great Britain in 1874, and awarded a royal charter in 1894).[45] Commercial dealers also played an increasingly prominent role in exhibitions, particularly over the second half of the nineteenth century, when they facilitated the rise of rotating exhibitions throughout the calendar year.[46]

## Epistemological Genealogy

Much of the press dedicated to art was concerned with the role of the arts in civic life—what would they contribute to British society? As Craig Hanson reminds us, epistemology was a critical strand in these discussions,[47] and it provided the genealogical framework for the "truth-to-nature" discourse of the Victorian age.

Hanson argues convincingly that the Royal Society "provided the first 'public' institutional basis for the arts in England." Home of the polymath virtuoso, the Royal Society catalyzed the formation of intellectual networks concerned with questions of methodology in the age of empiricism. In particular, Hanson explains, dilettanti in medicine and art debated similar issues regarding how to root their practices in reason and observation and how to establish authority and rational argument. This virtuosic tradition of intellectual exchange persisted well into the eighteenth century, as exemplified by the artists George Stubbs and Joseph Wright and the physician William Hunter, who became the first professor of anatomy at the Royal Academy of Arts in 1768.[48]

Hunter exemplifies Hanson's concept of the virtuoso, committed to investigating the natural world and advancing scientific enterprise as well as supporting the arts. Hunter, over his lifetime, formed an important collection that reflected his myriad interests in such fields as art, entomology, geology, medicine, numismatics, and zoology. Trained in anatomy at the University of Edinburgh, Hunter was responsible for the production of *The Anatomy of the Human Gravid Uterus*, published in 1774; its accuracy and evident close observation attracted leading artists who were seeking greater proficiency in human anatomy.[49] Hunter lectured at St. Martin's Lane Academy, as well as at the Royal Academy, using plaster casts of flayed figures that displayed considerable anatomical detail, and he also participated in the Society for the Encouragement of Arts (founded in 1754). His growing ambition as a collector and his intent to establish a museum gained him the attention of the Royal Society, who elected him a fellow in 1767; membership in the Society of Antiquaries followed in 1768.

As part of his collecting activities, Hunter commissioned the painter George Stubbs to paint an Asian antelope in 1769 and a North American bull moose in 1770; the latter was intended as visual evidence for Hunter's developing argument about the extinction of species.[50] Stubbs was a likely artist to take up such a task, as he had supplied engravings for Dr. John Burton's *Essay Towards a Complete New System of Midwifery Theoretical and Practical* (1751), a project analogous to Hunter's 1774 anatomy, and later dedicated himself to representing the anatomy of the horse thoroughly and accurately in his published treatise of 1766. Working closely from dissected specimens and attentive to detail, Stubbs was nonetheless keen to preserve the sense of the whole, as well as to express the vitality of the species, and he consistently represented the whole animal, often in implied motion.[51] His animal studies demonstrate not only the close visual observation of nature—as Oliver Kase points out, Stubbs emphasized that the figures were "drawn from nature"—but also experience, knowledge gained over time and through labor.[52]

In most narratives of artistic contributions to British Enlightenment scientific thought, Stubbs is linked closely to Joseph Wright of Derby, whose representations of scientific experiments and participation in the Lunar Society can be considered important precursors to the work of William Henry Fox Talbot. Born in Derby, Wright returned there after training with the painter Thomas Hudson in London, joining a lively intellectual community located in and around Birmingham composed of industrialists, philosophers, and medical practitioners. Later dubbed the Lunar Society because they met each month on the Monday closest to the full moon, they exchanged ideas and conducted experiments. The group included Erasmus Darwin and Joseph Priestley, as well as Josiah Wedgwood, who would make important contributions to the origins of photography. Their gatherings are believed to have inspired Wright's canvases *A Philosopher Lecturing on the Orrery* (1766) and *An Experiment on a Bird in the Air Pump* (1768).[53] An orrery is a device for demonstrating planetary motion, and the painting depicts a rapt audience that includes mesmerized children as well as a note-taking colleague who stands to the left of the red-jacketed philosopher. The scene is dramatic and striking, largely owing to Wright's handling of tenebrism. David Fraser has argued that the illuminated faces gathered around the device echo the mental enlightenment that science has provided them.[54] But Elizabeth Barker has suggested that audience's reactions to the display would have been more complicated than what Wright depicts. Drawing on contemporaneous historical evidence, she establishes that Wright has omitted a notable component of the orrery, which, as both she and Fraser note, was not a tool of scientific investigation but a device intended for popular demonstrations. Thus, while Wright's painting professes verisimilitude in its attention to physical detail and materiality, it actually deploys a dissembling facture, cunningly depicting "truths and fictions" to create an intriguing puzzle for viewers knowledgeable about the increasingly popular field of astronomy.[55] Likewise, *An Experiment on a Bird in the Air Pump* represents a plausible event, the demonstration of a scientific experiment, akin to those that Wright himself may have witnessed. It also inspires the viewer to ponder larger themes of life and death, as the bird flutters to the bottom of the glass globe to which the red-coated philosopher controls the airflow.[56] The complex iconography of Wright's paintings and the diverse strands of artistic and scientific representation from which he drew his references emphasize that art did not merely illustrate science but offered a powerful social commentary on science.

Indeed, Wright's painting *A Philosopher by Lamp Light* (1769) complicates the social value of natural philosophy. In the left foreground sits a hermit wrapped in a voluminous brown robe, one hand supporting his head, the other grasping the end of a bone as he tests the resilience of the string to which it is attached in order to understand physical movement. At right, two young pilgrims peer through the darkness, aided by the moon breaking through the clouds, to seek him out in his cave. Judy Egerton notes that Wright often referred to this hermit as Democritus, the ancient Greek philosopher regarded as the father of modern science.[57] Democritus was skeptical about whether true knowledge could be gained through the senses, and he exalted intellectual knowledge. By recognizing at least two different means by which knowledge could be gained, Democritus inferred that "when an object appears one way to one, and another to another, neither is nearer to the truth."[58] This acknowledgment of human fallibility and doubt concerning the scientific enterprise is a far cry from the resolute belief in the truth value of scientific knowledge trumpeted by James Stanley Little at the end of the nineteenth century. It is a reminder that science was an unstable field a century earlier, its cultural value not yet secured. Moreover, Wright represents the philosopher as a solitary figure, isolated in his own thoughts. This was precisely the image of the scientist that the generation of reformers in the 1830s sought to counter, insisting instead upon rigorous method checked through collective enterprise.[59]

## The Narrative

As noted earlier, my study begins in the 1830s, when efforts to define the field of science and to assert its cultural value began to gain considerable momentum. In chapter 1, I consider the origins of photography as manifested in the work of William Henry Fox Talbot. A polymath, Talbot was closely involved with leading scientists of the day, including Herschel and Brewster, and was a member of the Astronomical Society and the Royal Society. Keenly aware of debates then brewing about scientific method, he proudly announced his photographic process as proof of the inductive process, thus proclaiming his allegiance with the "Gentlemen of Science" of the 1830s. Concomitantly, Talbot inserted his photographs into artistic discourses through his exhibiting practice and rhetoric. This chapter considers these dual and overlapping contexts both synchronically and diachronically. While Talbot's pursuit of truth to nature in his photographs aligns him with developments in landscape representation associated with his predecessors and contemporaries Thomas Girtin, J. M. W. Turner, John Constable, and Cornelius Varley, Talbot also marks a distinct break from their efforts in his abnegation of self in the act of representation, deferring to the sun as the creator of the image.

Talbot's claim that his photograms and calotypes were nature's pencil, that is, that they were unmediated representations of nature on paper, is analogous to the advice of the writer and art critic John Ruskin, a leading voice in Victorian culture. Ruskin recommended that young artists, early in their studies, "go to Nature in all singleness of heart, and walk with her laboriously and trustingly, having no other thoughts but how best to penetrate her meaning, and remember her instruction; rejecting nothing, selecting nothing, and scorning nothing; believing all things to be right and good, and rejoicing always in the truth."[60] Furthermore, Ruskin counseled, these young students should adopt an "innocent eye," perceiving shapes and colors without attaching a priori knowledge or significance to them. Ruskin avidly engaged with contemporaneous developments in science, particularly those relating to geology (he joined the Geological Society in 1840), and he evinced great sympathy with natural theology.

The art critic F. G. Stephens, writing in 1850, reinforced the moral imperative of pursuing truth in art, turning to the physical sciences as a model. The new form of painting that Ruskin and Stephens envisioned is most closely associated with the Pre-Raphaelites. Chapter 1 analyzes artworks by painters associated with the movement: scenes of nature that bear the traces of strict mental discipline and intense physical labor in meticulously rendered details.

The methodology of direct, unmediated observation promoted by Talbot and Ruskin provoked resistance and carried its own contradictions. In concluding chapter 1, I examine a series of episodes—nature printing, Talbot's adoption of the picturesque, and several Pre-Raphaelite painters' struggles to follow Ruskin's dictates rigorously—that reveal the resilience of the "practiced eye," that is, visual perception trained to recognize significant patterns in nature.

The challenges posed by the pursuit of truth to nature are taken up in chapter 2, which focuses on two of the Pre-Raphaelite artists—John Brett and John Everett Millais—and traces their careers as landscape painters after the dissolution of the Pre-Raphaelite Brotherhood. In the 1850s and 1860s, a new generation of reforming scientists came to the fore in Victorian culture. Historians of science, particularly Frank Miller Turner, dubbed this community the scientific naturalists, who, as T. H. Huxley stated, argued for a "new Nature begotten by science upon fact."[61]

As Gowan Dawson and Bernard Lightman have recently observed, while the term *scientific naturalism* is often attributed to Huxley, who explored the concept in his *Essays upon Some Controverted Questions* (1892), it has a longer etymology that speaks to the complex landscape of science and religion in the nineteenth century and to changing understandings of professionalism. The term first emerged in the 1840s, in the United States, as a negative commentary on texts, such as *Vestiges of the Natural History of Creation* (1844), that attempted to provide rational explanations for physical phenomena. By the 1860s, the term had been taken up by the British religious press, and eventually it took on the more positive connotation "for the scientific rejection of all non-material phenomena." When Huxley adopted the term, he needed to cleanse it of "all the deleterious ethical and political connotations it had accrued since first coming into usage in the 1840s." He emphasized its associations with authority and trustworthiness and also deployed it to refer to a "specialized practitioner" possessing "intellectual independence." While recognizing the dynamic history of the concept of scientific naturalism and the problematic nature of labels, I nonetheless adopt the term *scientific naturalism* here "as a shorthand for a particular outlook among men of science and associated intellectuals in Victorian Britain," and in response to the considerable scholarly literature that has studied this phenomenon.[62]

The historical actors associated with scientific naturalism include John Tyndall, Joseph Dalton Hooker, George Henry Lewes, and Herbert Spencer. (Charles Darwin is often connected only loosely to this circle because his adherence to the inductive method was often disputed.)[63] This new generation of scientists laid claim to the public discourse of science, wresting it away from their gentlemanly predecessors, who had insisted that a divine presence guided nature. By contrast, the scientific naturalists eschewed religious explanations for natural phenomena and sought to both professionalize and popularize their discipline. In doing so, they argued not only that "they provided the best intellectual leadership of a modernized and industrialized Britain" but also that science was the best means of achieving knowledge and truth.[64] Most critically for this study, they did not concern themselves solely with scientific practice, but also believed that the scientific knowledge they promoted should be extended to culture and society.

In this context, the scientific naturalists frequently proposed a union between art and science, one that initially seemed well suited to the practices of Millais and Brett but became increasingly less so as their careers advanced. At first, it appeared that the painters' gradual disaffection with scientific naturalism could be accommodated within John Tyndall's new paradigm of the scientific use of the imagination. But eventually the rise of psychology, which questioned many of the principles of scientific naturalism, suggested new directions for landscape painting, provoking divergent responses from Millais and Brett. New investigations into mental perception led by the philosopher George Henry Lewes, the psychologist James Sully, and the physicians William Carpenter and Hermann von Helmholtz, among others, drove a wedge into scientific naturalism and also changed the criteria by which paintings should be judged. By the 1880s, a number of important intellectuals deeply familiar with scientific naturalism began to challenge its claims to cultural authority.[65] Frederic W. H. Myers, a founding member of the Society for Psychical Research, asserted that

> *the* emotional *creed of educated men is becoming divorced from their* scientific *creed; that just as the old orthodoxy of religion was too narrow to contain men's knowledge, so now the new orthodoxy of materialistic science is too narrow to contain their feelings and aspirations; and consequently that just as the fabric of religious orthodoxy used to be strained in order to admit the discoveries of geology or astronomy, so now also the obvious deductions of materialistic science are strained or overpassed in order to give sanction to feelings and aspirations which it is found impossible to ignore.*[66]

In this configuration of knowledge, science became associated with the intellect and art with feeling.

The conviction that art and science could productively be joined had begun to waver by the closing decades of the nineteenth century. Photography again became a lightning rod for these debates, particularly as represented by the photographer and writer P. H. Emerson, the subject of chapter 3. When Emerson began his career as a photographer in the 1880s, he was in many ways picking up the thread of photographic discourses established by Talbot. Emerson positioned photography as an unmediated, truthful expression of nature—what he called naturalism—and insisted that he had at last established a scientific basis for art. Within a decade, however, he had refuted his own tenet, denying the possibility that photography could be truthful to nature because of new scientific findings about perception that undermined the assumption that the eye, or the camera, could act as an impartial recorder. He now understood that the mind played a critical role in perceiving and interpreting the scene, no matter how unbiased the observer strove to be. Whereas, at midcentury, artists and scientists reveled in the possibility that the two disciplines could come together in the shared pursuit of truth through the close observation of nature, Emerson now declared that such a union was impossible.

Emerson confided many of his doubts about naturalism to the painter George Clausen, who likewise struggled with the paradigms of realism and naturalism that he had inherited. Chapter 3 considers how the changing nature of scientific naturalism and the rise of psychology affected the circle of British painters associated with the New English Art Club and impressionism, including Clausen, Philip Wilson Steer, and Walter Sickert. Rather than represent the visual field in a nonhierarchical manner that denies the role of the perceiving self, as had been attempted by the Pre-Raphaelites, Clausen and his colleagues strove for a new form of naturalism that required the artist to select what is most essential from the scene, an act of discernment that aligned with new understandings of human perception proposed by psychologists and physiologists.

Chapter 4 takes up the next generation of British modernists, particularly the painters associated with the Camden Town Circle, including Augustus John, J. D. Innes, Charles Ginner, and Harold Gilman, who likewise struggled to redefine realism and the task of representing nature for the modern age. Ginner and Gilman even coined the term "NeoRealism" to describe their painting practice and to underscore their difference from naturalism. Rather than draw from scientific naturalism—Innes dismissed geology and astronomy as fossilized dreamlands—they turned to philosophy, and in particular the writings of Henri Bergson. Informed by Bergson's concepts of intuition and instinct, they rejected the notion that the perceiving self should be suppressed in painted representations of natural scenes. Innes and John aimed to merge the creating subject (and implicitly the perceiving subject) with the landscape, and Ginner and Gilman devised means by which to assert visual mastery over external nature. Ironically, like the Pre-Raphaelites before them, they faced fierce criticism, though in this instance from within their own modernist camp, in the figure of the prominent critic T. E. Hulme.

While each successive generation, from Talbot in the 1830s to Ginner and Gilman in the 1910s, claimed to be pursuing truth to nature, their concept of truth was in fact historically contingent and inherently unstable. Moreover, these debates were not the exclusive purview of artists but were part of a broader cultural conversation that involved scientists, critics, and philosophers.

## Methodology

Central to my argument is the claim that artworks can be persuasively associated with changes in nineteenth-century scientific paradigms. While artworks necessarily occupy the foreground of this study, historical change is more than mere backdrop. If knowledge is always in formation and history constantly unfolding, then the art object and its social and historical context are always symbiotically interlocked. The task of my narrative, then, is not to extract or isolate the object from the dense network of relationships in which it was originally embedded, but to find the patterns of significance within that web of connections. I must therefore address questions of methodology and the archive.

Communities of both practice and reception were critical for nineteenth-century cultural production. As the discussion above suggests, both artists and scientists organized themselves into societies and other groups, and such "clubbiness" was shared across Victorian culture. Therefore, in many cases, links can be established between art and science through personal networks. This is most strongly evident in the case of William Henry Fox Talbot.[67] Talbot's extensive correspondence network establishes his close ties to the other leading scientists of his day. Fragments of correspondence and diaries preserved for John Brett, William Holman Hunt, John Everett Millais, P. H. Emerson, George Clausen, Augustus John, and J. D. Innes occasionally reveal personal ties between artists and scientists—such as John Tyndall's invitation to Millais to attend his lecture on rainbows—but more often reveal an exchange of ideas about scientific knowledge, particularly as it related to artistic practice. These dialogues were often mediated by texts, as when Brett recorded his enthusiastic reaction to Ruskin's *Modern Painters* and "Pre-Raphaelitism," or when Augustus John and his patron John Quinn recommended Bergson's books to each other.

Print culture fostered even broader communities of affiliation and interest, and the analysis of discourse is at the heart of this study. Discourses are critical means by which knowledge and power take shape and are situated within, and constitutive of, disciplines and institutions.[68] Daily, weekly, and monthly newspapers, journals, and books were public spaces in which ideas about art and science were debated. They were points from which new ideas were launched into the public sphere, where they were received and reviewed; the conversations that emerged from these pages sometimes reached consensus but more frequently were conflicting, even paradoxical, as positions were staked out, withdrawn, and reestablished. Heated rhetoric suited the economics of Victorian periodical publication. As Laurel Brake observes, these texts cannot be separated from publishers' need to generate revenue through the consumption of print, which fueled strategies to attract and retain the reader's attention and pocketbook.[69]

Print culture produced a circulatory network, with branches reaching far beyond the nodes of production and reception in metropolitan centers. It also helped to form what Pierre Bourdieu has described as a field of cultural production, "the space of positions and the space of position-takings" created by both "the structured set of the manifestations of the social agents involved in the field—literary and artistic works . . . [and] also political acts or pronouncements, manifestations, or polemics, etc." Critically, Bourdieu observes that the "literary or artistic field is a *field of forces,* but it is also a *field of struggles* tending to transform or conserve this field of forces."[70]

One way to approach this "field of struggles" is to consider how neither producers nor receivers are universal or ideal but are historically constituted and locally situated. The task of the historian, then, is to find patterns of significance within the welter of positions and acts of interpretation. The historians of science Aileen Fyfe and Bernard Lightman, drawing on reception theory, offer a useful model, one well suited to their project of investigating how science was popularized in nineteenth-century British consumer culture. They describe how their "efforts to place some order onto histories of those interpretations have led to the ideas of *horizons of expectation* (the idea that readers' education, experiences, etc. will help determine how they react to particular texts) and of *interpretative communities* (groups of readers who see things in broadly the same way)." Furthermore, they argue that we should "extrapolate from texts to include images, exhibits, and lectures and expand the notion of readers to include viewers and listeners."[71]

While the model that Fyfe and Lightman offer aligns art with other forms of cultural production and discourse, I also want to attend to the particular materiality of art. This includes the physicality of the artifact itself, such as patterns of oxidized silver on paper, or glazes of paint on prepared canvas, as well as the ways in which the material object was made visible in the public sphere. I deliberately begin each chapter with a close description of an artwork in order to ground my questions and subsequent analysis in relation to a work of art.

In giving this close description, I am conscious of the work of Michael Baxandall, who wisely advised that "the description is less a representation of the picture, or even a representation of seeing the picture, than a representation of thinking about having seen the picture. To put it another way, we address a relationship between picture and concepts." Thus I acknowledge that my descriptions are forms of argument, ways in which I can begin to weave together the connections between image and contemporaneous ideas about science. In building my case about the relationship between artworks and their context, I am also mindful of Baxandall's point about "intention," which he uses in reference "to pictures rather more than to painters. In particular cases it will be a construct descriptive of a relationship between a picture and its circumstances." To elucidate this point, Baxandall explains that "the painter's role has been to make marks on a plane surface in such a way that their visual interest is directed to an end."[72] In analyzing J. S. Chardin's *A Lady Taking Tea* (1735), a painting acquired by the physician William Hunter and now in the collections of the Hunterian Museum and Art Gallery in Glasgow, Baxandall employs a set of constraints or guidelines that I adopt here.

> *I shall make two limiting demands of the connection between ideas and painting I want to pursue. First, the science or philosophy invoked must be made to entail fairly directly a particular thing about visual experience and so about possible pictorial character. I shall hope for a pictorial corollary, as it were, such that the scientific thought could enter the painter's self-Briefing. Secondly, I shall demand some indication that it was conceivable, in the period, for the two universes to be brought into this sort of relation. In effect this means I need to produce men capable of reflecting on the relation of painting and science that I claim existed.*[73]

In sum, the archive from which I derive my arguments is composed of the artworks themselves, the surviving personal correspondence of the makers, and the larger print culture in which the production and reception of these works of art was situated and through which contemporaneous debates about the nature of both art and science were waged. I begin my pursuit of the pictorial brief that conjoined art and science in the 1830s, when new models of scientific thought emerged and scientific method was rigorously debated. At the same moment, a new mode of art making—the photograph—emerged, and painters and critics began to rethink both theories of representation and the physical craft of painting. I then map out how artistic debates concerning truth to nature changed in dynamic relationship to scientific paradigms through the ensuing decades until the birth of a new century.

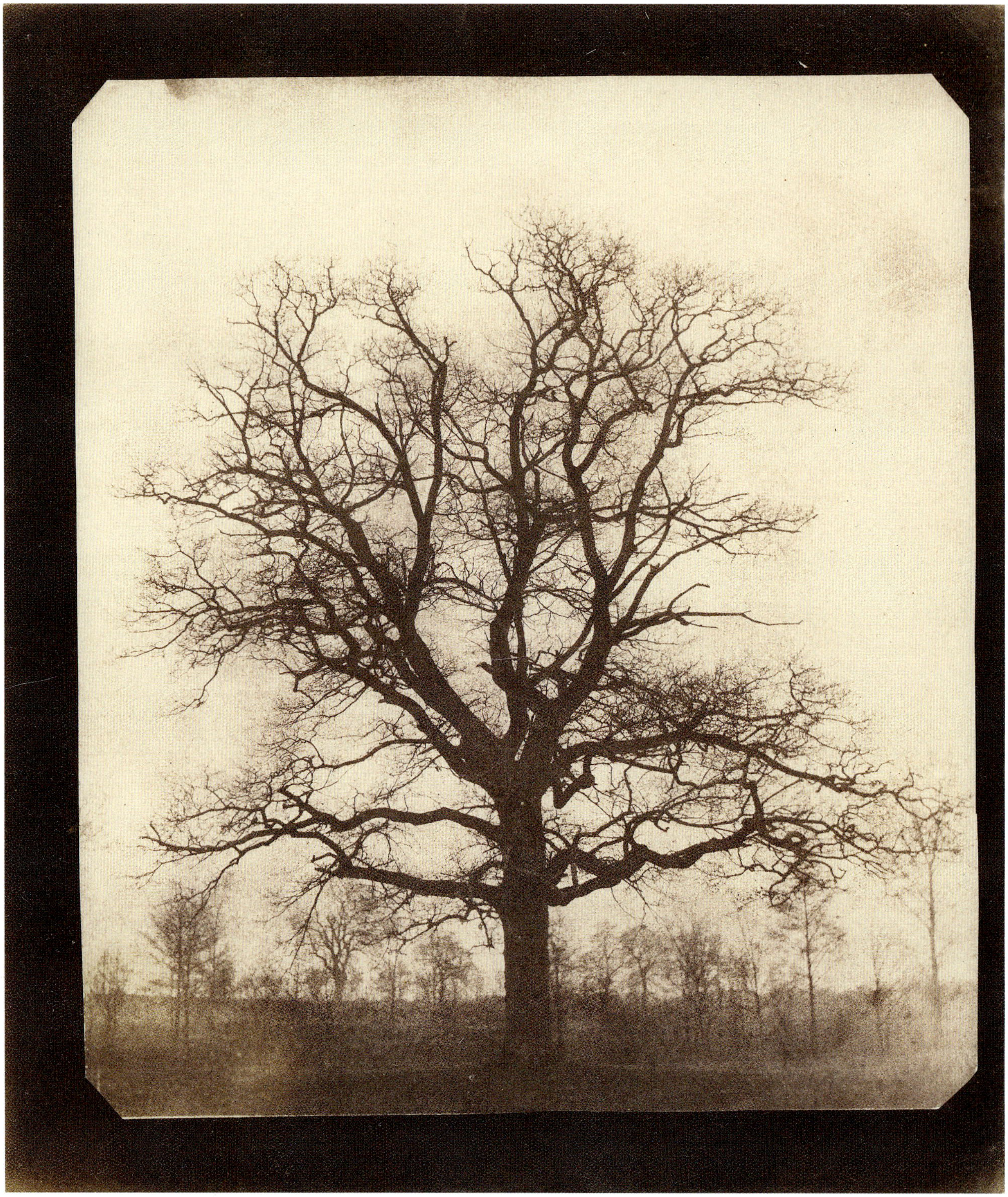

FIG. 1

William Henry Fox Talbot (1800–1877), *An Oak Tree in Winter*, probably 1842–43. Salted paper print, image 19.4 × 16.6 cm. The J. Paul Getty Museum, Los Angeles.

{ CHAPTER ONE }

# Truth to Nature and the "Innocent Eye"

My first object of inquiry, a photograph, could be regarded as an experiment, a demonstration of a chemical process, an illustration of a place, or a testament to the noble dignity of nature (fig. 1). The paper print portrays a stately oak tree in winter at Lacock Abbey, the estate of William Henry Fox Talbot. The central tree's grand height is accentuated by the low horizon line, broken in places by slim neighboring trees that pay homage to their great predecessor, their crowns merely reaching its lower branches. Both ground and oak tree are rendered in varying purplish tones, the darkest of which are found in the tree's trunk. This coloration conveys a sense of the tree's solidity at the same time that the darkened shape is silhouetted and flattened against the gray sky. A few splotches of brighter tones appear in the left and right corners of the image, accidents of chance and reminders of the chemical processes used to make the image. The mighty tree, splayed out on the paper and depicted on a portable scale, is reminiscent of a botanical specimen to be shared among explorers, gardeners, nurserymen, and cognoscenti. The photograph, however, lacks any notation of species, genus, and so on. It is not the thing itself; it is a reduced, flat, mottled-purple representation. Yet the photograph partook in conventions that allowed it, with all its idiosyncrasies, to be read as truthful to nature.

My second object of inquiry, which I wish to bring into dialogue with my first, is a small, oval-shaped oil painting depicting a woodland subject (fig. 2). Near the center of the scene lies the key point of visual interest, a patch of white and blue sky reflected on the surface of a stream. Only once our gaze is fixed there do we notice the woman to the left, seated at the foot of the tree that arches its branches over the meandering stream, merging with the foliage at the far side of the bank. The woman, her contemplative visage framed by her hat, devotes her gaze to the book she holds. Time must be taken to make sense of this image; its densely organized composition makes quick summation difficult. Close study is repaid by glimpses of shimmering reflections on still pools of water and skeins of underwater plants dragged by watery currents.

The painting, approximately ten inches in length, is comparable in size to a photograph and, like many photographs of rural scenes from around this time, is preoccupied with the dappling of light and shadow.

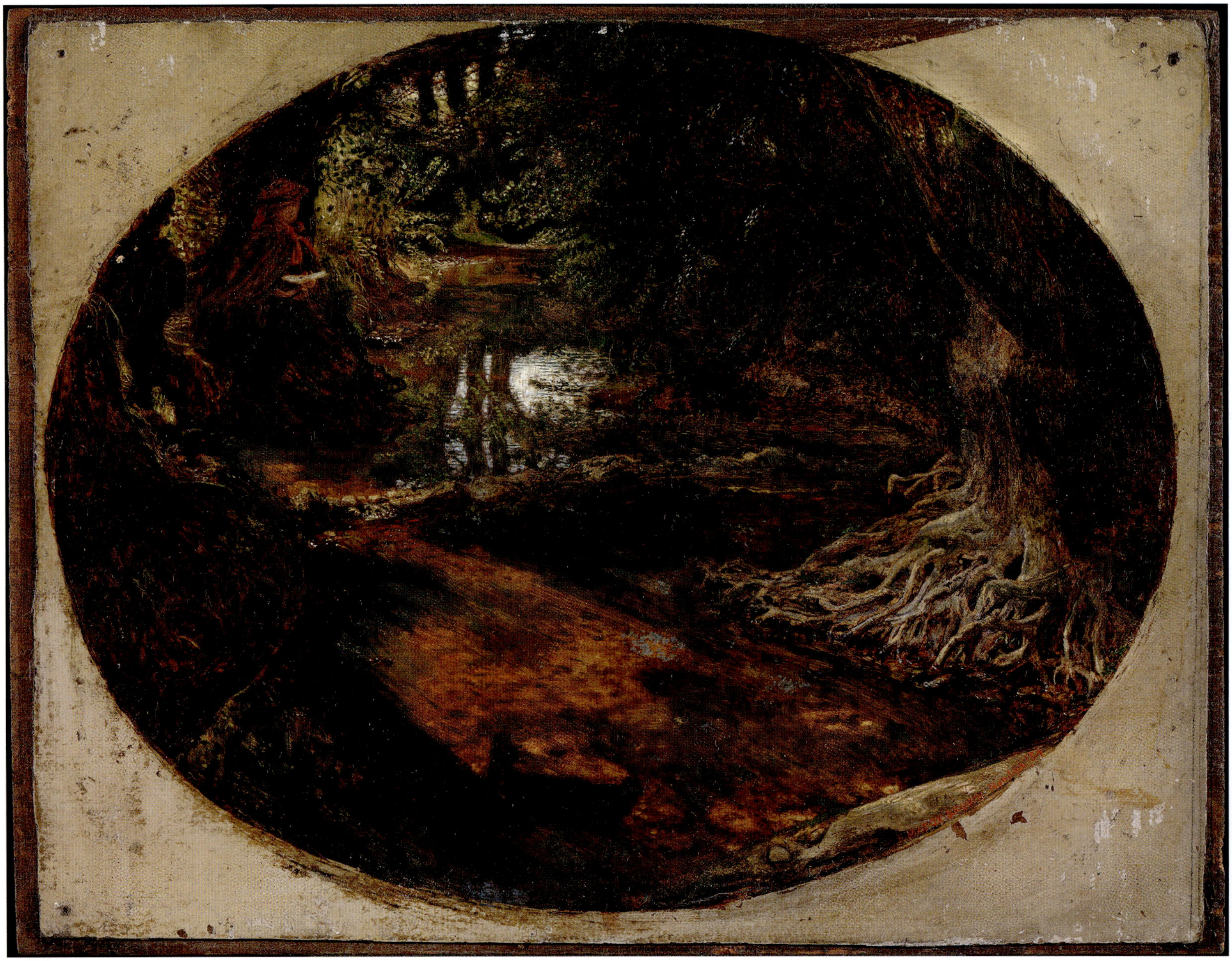

FIG. 2
Ford Madox Brown (1821–1893), *The Brent at Hendon*, 1854–55. Oil on board on mahogany, support 203 × 248 mm. Presented by F. Hindley Smith, 1920. © Tate, London 2015.

It is these analogies that first drew me to the question of the relationship between photography and painting in this period—a question that had struck a number of scholars before me, including Diane Waggoner, whose exhibition *The Pre-Raphaelite Lens* revealed the extensive iconographic and formal links between photography and Pre-Raphaelite painting. Yet I am also conscious of the ways in which painting and photography were profoundly different. Paper photography, with its sepia tones and delicate fibrous surfaces, is obviously distinct from the early painted canvases of the Pre-Raphaelites, who laid down paint in thin glazes, leading to jewel-like tones, sharply demarcated forms, and smooth, glossy surfaces. Nonetheless, in sifting through the vast textual archive created by the extensive commentary on both painting and photography produced around the middle of the nineteenth century, I was repeatedly struck by the ways in which the aims of early photography and Pre-Raphaelitism were described in analogous terms that pointed to a shared intellectual project. For example, Talbot announced that one of the chief benefits of photography was that "it will enable us to introduce into our pictures a multitude of minute details which add to the truth and reality of the representation, but which no artist would take the trouble to copy faithfully from nature."[1] When the Pre-Raphaelite painters began to exhibit publicly, their apologist F. G. Stephens asserted that their work ushered in "a great change in the character of the productions of the modern school, a marked attempt to lead the taste of the public into a new channel by producing pure transcripts and faithful studies from nature."[2]

In particular, the discourses of modern painting and photography shared a concern with truth to nature, and the foundation of this discourse at midcentury is the subject of this chapter. I ask how this interest came about, how it was visually manifested, and what its critical concepts were. Much of the archival evidence that I investigate is in the form of treatises. Their authors aimed to shape a community of practice, and their advice reflects their inherent belief that with sufficient guidance, other practitioners could produce replicable results that would point to a universal truth. Indeed, this belief continued to propel scientific and artistic activity well into the closing decades of the nineteenth century. For many of the authors and artists with whom I am concerned, the stakes were quite high, because they aimed to define pictorial representation for the modern age. Their leading strategy was to claim truth to nature—but how this truth could be achieved was a matter of vigorous debate.

## The Invention of Photography in the Field of Science: Photogenic Drawing

William Henry Fox Talbot easily meets the criteria of the virtuoso that Craig Hanson has traced from the early seventeenth to the mid-eighteenth century in *The English Virtuoso*. As a young man at preparatory school, Talbot was interested in both botany and chemistry, fields of study not then included in the formal curriculum. He continued his training at Trinity College, Cambridge, focusing on mathematics. In 1818, William Whewell became his lecturer. Whewell had studied at Trinity, became a fellow there in 1817, and assumed the professorship of mineralogy from 1828 to 1832. He then became professor of moral philosophy from 1838 to 1855, served as college master from 1841 to 1866, and was a major intellectual force in shaping the emerging discipline of science. By the time Talbot began his studies at Cambridge, the previous generation of undergraduates—namely, George Peacock, Charles Babbage, and John Herschel, with whom Talbot would go on to form important relationships—had shaped the mathematics curriculum by arguing successfully for the inclusion of treatises by non-English authors.[3]

After taking his degree and a period traveling in Europe, Talbot returned to England and, while attending to the affairs of his estate, began to develop his extensive network of correspondence with like-minded colleagues such as Herschel, David Brewster, and Michael Faraday. Faraday was closely associated with the Royal Institution, where he had begun his career as a chemical assistant in 1813; in 1821 he became the supervisor of the institution and in 1825 the director of the laboratory. In that capacity, Faraday worked with Talbot in the 1820s and 1830s on experiments concerning chemistry and optics.[4] In 1831, Talbot was elected to the Royal Society, sponsored by Faraday, Peacock, and Whewell and by his uncles Charles Lemon, a member of the Royal Society who eventually became the second president of the Statistical Society, and William Thomas Horner Fox-Strangways, a diplomat. At this time, Talbot was best known for his work in optics and mathematics; in 1838, he was awarded the society's Royal Medal for two papers on elliptical integrals.[5] His notebooks from this period, recently analyzed by scholars Mirjam Brusius and Chitra Ramalingam, also reveal an interest in botany, classics, philology, and etymology.[6]

Talbot joined the Royal Society a year after the institution had become caught up in acrimonious debates over leadership succession, pitting John Herschel against the duke of Sussex, with the duke emerging victorious. Babbage's diatribe against the institution was published around the same time.[7] These circumstances energized many members of the British Association for the Advancement of Science, with which Talbot became involved. In 1836, when the group was slated to meet in Bristol, Talbot invited Whewell, along with Babbage, Brewster, and other scholars, to his home at Lacock Abbey.[8]

The circle that Talbot gathered about him that August included some of the key figures in debates regarding optics, among them Charles Wheatstone, a physicist who became a fellow of the Royal Society in 1836 and presented his theory of stereoscopic vision to the society in 1838, and David Brewster, appointed to the Royal Society in 1815, who also worked on the question of stereoscopic vision in connection with his long-standing interest in the behavior of light. Talbot's friendship with Brewster was facilitated by Herschel, whom Talbot had met in Munich in 1824. Talbot kept up a vigorous correspondence with this circle, as their scientific interests overlapped considerably, and they often exchanged advice on conducting experiments.

The state of the British Association provides a useful window onto concepts of modern science at this time. The *British Critic*, in 1839, reviewed the meetings of the association up to that date. Comparing the present moment to the past, the author found that present-day scientists were concerned with tangible objects as opposed to abstract concepts. Science, rather than comprehending the entire field of knowledge, now focused upon "physical science, the science, in other words, of matter and material things." The present-day philosopher

> *presents not the image of one who by dint of deep reflection, or of converse with the past, has opened up new springs of moral truth, or obtained a deeper*

> *insight than his fellows into the mysteries of our spiritual nature, but of one who by a series of skillful manipulations has succeeded in producing some visible effect of a striking kind,—who by some practical discovery has enabled us to dive deeper into the earth, or to skim more rapidly along its surface than we could do before,—who can astonish us by dissolving the insoluble, by burning the incombustible, or dazzle us by some light more brilliant than has yet shone upon human eyes.*[9]

But this new science should not, the author cautioned, be spectacle for spectacle's sake.[10] The task of science was to benefit the nations "in a spirit of dutiful submission to the revealed councils of Heaven," that is, the Christian God.[11]

While Talbot's work in mathematics, classics, and philology link him to the generation of philosophers that was passing away, his work in photography—which was, in essence, making visible and tangible the passage of light—demonstrates that he kept pace with scientific change. Indeed, Larry Schaaf has argued persuasively that the close relationship between Talbot and Herschel explains why photography developed in the late 1830s—namely, because each man was able to advance the work of the other.[12]

Herschel, son of the astronomer William Herschel, had grown up surrounded by the spirit of scientific investigation and trained in mathematics at Cambridge. He would go on to make major contributions in astronomy, optics, the study of light, and botany (he and Talbot exchanged plants). In 1830, Herschel opened wide the debate regarding scientific methodology when he published his *Preliminary Discourse on the Study of Natural Philosophy*. As the historian of science Richard Yeo has observed, behind many of these debates was the legacy of Francis Bacon and the concepts of induction and hypotheses, as well as "moral connotations about the proper use of reason."[13] Above all, the question was how to arrive at verifiable truth.

Herschel's treatise was aimed at the emerging public for scientific literature, but it was also taken up by his peers.[14] The author announced his intellectual debt and lineage on the title page, which was illustrated with a half-length bust of Bacon, its plinth ornamented with a telescope and other instruments. Bacon was regarded as the "father of experimental science" and was revered in particular for his formulation of the inductive method.[15] Herschel argued that the task of science was to ascertain the laws of nature, with the expectation that "a knowledge of natural laws enables us to improve our condition."[16] Herschel observed that Bacon had provided the framework by which such laws could be derived, that is, a system of classifying "particular facts" into "general facts," which point to laws, and then, finally, "we arrive at *axioms* of the highest degree of generality of which science is capable."[17]

Observation of external nature was critical for accumulating facts, as was experiment, which allowed cause and effect to be tested and verified.[18] Herschel anchored his discussion of critical concepts in successful examples of scientific practice. In reading through both Herschel's experiments and theoretical discussions, it becomes clear that he conceived knowledge as obtained visually by observing phenomena from which facts, and then laws, could be derived.

The correspondence between Talbot and Herschel is filled with examples of visual observation and experimental investigation, and with a desire to share knowledge in ways that would allow others to partake in their inquiries. Since Larry Schaaf has amply discussed how the dynamic relationship between Talbot and Herschel informed Talbot's investigations into permanently fixing the passage of light, I will not rehearse all of their critically important exchanges here. But it is important to establish the two scholars' shared interest in studying the properties of the light spectrum and other optical issues, as well as chemistry, which Herschel encouraged Talbot to pursue.[19] One result of their informal collaboration was Talbot's invention of photogenic drawing, "forming pictures and images of natural objects by means of solar light," as Talbot described it to the editor William Jerdan.[20] Photogenic drawing, or what is often called a photogram today, is based on the principle that paper, chemically prepared with a solution of silver nitrate to become light-sensitive, will blacken when exposed to light. Any object on the paper that blocks the passage of light will leave a "shadow" of itself in the form of a white space, establishing its outline (fig. 3). On 25 January 1839, Talbot wrote to Herschel that he was about to present to the Royal Society "a new Art of Design which I discovered about five years ago, viz. the possibility of fixing upon paper the image formed by a Camera Obscura; or rather, I should say, causing it to *fix itself*."[21]

On 31 January 1839, Talbot read his paper "Photogenic Drawing: Some Account of the Art of Photogenic Drawings" to the Royal Society, introducing the results of his experiments to his colleagues.[22] But in fact examples of his photogenic drawings had already been shared with the public on 25 January, as Faraday had included them as part of his popular Friday lecture at the Royal Institution that day.[23] Subsequently, Talbot published a synopsis and then a full account of his paper in the *Athenaeum*.[24] He was clearly taking advantage of new publishing opportunities; yet, when he submitted his article to the Royal Society's *Transactions*, the council voted against accepting his article, finding that he did not sufficiently explain his process.

This turn of events sparked a series of letters between Talbot and Herschel about the announcement and publication of their respective findings. These commentaries were interwoven with further observations and advances on the chemical preparation for photography (particularly with respect to fixing the image); they also sent samples of their prints to each other.[25] In this way, they grounded their experimental discussions in successful examples, as Herschel had done in his treatise. They also couched their results in terms of national benefits, comparing their methodology favorably with that of L. J. Daguerre, whose daguerreotype was a singular image on metal, whereas the photogenic drawing could be multiplied by using the initial paper print, rendered semitranslucent, as a negative for a subsequent print in which the lights and darks would be reversed.[26]

I want to attend further to Talbot's published accounts of his photogenic process here because they are early expressions of the discourse of truth to nature. In his publications, Talbot largely eschewed the topics of chemistry, paper preparation, and the intricate steps of his experiments that he pursued with Herschel. Instead, he focused on methodology and viable applications. As mentioned earlier, Talbot argued in his first publication on photogenic drawing that

> *this remarkable phenomenon, of whatever value it may turn out in its application to the arts, will at least be accepted as a new proof of the inductive methods of modern science, which by noticing the occurrence of unusual circumstances (which accident*

*perhaps first manifests in some small degree), and by following them up with experiments, and varying the conditions of these until the true law of nature which they express is apprehended, conducts us at length to consequences altogether unexpected, remote from the usual experience, and contrary to almost universal belief. Such is the fact, that we may receive on paper the fleeting shadow, arrest it there, and in the space of a single minute fix it there so firmly as to be no more capable of change, even if thrown back into the sunbeam from which it derived its origin.*[27]

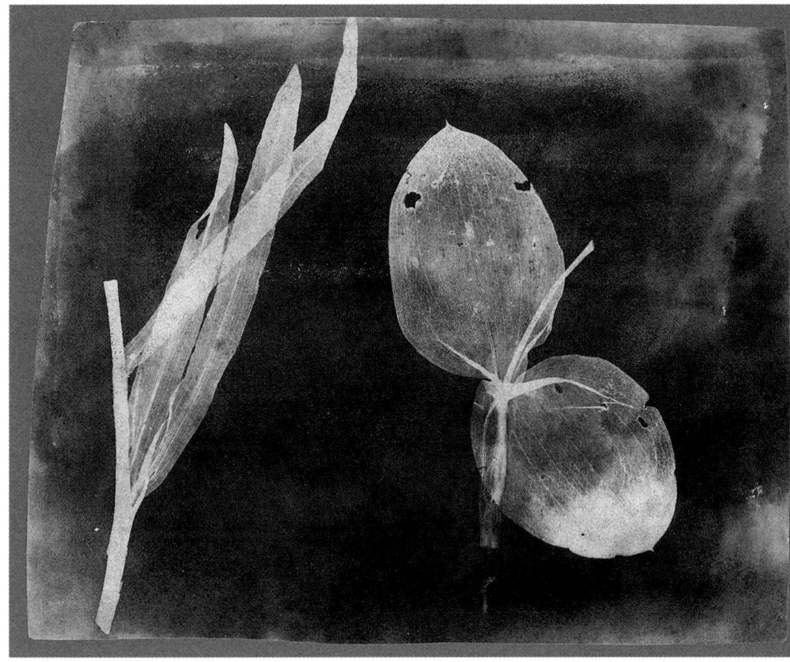

FIG. 3
William Henry Fox Talbot, *Leaves of Orchidea*, April 1839. Photogenic drawing negative, image 17.1 × 20.8 cm. The J. Paul Getty Museum, Los Angeles.

In this passage, Talbot positioned modern science, and in particular Herschel's inductive methodology, as his project's metanarrative, and photogenic drawing as the supporting evidence. Furthermore, in the subtitle of his essay—"Or the process by which Natural objects may be made to delineate themselves without the aid of the artist's pencil"—Talbot suggested that photogenic drawing was the result of nature's revealing itself to the human observer. Such assertions, as Herta Wolf observes, were in keeping with contemporary notions of chemistry that depicted "nature as active and operative."[28]

The absence of human agency was integral to the practical applications of photogenic drawing that Talbot proposed. Portraits, specifically silhouettes, executed as photogenic drawings would be far better than those drawn by hand because "the hand is liable to err from the true outline, and a very small deviation causes a notable diminution in the resemblance," whereas the portrait produced by "means of solar light" possessed greater "truth and fidelity." Botanical specimens would also be more accurately recorded: "with the utmost truth and fidelity, exhibiting even the venation of the leaves, the minute hairs that clothe the plant, &c."[29]

Over the course of 1839, Talbot continued to disseminate his process, showing a body of his work at the meeting of the British Association for the Advancement of Science held in Birmingham. The following year he exhibited a collection of his photogenic drawings at the Graphic Society, which suggests that he wanted to convince not only scientists of the value of his discovery but also artists. The Graphic Society had been established in 1833 by William Brockedon, who had also been instrumental in forming the Geographical Society. The membership of the Graphic Society was composed of professional artists, but, like the scientific societies, the society organized conversations, public discussions, and displays to which guests were invited. Scientists such as Charles Babbage and Michael Faraday regularly attended. Talbot's collection of photogenic drawings was favorably reviewed by the *Literary Gazette*, which had already demonstrated keen interest in the process:

> *The representation of objects is perfect. Various views of Lacock Abbey, the seat of Mr. Talbot; of Bowood; of trees; of old walls and buildings, with implements of husbandry; of carriages; of tables covered with breakfast things; of busts and statues; and, in short, of every matter from a botanical specimen to a fine landscape, from an ancient record to an ancient abbey, are given with a fidelity that is altogether wonderful. . . . Indeed, there is nothing in these pictures which is not at once accurate and picturesque.*[30]

This notion of accuracy was echoed by the artist and scientist Arthur Parsey, who claimed in his treatise *The Science of Vision* (1840) that the photographs of Talbot and Daguerre, produced in the camera obscura, were the first instances of "natural images," meaning that the images appeared just as they would to the eye, untainted by conventional rules of perspective as taught in art schools. Shorn of convention, these images were, according to Parsey, inherently truthful.[31] Talbot himself reinforced such ideas with an anecdote of a viewer who mistook his photogenic drawing of lace for actual lace.

### The Genealogy of Natural Images

Parsey's announcement of the first instance of "natural images" points to a larger inquiry then taking place regarding the relationship between nature and visual representation. A lightning rod for this discussion was the painter John Constable, who was deeply engaged in the question of how to represent landscape in ways faithful to nature; in 1833 he issued his print portfolio *Various Subjects of Landscape, Characteristic of English Scenery, Principally intended to display the Phenomena of the Chiar'oscuro of Nature, from Pictures Painted by John Constable, R.A., Engraved by David Lucas*. The collection, Constable explained to readers, represented "home scenery . . . taken from real places . . . the effects of light and shadow being transcripts only of such as occurred at the time of being taken."[32] Constable shared with Talbot a concern with "indexicality," that is, a mode of representation based directly upon the observed object, and with temporal equivalence, a mode of representation that signifies a single moment of time. The two men also shared a desire to bring together the practices of art and science.

Constable's series of public lectures on landscape, delivered to the Literary and Scientific Society of Hampstead, the Worcester Literary and Scientific Institution, and the Royal Institution, clearly demonstrate his commitment to marrying art and science.[33] In concluding his fourth lecture, the painter proposed that "painting is a science, and should be pursued as an inquiry into the laws of nature. Why, then, may not landscape painting be considered as a branch of natural philosophy, of which pictures are but experiments?"[34] The artist's earlier meteorological studies of the sky, accompanied by careful notation of weather conditions, exemplify this concept of painting as experiment.[35] Constable's proposition offers an intellectual framework for landscape painting equivalent to the one Talbot constructed for photogenic drawing. Both painting and photogenic drawing are experimental means of pursuing the inductive agenda of discovering the laws of nature, which Herschel believed should be the primary concern of natural philosophy, or science.

To understand how Constable arrived at this proposition, which was ideally suited to his audience at the Royal Institution, it is necessary to understand more fully the genealogical context in which his work appeared. In his fourth lecture, Constable voiced his admiration for the work of both Cozens and Girtin.[36] R. B. Beckett, who compiled Constable's lectures, believes that Constable was referring to John Cozens, but he admired both the father, Alexander Cozens, and his son, John. Alexander Cozens was a drawing master, and that experience informed his series of treatises on landscape composition. In 1759, he published his *Essay to Facilitate the Inventing of Landskips, intended for Students in the Art*, which offered a system by which to represent landscapes. Working from "blots" or "rude black sketches," the would-be artist was advised to shape these into landscape compositions that emulated the great masters.[37] Crucially, the success of a drawing was measured not by its resemblance to the natural scene but by its resemblance to other works of art.

Constable's great patron and mentor, Sir George Beaumont, studied with Alexander Cozens and in the early 1770s had the opportunity to work from Cozens's incomplete treatise *The Various Species of Landscape, &c. in Nature*. Beaumont later shared his notes from this experience with Constable. Cozens's treatise was another attempt to systematize the representation of nature, broken down into the three categories of composition, objects, and circumstances. It also explored the potential emotional resonance of particular landscape motifs; for example, "a hollow or chasm may be horribly great," giving rise to feelings of "greatness, awe, surprize [*sic*], danger."[38] This associationist methodology fits comfortably within contemporaneous aesthetic theories that generally divided landscape into broad categories, namely, the beautiful (closely related to the pastoral), the sublime, and the picturesque. Edmund Burke, in his *Philosophical Enquiry into the Origin of Our Ideas of the Sublime and Beautiful*, first published in 1757, explored how individual sensibilities and imaginations could participate in shared aesthetic responses, or what he called "taste." The trigger was the physical landscape and the perceptions or associations it engendered. For example, "Whatever is fitted in any sort to excite the ideas of pain, and danger, that is to say, what is in any sort terrible, . . . or operates in a manner analogous to terror, is a source of the *sublime*."[39]

William Gilpin, in his three essays *On Picturesque Beauty*, *On Picturesque Travel*, and *On Sketching Landscape*, clarified for readers that in pursuing the picturesque, they were seeking that "which please[s] from some quality, capable of being *illustrated in painting*." Gilpin's task was to disinter the source of that pleasure, which he argued stemmed from qualities of roughness and ruggedness, a conclusion he then tested against experience, including artistic practice. For example, he claimed that the artist sought the picturesque because his composition demanded it: "in *landscape painting* smooth objects would produce no composition at all." He added that "picturesque composition consists in uniting in one whole a variety of parts; and these parts can only be obtained from rough objects." The picturesque also entailed variety and contrast, conveyed largely by the treatment of light, shade, and color (figs. 4, 5).[40]

While Constable greatly admired his predecessors and studied their work closely, particularly with the assistance of Beaumont, a great intellectual distance

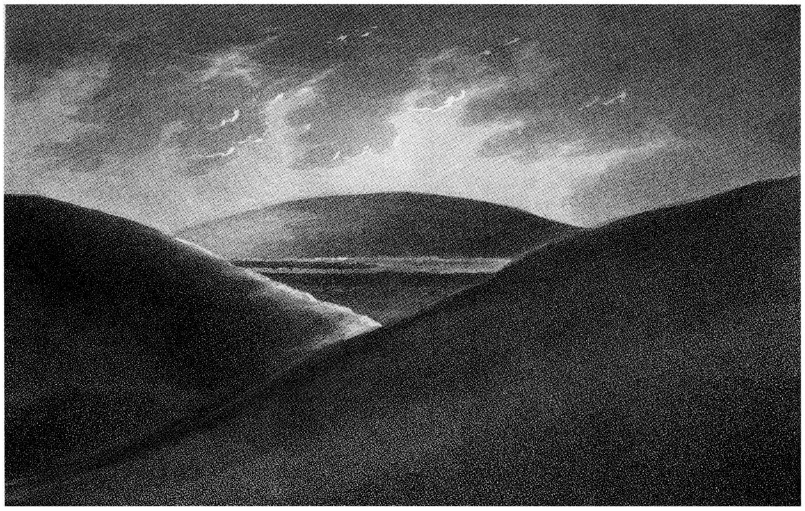

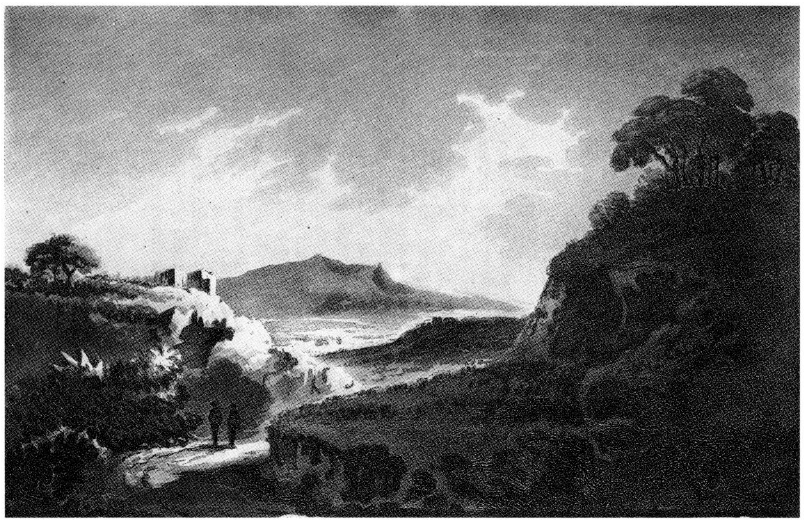

FIGS. 4, 5

"It is the intention of these two prints to illustrate how adverse the idea of *smoothness* is to the *composition* of landscape. In the second of them the *great lines* of the landscape are exactly the same as in the first; only they are *more broken.*" William Gilpin, "Explanation of the Prints," in *Three Essays: On Picturesque Beauty, On Picturesque Travel, and On Sketching Landscape*, 3rd ed. (London: T. Cadell and W. Davies, 1808), facing pp. 18, 19. The Getty Research Institute, Los Angeles (92-B24135).

nonetheless separates the productions of the two Cozens from those of Constable. The art historian Charlotte Klonk adroitly summarizes this distance and the transition "from the belief that the representation of nature must be framed within some objective ordering principle, to an acceptance of the receptivity of the individual subject as the sole point of unity for artistic activity."[41] Klonk's analysis of how and why this change took place is critically important to this study. She asserts that these changes cannot be explained solely by art's internal logic and argues that the frame of reference must be broadened to include contemporaneous understandings of nature, which were largely shaped by natural philosophy and science. More precisely, Klonk is interested in "phenomenalism," a term she deploys "to denote the complex of attitudes concerning the relationship between mind and nature. . . . According to phenomenalism, artists or scientists must confine themselves strictly to what is given to the perceiving subject, without making any prior suppositions concerning the underlying mechanisms by which what is observed is connected. For phenomenalism, generalization is permissible only on the basis of the painstaking accumulation of particular instances."[42]

Phenomenalism emerged, Klonk explains, at the end of the eighteenth century in philosophical and scientific circles and gained traction in visual culture in the early nineteenth century. Phenomenalism was critical, for example, to debates in the Geological Society; representations of "stratigraphic appearance" were considered important contributions to the accumulation of data required to lift the discipline out of the quagmire of competing theories of causation, i.e., volcanoes, floods, the deity, etc. Yet, as Klonk observes, "this attempt to unite the phenomenalist conception of the primacy of observation with the scientific ideal of objectivity is inherently unstable." Observation is rooted in the discerning individual, opening the door to "a more subjectivist mode of perception," as in the case of J. M. W. Turner's *Staffa, Fingal's Cave* (1832), in which the artist obscures the geological formation in swaths of mist and focuses visual interest upon the smoke-belching steamboat.[43]

"Strict phenomenalism was only a transitional phase" for the artists Klonk discusses. She argues, for example, that Constable's work was the most phenomenal around 1815, when he strove for a nonhierarchical representation of the landscape that gave equal attention to all observed phenomena. But such an achievement rested upon the perceiving subject, who became an increasingly registered presence in Constable's art.[44]

Positioning Constable as a subjective artist must be balanced with understanding his contributions to the systematization of landscape representation. His formulation of guidelines and advice for other artists indicates that Constable relied on more than just a personal response to nature. In his *Various Subjects of Landscape*, for example, he considered how to impart "Expression, Taste, and Sentiment."[45] He explained to readers that the artist must first define his desired impact on the viewer, and then select the appropriate components of the natural scene to generate that outcome. Even paintings tied closely to Constable's period of intense concern with naturalism, such as *The Haywain* (1821), reveal his engagement with referents outside the scene, such as past masters of landscape painting and georgic poetry; his later work demonstrates an engagement with the picturesque.[46]

J. M. W. Turner, who frequently competed with Constable for attention on the walls of the Royal Academy exhibitions, likewise had a scientifically informed practice that partook of phenomenalism and yet was concerned with both systematizing and theorizing landscape representations and provoking emotional responses to scenes of nature. Turner first emerged as a topographical artist and then advanced his training at Dr. Monro's Academy, Thomas Munro's informal gatherings for artists. Shortly thereafter, in 1807, Turner was appointed professor of perspective at the Royal Academy of Arts (a position he held until 1837), and it is worth remembering that geometry and mathematics were regarded as core disciplines within the sciences at this time.[47] This experience contributed to Turner's *Liber Studiorum* (first issued in 1807), a collection of landscape prints designed to introduce the reader to different typologies of landscape, such as pastoral or marine.[48]

Later in his career, Turner circulated within a community of scientists: he met the chemist Humphry Davy (professor at the Royal Institution, president of the Royal Society, and founding member of the Geological Society) in 1819 and Brewster in 1834; he knew Faraday, and he was friends with Mary Somerville and her husband.[49] As James Hamilton explains, Somerville's experiments in magnetism and those of Faraday were of great interest to Turner and informed his painting *Snow Storm—Steam-boat off a Harbour's Mouth Making Signals in Shallow Water, and Going by the Lead. The Author was in this Storm on the Night the Ariel left Harwich* (1842), which shows an iron ship subjected to the forces of the energized water.[50] This painting also evokes the sublime. It, like *Staffa, Fingal's Cave*, is also remarkably indistinct in the portrayal of forms; this technique, the art historian Kay Dian Kriz explains, drew attention to the imaginative power, even genius, of the artist. Kriz proposes that Turner's indistinctness was so troubling for his contemporaries because it posits "an artistic subject which is so autonomous that it has no longer the need or capacity for social engagement." The significance of Turner's paintings, according to this paradigm of genius, was measured by their ability to provoke strong feelings, fostered by his bravura effects.[51]

The pronounced presence of the subjective artist in representations of nature drew to a close the convergence between science and art fostered by phenomenalism, according to Klonk. Indeed, she argues that this moment of convergence was exceptional.[52] But that commingling of art and science, as Talbot's photogenic drawings and accompanying rhetoric make clear, persisted well into the nineteenth century and, as this study will reveal, was episodically debated and reconfigured.

Certainly, photogenic drawing and phenomenalist works of art demonstrate a shared engagement with the close observation of nature expressed in an evenhanded and dispassionate manner. Their makers were also driven by similar inquiries into how mechanical devices might aid in rendering nature. Both Talbot and the artists John and Cornelius Varley, eager participants in phenomenalism, utilized the camera lucida, an instrument for transmitting, on a diminished scale, a scene of nature onto a drawing surface. Talbot's frustrations with the instrument led him to seek an alternative means by which to fix natural scenery upon paper.

But unlike the phenomenalists, Talbot was untroubled by the figure of the perceiving artist; he suppressed the individual. Nature, in his view, rendered the images without the aid of the artist.[53] His confidence in the process itself as a guarantee of truth echoes Babbage's arguments regarding the value of scientific instruments, such as the telescope or the barometer. Babbage argued that they place "Nature in such circumstances, that she is forced to record her minutest variations on so magnified a scale, that an observer, possessing ordinary faculties, shall find them legibly written."[54] Experimental methodology—repeatable and thus verifiable—was also a guarantee against the undue influence of, as Herschel put it, individual "prejudices."[55] Recognizing the danger of the perceiving observer's or artist's subjectivity, the scientific community determined a means of patrolling and controlling that observer/artist. And Talbot arrived at a visual method that could participate in these efforts.[56]

## Debates About Inductive Methodology

But the problem of the perceiving subject persisted within Talbot's intellectual network, emerging quite strongly in Whewell's *Philosophy of the Inductive Sciences* (1840), which built upon his earlier *History of the Inductive Sciences* (1837) (dedicated to Herschel). Talbot's Trinity colleague Thomas Worsley reported eagerly to Talbot on the publication of Whewell's *Philosophy*. Like Herschel, Whewell acknowledged an intellectual debt to Francis Bacon, and in particular to the methodology of induction.[57] But, unlike Herschel, Whewell wrote from the point of view of a philosopher rather than a hands-on experimenter.[58] Whewell argued strenuously on behalf of the unity of the sciences and challenged the belief that the strategy of induction would lead to ultimate truth.

Whewell questioned the efficacy of the method of working strictly from an accumulation of observed facts to general laws.[59] He charged that the pursuit of facts was always informed by an idea that existed a priori. Knowledge was advanced, the professor argued, by two intertwined processes: "by which *conceptions* are *made more clear* in themselves, or by which the conceptions more strictly *bind together the facts*."[60] One could not recognize the true from the false, Whewell insisted in a rebuke to Bacon, without a preexisting theory or hypothesis: "The necessity of a *conception* which must be furnished by the mind in order to bind together the facts, could hardly have escaped the eye of Bacon, if he had cultivated more carefully the ideal side of his own philosophy."[61]

Like his intellectual predecessors, Whewell recognized the value of observation, as he emphasized in the second edition of his *Philosophy of the Inductive Sciences*: "knowledge is obtained from the observation of the external world. Nature is the Book, and Man is the *Interpreter*. The facts of the external world are marks, in which man discovers a meaning, and so reads them. Man is the Interpreter of Nature, and Science is the right Interpretation." But he also acknowledged how difficult it was to rely exclusively on sight for knowledge: "Even in the case in which our perceptions appear to be most direct, and least to involve any interpretations of our own,—in the simple process of seeing,—who does not know how much we, by an act of mind, add to that which our senses receive?" Whewell emphasized how the observer often failed to grasp the unconscious acts of interpretation exercised by the mind in viewing visual representations:

> *Does any one fancy that he sees a solid cube? It is easy to show that the solidity of the figure, the relative position of its faces and edges, are inferences of the spectator; no more conveyed to his conviction by the eye alone, than they would be if he were looking at a painted representation of a cube. The scene of nature is a picture without depth of substance, no less than the scene of art; and in the one case as in the other, it is the mind which, by an act of its own, discovers that colour and shape denote distance and solidity. Most men are unconscious of this perpetual habit of reading the language of the external world, and translating as they read. The draughtsman, indeed, is compelled, for his purposes, to return back in thought from the solid bodies which he has inferred, to the shapes of surface which he really sees. He knows that there is a mask of theory over the whole face of nature, if it be* theory *to infer more than we* see. *But other men, unaware of this masquerade, hold it to be a fact that they see cubes and spheres, spacious apartments and winding avenues. And these things are facts to them, because they are unconscious of the mental operation by which they have penetrated nature's disguise.*[62]

But it is important to recall that Whewell's conception of the perceiving subject existed within the framework of Christianity. Whewell had demonstrated his commitment to natural theology in his *Bridgewater Treatises* (1839), and he reiterated this point in the *Philosophy*, stating that "the Ideas which we necessarily employ in the contemplating of the world around us, afford us the only natural means of forming any conception of the Creator and Governor of the Universe."[63] In other words, an ultimate authority governed the universe, although that authority might be viewed or understood only dimly via the workings of nature.

Whewell's first edition of *The Philosophy of the Inductive Sciences* provoked a lengthy, largely respectful review by Herschel in the *Quarterly Review*. Herschel did register his dissatisfaction, however, with Whewell's emphasis on "abstract truth." Herschel insisted that ideas could only derive from human experience and that "*consistency of suggestion* by many particular events and objects . . . leads us to make any abstract propositions at all." Experience, in other words, trained and constrained the mind, and repetition achieved by a community of practitioners verified results.[64] Thus Herschel diminished the potential destabilizing influence of individual subjectivity.

David Brewster, writing in the *Edinburgh Review*, was far harsher in his criticism of Whewell's text, which he regarded as a "*Philosophy* of scholastic *Metaphysics* rather than of physical *Science*."[65] While Brewster, like Whewell, believed in the power of imagination, he nonetheless objected that Whewell's methodology was not sufficiently engaged with the actual encounter with phenomena.[66]

> *Whatever the methods may be by which truth has been sought, whether by the imagination or the judgment, whether by the use of hypotheses, or the torture of facts, we are persuaded that there is but one way by which it can be found— . . . by a lofty estimate of its value; by an abiding conviction that it is within our reach; by patience and single-mindedness in its pursuit; by an indomitable perseverance amid obstructions and defeats . . . and, by the application of all means and instruments at our disposal, to lay hold of the prize, insulate the gem from the matrix, and exhibit the diamond in its native purity and lustre.*[67]

In short, laborious, systematic inquiry was the best insurance for valid conclusions.

Arguably, the most pointed critique of Whewell came from John Stuart Mill, who was finishing his *System of Logic, Ratiocinative and Inductive* (1843) when Whewell's *Philosophy* appeared and offered Mill a foil. As a philosopher, Mill is perhaps better known for his writings on political theory and economy; it is therefore worth remembering that science, and induction in particular, was promoted in this period as a method applicable to social concerns as well as to the physical sciences. Mill argued that Whewell's philosophy was, in essence, noninductionist. As Laura Snyder succinctly states, Mill wished to foreclose on the role of the mind in induction; "he thought that leaving room for any such contribution was to open the door to intuitionism."[68] In his introductory chapter, for example, Mill asserted that what observers often attributed to intuition, as in the perception of objects in the distance, was actually inference based on experience. Such inferences, which led to particular facts or general truths, composed the "greatest portion of our knowledge," and Mill proposed that logic provided the "principles and rules" by which these inferences or evidence could be judged.[69] In short, as Snyder summarizes, Mill "endorsed an ultra-empiricist epistemology . . . [that] led to a scientific methodology that denied the possibility of inductive discovery of theories postulating unobservable entities or events." What was observed, in short, was a valid pathway to truth. Significantly, Mill's book was far more successful than Whewell's, becoming, as Snyder notes, the "standard textbook on logic at Oxford until the end of the nineteenth century."[70] Mill had helped lay to rest, albeit temporarily, concerns about the perceiving subject.

## The Invention of Photography in the Field of Science: The Calotype

Talbot's further refinements and dissemination of the photographic processes benefited from these debates about scientific methodology. Continued experimentation, spurred on by Herschel, led Talbot to refine the process of developing a negative—in which lights and darks are reversed—that could be used to print a positive that mirrored lights and darks as found in nature.[71] His new process, which he called calotype, coupled with the use of improved paper and chemical formula, was much faster in recording the scene. Talbot published an account of his calotype in the *Literary Gazette*, positioning it as a significant advance over photogenic drawing in representing external nature accurately. For example, his new process allowed for "increased sharpness and distinctness in the outlines of objects."[72] Talbot registered a patent for his new process, a decision that initially meant that he did not publish technical details, considering these proprietary information.[73] He was nonetheless eager to make the larger scientific community aware of his advances, and on 10 June 1841 he delivered a paper on the calotype process before the Royal Society. He also shared examples of his calotypes with David Brewster, who presented them to the St. Andrews Literary and Philosophical Society in July 1841; at least two of these calotypes are believed to be preserved in the Brewster Album.[74]

Talbot offered a more public demonstration of the medium in his book *The Pencil of Nature*, issued in six fascicles (1844–46). Produced at his newly established business at Reading, this work required intense labor by Talbot and his assistant, Nicholas Hennemen, as each print had to be produced by hand. This limited-edition collection showcased the calotype process as well as its possible applications, such as recording collections, depicting sculpture, facsimile reproduction of manuscript pages, copying prints and drawings, and recording antiquities.

This landmark publication, the first book to be illustrated with photographs, has received considerable scholarly attention; here, I want to focus on how Talbot fashioned the perceiving subject and the rhetoric of truth to nature in his text.[75] In the chapter "Brief Historical Sketch of the Invention of the Art," Talbot recounts how his experiments, readings, and discussions with scientists like Humphry Davy led to a "labyrinth of facts" through which he had to wind in order to find the underlying laws. The photograph was, in essence, a means of discovering those laws, an inductive tool of investigation as imagined by Herschel. But Talbot suggested at the same time that his investigations were fed by his theory regarding the sensitivity of silver nitrate to light. This "theory," he wrote, "impelled me to explore a path so deeply hidden among nature's secrets"; it was a mode of intellectual inquiry more analogous to Whewell's than to Herschel's.[76]

But despite the dominance of the first-person narrator in the first section of *The Pencil of Nature*, Talbot's text again suppressed human agency in favor of that of nature. In his introductory remarks, for example, Talbot described his calotypes as "impressed by Nature's hand; and what they want as yet of delicacy and finish of execution arises chiefly from our want of sufficient knowledge of her laws." He insisted that the illustrative plates had been "formed or depicted by optical and chemical means alone, and without the aid of anyone acquainted with the art of drawing." Later in the narrative, in connection to a photograph of chinaware displayed in tidily arranged rows, he celebrated the camera's ability to capture the whole scene, without imposing a hierarchy. "The Camera depicts them all at once. It may be said to make a picture of whatever *it sees*. The object glass is the *eye* of the instrument—the sensitive paper may be compared to the *retina*."[77] Here, the instrument takes over the act of perception typically performed by the individual.

The photograph could not only perform the work of an individual, but it could do it better, with more strict discipline and close observation. In connection to his photograph *A Scene in a Library*, showing books lined on two shelves, Talbot argued that "the eye of the camera would see plainly where the human eye would find nothing but darkness." The photograph *The Haystack* prompted Talbot's claim that "one advantage of the discovery of the Photographic Art will be, that it will enable us to introduce into our pictures a multitude of minute details which add to the truth and reality of the representation, but which no artist would take the trouble to copy faithfully from nature."[78] Talbot added, in a dig that responded to the rhetoric that indistinctness was a marker of artistic genius, that the average artist, "contenting himself with a general effect, . . . would probably deem it beneath his genius to copy every accident of light and shade" (fig. 6).[79] Talbot projected an alternative, moralistic vision of a producer of images dedicated to the pursuit of truth to nature and rigorously following the methods of inductive, experimental philosophy.

The value of photography, in Talbot's narrative, rests upon its origins in this inductive methodology, its suppression of the human agent in favor of instrumentation, and its ability to capture minute details. The critical reviews of Talbot's book repeated this rhetoric: the *Morning Herald*, for example, argued that the photographs were "not simple imitative transcripts, but *book pictures* executed by nature herself: and it is almost beyond belief that they emanate from mere chemical sources and exist in total independence of either pencil or burin."[80]

It is important, at this juncture, to draw attention to scholar Jennifer Green-Lewis's point that photography's supposedly indexical relationship to the real was the result of "an ostensibly consensual mode of representation . . . a shared agreement, a complicity."[81] When Talbot's generation read the photograph—a two-dimensional patterning of shades of color, from white to black and brown—as a transcript of nature, they suspended explicit acknowledgment of the monochrome nature of the photograph, its diminished scale when compared to the originating scene, its fixed point of view, and its defined frame.[82] But the rhetoric of Talbot and his reviewers helped to forge the consensus that Green-Lewis describes.

As already suggested, one of the communities in which Talbot sought to exert influence was the artistic, despite his rhetorical insistence that photogenic drawing and the calotype offered a way to dispense with the artist. When he announced his calotype process in the *Literary Gazette*, Talbot anticipated the charge that photogenic drawing might "prove injurious to art, as substituting mere mechanical labour in lieu of talent and experience." Photography, he insisted, provided "ample room for the exercise of skill and judgment"; for example, different effects could be achieved by controlling the exposure of the paper to light, or manipulating the fixing process. If, as a result, the artist became "a chemist and an optician," Talbot felt that both art and science would be improved through such an alliance.[83]

While Talbot became increasingly preoccupied with other interests, particularly Assyriology, in the decades immediately following the publication of *The Pencil of Nature*, the field of photography expanded rapidly. In 1853, a professional society of photographers formed in London. The community of photographers began to be served by its own literature, such as Robert Hunt's treatises and manuals, the successive editions of which map the contours of the growth of knowledge about photography.[84] Photography journals emerged, allowing photographers to share knowledge and stage debates.

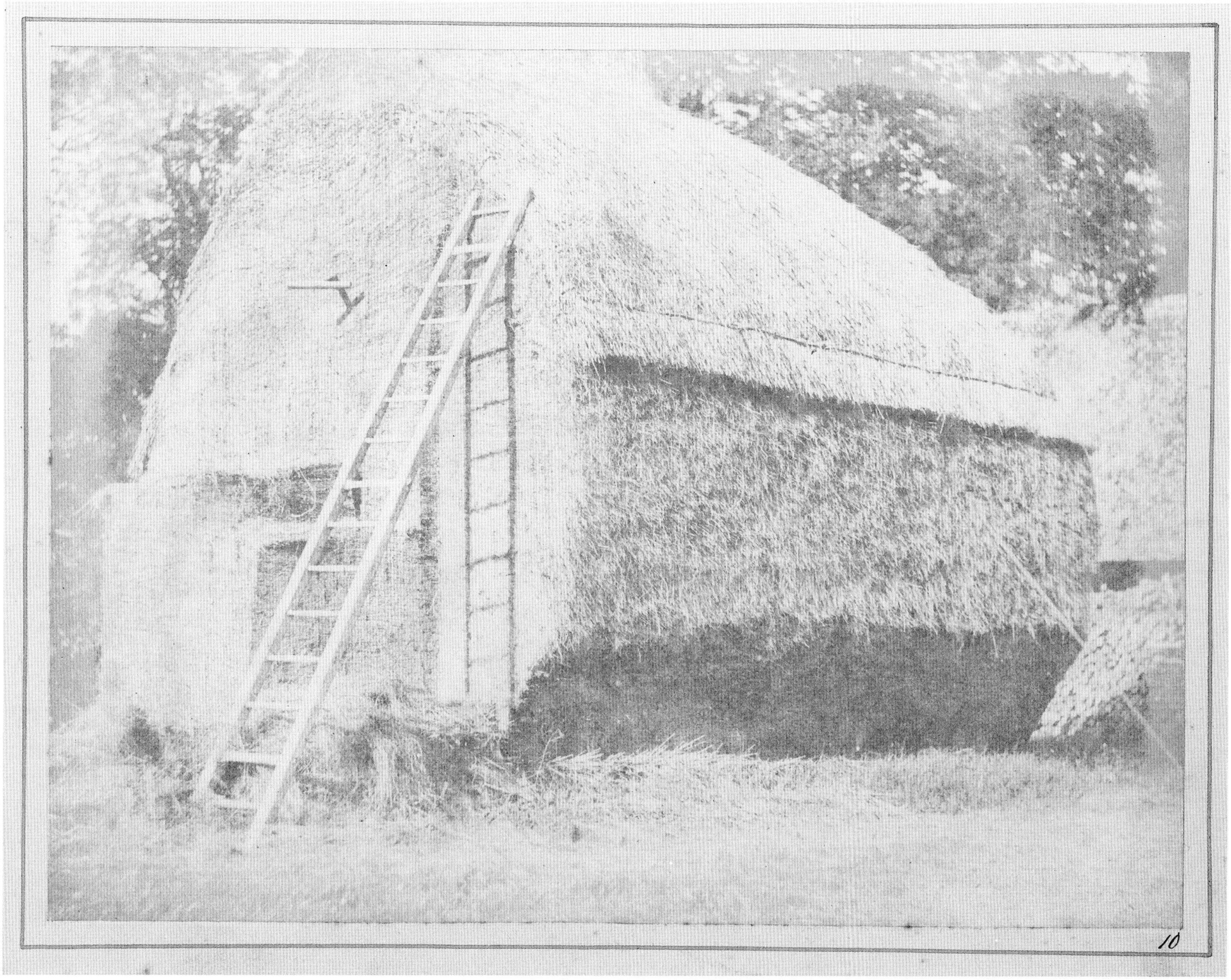


FIG. 6
William Henry Fox Talbot, *The Haystack*, 1844. Salted paper print, 16 × 20.7 cm. In Talbot, *The Pencil of Nature*, plate X. Gernsheim Collection, Harry Ransom Center, The University of Texas at Austin.

These journals show clearly that the relationship between photographic and artistic communities of practice was a major point of concern. I will mention only one example here, but this textual archive reveals the high degree of discursive interaction between these communities. In 1856, an anonymous article titled "Photography and the Elder Fine Arts" in the *Journal of the Photographic Society of London* proposed that photography was not only an art—"a matter-of-fact art, an art that cannot idealize or flatter"—but an art that could improve painting: "it is calculated to exercise a useful corrective influence upon the contemporaneous schools of painting." The author continued, "It shows what Nature is and was before pre-Raphaelites magnified her minutiae, because they knew no better how to express them, or their imitators gave to her minor details a prominency which she only imparted to her important features; and by this it is to be hoped it will correct a foolish vagary."[85] This commentary reflects an uneasy relationship between Pre-Raphaelite painting and photography. Furthermore, it undermines claims that Pre-Raphaelite painting was a response to the influence of photography. The author suggests that the Pre-Raphaelites ignored the visual vocabulary that photography offered them. The relationship between painting and photography, as I argue in this study, was configured not by influence but by shared interests and intersecting discourses shaped by the rise of scientific knowledge.

### Pre-Raphaelite Painting and Truth to Nature

The structure of my narrative, which begins with photography and then turns to painting, carries with it the undesired implication that photography influenced painting. This oft-repeated trope informs the poet William Bell Scott's account of visiting the Pre-Raphaelite painter William Holman Hunt's studio around Christmas 1847:

> *Every movement has its genesis, as every flower its seed; the seed of the flower of Pre-Raphaelism* [sic] *was photography. The seriousness and honesty of motive, the unerring fatalism of the sun's action, as well as the perfection of the impression on the eye, was what it aspired to. History, genre, mediaevalism, or any poetry or literality, were allowable as subject, but the execution was to be like the binocular representations of leaves that the stereoscope was then beginning to show.*[86]

I contend that the question is not whether, or when, photography influenced painting but how both painting and photography participated in the processes of knowledge formation shared with science. William Bell Scott's commentary, when read closely, indicates that photography and painting shared an interest in human perception, an interest also shared with scientists.[87] The self-discipline and probity admired in photography and inductive philosophy were also adopted by the Pre-Raphaelite painters. My argument is that the Pre-Raphaelite painters pursued their distinctive painting practice not because of the presence of the camera but because of the larger cultural forces that likewise stimulated Talbot and his colleagues to pursue mechanical means by which to achieve truth to nature.

The Pre-Raphaelite group formed in the late 1840s, based on friendships that had begun, for several of the artists, in art school. John Everett Millais and William Holman Hunt met at the Royal Academy schools. Millais, born in 1829, had demonstrated artistic talent at a young age, and his family's wealth allowed him to pursue a career in art. After attending Sass's Academy, a private school that helped train students for the Royal Academy entrance exam, he gained entry to the academy schools. Hunt's family background was quite different; his father had been a warehouse manager, and Hunt became an office clerk in 1839 at age twelve. He finally gained the family's permission to study art, and upon his third attempt, in 1844, he passed the entrance exam for the academy schools. There, Millais and Hunt befriended Dante Gabriel Rossetti, who entered the Royal Academy schools as a probationary student in the summer of 1844 after a spell at Sass's Academy. Rossetti's Italian immigrant father was a professor of Italian at King's College, and his keen interest in poetry and great works of literature was shared with his son.

Rossetti was drawn into the friendship between Millais and Hunt when he responded positively to Hunt's submission to the Royal Academy's annual exhibition of 1848, *The Flight of Madeline and Porphyro During the Drunkenness Attending the Revelry* (now known as *The Eve of St. Agnes*, Guildhall Art Gallery). Rossetti introduced them to fellow academy student Thomas Woolner, who focused on sculpture. The circle widened to include two additional academy students, James Collinson, the son of a bookseller and stationer, and F. G. Stephens, whose parents were the master and mistress of the Strand Union Workhouse. Together the group formed a sketching club, the Cyclographic Society, which became the basis for the Pre-Raphaelite Brotherhood, officially founded in September 1848. They were joined by Dante Gabriel's brother, William Michael Rossetti, who had accepted a post as excise clerk while pursuing a career as a writer and critic. In 1850, William became the art critic for the *Spectator*, a position he held for eight years; he also contributed reviews to such journals as the *London Review*, *Fraser's*, and occasionally more specialist art periodicals such as the *Fine Arts Quarterly*. He also helped to disseminate the ideas of the Pre-Raphaelites in the United States via his contributions to the American journal the *Crayon*.[88]

The formation of the Brotherhood fostered a like-minded community. Harking back to the founding days, William Holman Hunt recalled to the French critic Ernest Chesneau:

> *We in a sort of youthful bravado adopted the letters P.R.B. as our insignia, we accepted the pre-Raphaelite designation because in arguing against the conventionalities of current art we were told that by our principles we repudiated Raffael's teaching. We agreed perhaps only in hatred of Academic traditions and in the need of referring every rule of Art to its use in rendering Nature. High finish was in no way a necessary part of our doctrine altho' by the way of exhibiting love for the features of Nature we had pleasure—and we thought to find instruction in lingering over their parts.*[89]

While Hunt's desire to mythicize the founding of the group is now well recognized, and while the Pre-Raphaelite opposition to the Royal Academy has been too often overemphasized, it is clear that the group of young men did not want simply to accept the conventions of art making associated with the academy schools, which stressed acquiring skills and knowledge through drawing from the antique.[90] Turning to the art before Raphael, particularly fourteenth- and fifteenth-century Italian frescos, the artists fashioned an alternative genealogy for their practice. They complemented their

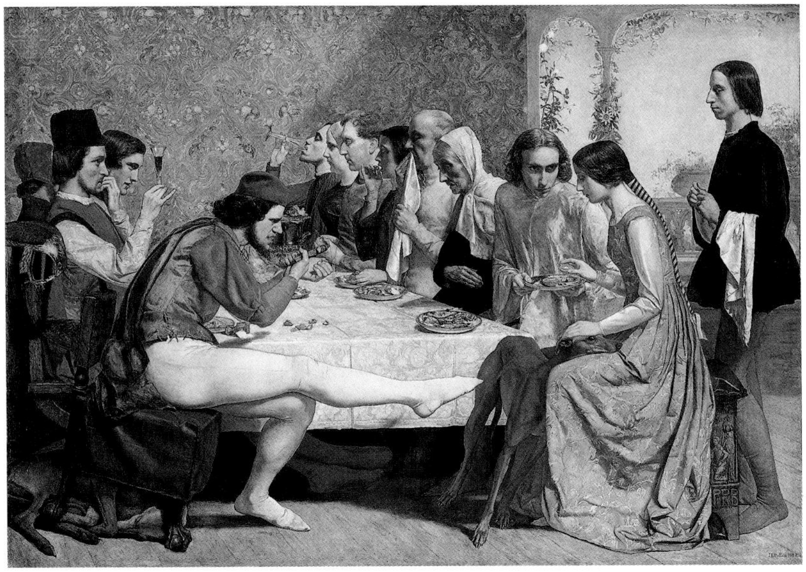

FIG. 7
John Everett Millais (1829–1896), *Isabella*, 1848–49. Oil on canvas, 103 × 142.8 cm. Purchased by the Walker Art Gallery in 1884. Courtesy of National Museums Liverpool (Walker Art Gallery).

internal discussions with a public campaign of displaying their work in important exhibitions: Millais's *Isabella* and Hunt's *Rienzi Vowing to Obtain Justice for the Death of His Young Brother, Slain in a Skirmish Between the Colonna and the Orsini Factions* were both shown at the Royal Academy in 1849, and Rossetti sent *The Girlhood of the Virgin Mary* to the Free Exhibition that year. All three works received considerable critical attention.[91] Millais's *Isabella* (1848–49) clearly manifests the Pre-Raphaelites' desire to dispense with previous painterly conventions in its rejection of linear or Cartesian perspective (fig. 7).[92] This disavowal is admittedly incomplete, as Millais retains many markers of recession into space, such as the overlapping of forms and diminution; nonetheless, the absence of converging orthogonals and the use of hard outlines for forms, even as they recede in space, were sufficient to disrupt prevailing modes of representation.

Press notices for Pre-Raphaelite paintings exhibited in the years immediately following became increasingly bitter and antagonistic with respect to the artists' choice of subject matter and manner of painting, as well as their nomenclature. This was particularly true of Millais's Royal Academy submission of 1850, *Christ in the House of His Parents*, which imagined Christ as a young child in his father's carpentry shop. William Michael Rossetti and, eventually, Ruskin came to the defense of the artists.

## John Ruskin, Pre-Raphaelitism, and Inductive Methodology

John Ruskin by this date was a well-established art critic; in 1843, he published the first volume of *Modern Painters*, an exegesis on the work of J. M. W. Turner to which I shall return shortly. Ruskin's father, a sherry and wine importer, had encouraged his son's education and travel in Britain and Europe; his mother's evangelical Christianity also had a strong impact on the young Ruskin. One of Ruskin's first intellectual interests was natural history, and among his earliest publications was a series of articles for *Loudon's Magazine of Natural History* (1834) delineating the physical causes of visual phenomena he had observed in nature, such as the colors of the water of the Rhine and the strata of Mont Blanc in Switzerland.[93] He maintained a lifelong engagement with geology and attended meetings of the Geological Society as a young man. In 1847, he served as secretary of the Geological Section on the occasion of the Oxford meeting of the British Association for the Advancement of Science.[94] Ruskin later reminisced, reflecting on his abiding fascination with botany and geology, "that my natural disposition [f]or these sciences would certainly long ago have made me a leading member of the British Association for the Advancement of Science; or—who knows?—even raised me to the position which it was always the summit of my earthly ambition to attain, that of President of the Geological Society."[95]

As a young man, Ruskin added meteorology to his repertoire of intellectual interests, writing a study for the *Transactions of the Meteorological Society* in 1839 in which he argued that if the discipline were to advance, it could not be practiced in a solitary fashion but required many observers "contributing to the knowledge, and aiding the efforts, which will be capable of solving the most deeply hidden problems of Nature, penetrating into the most occult causes, and reducing to principle and order the vast multitude of beautiful and wonderful phenomena, by which the wisdom and benevolence of the Supreme Deity regulates."[96] Such passages demonstrate Ruskin's adherence to natural theology and his commitment to the inductive processes of observation, which would lead to the recognition of the laws and principles of nature.

In 1836, Ruskin took up formal studies at Oxford University. There, he met the geologist William Buckland, who admired his landscape views for their attention to geological features.[97] Through Buckland, Ruskin met other leading scientists, including Herschel.[98] After a period of ill health and travel to Europe with his parents, he completed his master's degree in 1843. After another stint of extensive travel, he completed the second volume of *Modern Painters* (1846), followed by *The Seven Lamps of Architecture* (1849), based on his study of Gothic architecture in Britain and Europe, and the three-volume *Stones of Venice* (1851–53), on Venetian architecture. *The Seven Lamps of Architecture* drew the attention of Whewell, who invited Ruskin to visit him in Cambridge, and together the two visited Ely Cathedral.[99]

In 1851, Ruskin stepped forward to defend the Pre-Raphaelites, triggered by critics' negative comments on the Pre-Raphaelite works then on view at the Royal Academy exhibition, including Millais's *The Woodman's Daughter* (1850–51), *The Return of the Dove to the Ark* (1851), and *Mariana* (1850–51), Hunt's *Valentine Rescuing Sylvia from Proteus* (1850–51), and Collins's *Convent Thoughts* (1850–51). Ruskin's letter to the *Times*, which appeared on the same page as the announcement of the opening of the Museum of Practical Geology in London, defined new terms of visual objectivity by which art should be judged.

Noting that he had no prior acquaintance with Millais or Hunt, the two artists he discussed most extensively in this letter, Ruskin explained to readers that the Pre-Raphaelite artists "will draw either what they see, or what they suppose might have been the actual facts of the scene they desire to represent, irrespective of any conventional rules of picture-making." He called attention to the botanical accuracy of the plant material represented and the "truth, power, and finish" of the drapery depicted by Millais and Hunt, and he noted only one error in perspective, in Millais's *Mariana*—"that the top of the green curtain in the distant window has too low a vanishing point." The ultimate value of their paintings, he insisted, lay in "their fidelity to a certain order of truth (labour and fidelity which are altogether indisputable)."[100]

Here, Ruskin has transferred processes of inquiry associated with the natural sciences to art. He notes that the Pre-Raphaelites observed nature closely, and laboriously represented their findings on canvas, not manipulating them to fit codified principles of art making. Even for subjects based on literature, they created circumstances in which they could paint directly from the scene. For example, for *The Woodman's Daughter*, based on Coventry Patmore's poem, Millais temporarily moved from London to Oxford, where he found a suitable wood for the background scene. Collins joined Millais there and based the garden shown in *Convent Thoughts* on a garden cultivated by Thomas Combe behind his house at Clarendon Press. Combe's garden was also used for the view through the window before which Mariana stitches her embroidery; in turn, the seamstress models her threaded leaves upon those visible through the glass panes. Once the backgrounds were completed from life, the artists added the human figures, a process typically undertaken in the studio, again painted from life. Millais, for example, was concerned that Maude, the woodman's daughter, be suitably dressed for his rendering and wrote to Mrs. Combe asking if she could help him procure a used pinafore and an old pair of walking boots that he had observed a young girl in the neighborhood wearing.[101]

The links between Pre-Raphaelite artists and scientists were largely mediated, rather than participatory, with the exception of John Brett, who became a member of the Astronomical Society. But the artists associated with Pre-Raphaelitism traveled in the same circles as many scientists. The physicist John Tyndall, for example, extended an invitation to Millais to attend his lecture on rainbows.[102] Both Millais and Hunt knew Thomas Combe, the superintendent of Oxford University Press, who was responsible for publishing scientific treatises issued by the Ashmolean Society.[103] Combe was the publisher of Reverend Baden Powell's *Essays on the Spirit of the Inductive Philosophy* (1855), which urged long training, deep experience, and accurate observation in order to reach secure conclusions.[104]

The artists' own writings offer intriguing evidence of their engagement in science. Dante Gabriel Rossetti, although avowedly uninterested in recent research into the material world, nonetheless "proclaimed that the radical want in modern art was a stricter study of Nature." Hunt, in his autobiography, insisted that the Pre-Raphaelites marked a new path in art by drawing upon "nature and scientific knowledge." "The additional knowledge and longings of the modern mind," Hunt added, required "more exact truth," which was incumbent upon artists to supply.[105] Such comments give rise to the question that informs my discussion here: Why and how did these artists give such authority to science?

One answer lies in the increasing visibility of science, with the formation of professional societies and the emergence of popular literature on the topic; another answer lies in scientists' emphatic claims that their methods, particularly induction, could be extended from the study of the natural world to other aspects of cultural production. Most critical for my argument is the interest in perception and the natural world shared by artists and scientists. Crucial for the interaction and intersections between these two communities were figures such as Ruskin, who moved between the artistic and scientific communities and was interested in reforming artistic practice. Ruskin found a model in scientific practice, particularly induction, which was currently being argued with a rigor and intensity that he clearly admired. Like his scientific peers, Ruskin was committed to the pursuit of truth and indeed believed that this was the artist's and the critic's highest calling.[106]

Truth was the leading leitmotif of the first volume of Ruskin's *Modern Painters*, which became a catechism of sorts for the Pre-Raphaelites.[107] Chapter headings include "Of Ideas of Truth in Their Connection with Those of Beauty and Relation," "That the Truth of Nature is Not to Be Discerned by the Uneducated Senses," "Of Truth of Space:—First, as Dependent on the Focus of the Eye," and "Of Truth of Clouds:—First, of the Region of Cirrus." Like treatises on natural philosophy written by his contemporaries, Ruskin provided principles to guide the work of his colleagues, critiqued the work of previous theorists, and supported his assertions with "demonstrations," or artistic examples. Turner was his leading exemplar.

In many ways, Ruskin was battling wits with Sir Joshua Reynolds, the first president of the Royal Academy and the author of *Discourses on Art* (based on lectures delivered between 1769 and 1790). While this debate can be handily summarized as between realism and idealism, the specific terms of Ruskin's disagreement with Reynolds reinforce the critic's links to the practices of inductive science. Ruskin charged that the aphorism "general truths are more important than particular ones," associated with Reynolds, was "altogether false." Rather than general truths and broad effects of tone, color, space, and light, Ruskin preferred distinctive truths, which distinguish, characterize, and separate each object or phenomenon from another. Therefore, Ruskin explained, "real truthfulness of the painter is in proportion to the number and variety of the facts he has so illustrated." Ruskin's pictorial logic is firmly embedded within the framework of inductive philosophy. Indeed, Ruskin insisted that the artist select those facts that realize "a general principle," that is, nature's laws.[108]

Ruskin did not wish to dispense with general effect altogether; what he deemed a "complete picture" had "both general wholeness and effect of nature, and the inexhaustible perfection of nature's details." However, given that Reynolds and his generation had so forcefully emphasized breadth of effect, it is no wonder that Ruskin's contemporaries commented so extensively on attention to nature's details. Moreover, Ruskin invested the pursuit of the detail with moral value, seeing detail as indicative of "the inestimable beauty which exists in the slightest and least of God's works."[109] It is on this point about the value of the detail that the sharpest distinction between Reynolds and Ruskin can be drawn.

Reynolds had recommended that young artists pursue breadth "so that the general effect and power of the whole may take possession of the mind, and for a while suspend the consideration of the subordinate and particular beauties or defects." Details were distracting, according to Reynolds, because they detracted from perception of ideal beauty and also from "expression of the main characteristick [*sic*]" of a pictorial subject.[110] Ruskin, by contrast, argued that it was

> *the imperative duty of the landscape painter, to descend to the lowest detail with undiminished attention. Every herb and flower of the field has its specific, distinct and perfect beauty; it has its peculiar habitation, expression, and function. The highest art is that which seizes this specific character, which develops and illustrates it, which assigns to it its proper position in the landscape, and which, by means of it, enhances and enforces the great impression which the picture is intended to convey. Nor is it of herbs and flowers alone that such scientific representation is required. Every class of rock, every kind of earth, every form of cloud, must be studied with equal industry, and rendered with equal precision.*[111]

The two theorists are offering very different ways of both producing and perceiving the work of art. For Reynolds, the success of the final work should be judged by its ability to achieve a general effect or impression. Only those details necessary to help the mind build its associations should be included. Recalling Burke, these associations should lead to universally recognized aesthetic categories, such as the sublime or the beautiful. Ruskin, by contrast, emphasizes that it is the artist's task to capture every particularity of nature, and that the resulting painting should provide viewers with all that they need to know and perceive. The painting, in and of itself, is complete. Visual observation is the source of knowledge for both artist and viewer.

Ruskin thus argued that young artists should abandon the common practice of emulating past masters and building up a mental map of past artistic compositions, and should instead develop their "powers of observation."[112] In a passage that Ruskin himself later reappropriated for an essay on Pre-Raphaelitism, he advised young painters to "go to Nature in all singleness of heart, and walk with her laboriously and trustingly, having no other thoughts but how best to penetrate her meaning, and remember her instruction; rejecting nothing, selecting nothing, and scorning nothing; believing all things to be right and good, and rejoicing always in the truth." Only then, "when their memories are stored, and their imaginations led, and their hands firm" (a reversal in the order of steps that Reynolds prescribed), could students "take up the scarlet and the gold, give the reins to their fancy, and show us what their heads are made of."[113]

Ruskin's proposition is clearly rhetorical, in that any act of representation entails an act of selection; no work of art can encompass the universe. His language is intended to emphasize and enforce. Translating Ruskin's project into the language of inductive philosophy, he has proposed that the young artist observe nature closely without prejudgment, acquire facts, and only once these facts have been ascertained allow the imagination or hypotheses to shape the artist's interpretation. This act of imagination is grounded in the artist's accumulated experience of nature, and the final work of art should attend to all the individual particularities of nature. By contrast, Reynolds argued that experience should produce an artwork that transcends individual specificity and particular details, and corrects blemishes and imperfections, in order to achieve ideal beauty.

The point at which imagination could be liberated and exert influence over observations was an important concern for Ruskin, as it was for his scientific colleagues. In the essay "Pre-Raphaelitism" (1851), written to rebut misperceptions of their work, just as *Modern Painters* was written to refute critics' swipes at Turner, Ruskin focused on this question, using Millais and Turner as case studies—polar opposites—to illustrate his points. According to Ruskin, both painters were motivated by "a humble desire to render some part of what they saw in nature faithfully," yet, encountering the "same field in a mountain valley," Millais "sees everything, small and large, with almost the same clearness . . . he can remember nothing, and invent nothing," whereas Turner "has been watching the change of the clouds, and the march of the light along the mountain sides; he beholds the entire scene in broad, soft masses of true gradation." Millais then "patiently . . . sets himself to his mighty task; abandoning at once all thoughts of seizing transient effects, or giving general impressions of that which his eyes present to him in microscopical dissection, he chooses some small portion out of the infinite scene, and calculates with courage the number of weeks which must elapse before he can do justice to the intensity of his perceptions, or the fullness of matter in his subject."[114] In his labor and attention to visual observation, Millais humbles himself before nature, understood as the final arbiter of truth.

Turner, by contrast, would commit the transient scenes he witnessed to memory so that "thousands and thousands of such images, of older scenes, remain congregated in his mind, each mingling in new associations with those now visibly passing before him, and these again confused with other images of his own ceaseless, sleepless imagination, flashing by in sudden troops"; "he may take one of them out perhaps, this day twenty years, and paint it in his dark room, far away." Here is bravura genius, capable of capturing transient effects of nature even without the model before his eyes. Ruskin attributes to Turner both a memory and an imagination far beyond the capacities of his peers.[115]

These two artists thus proved Ruskin's larger point about the role of individual perception in deriving truth to nature.[116] It is as though Ruskin has projected onto Millais and Turner the various possibilities offered by inductive philosophy in the early nineteenth century: working strictly from visual observations to discern facts and slowly assembling those facts into a whole, or, with the accumulated knowledge gained from visual observation and facts, perceiving a whole at a glance. This distinction between Turner and the Pre-Raphaelites was clearly not lost upon readers. The art critic for the *Athenaeum* made precisely this argument in a review of the Royal Academy exhibition of 1852, which included several examples of Pre-Raphaelite painting, including Millais's *Ophelia* (1851–52), Hunt's *Hireling Shepherd* (1851–52), and Charles Allston Collins's *May, in the Regent's Park* (1851). In this notice, which followed a summary of recent meetings of scientific societies, the author observed that the Pre-Raphaelites, with their exaggerated methodology of strict observation—"an atomic analysis which demands the microscope to examine and the leisure of monastic illuminators to execute it"—and their desire to reform art in pursuit of truth, were a "reaction against ultra-Turnerism; which left too much to the imagination, and only shadowed forth what might and ought to have been expressed by outline and detail."[117]

In his essay on Pre-Raphaelitism, Ruskin also made clear that science should serve as the midwife to the rebirth of art in the modern age. If the Pre-Raphaelites "adhere to their principles, and paint nature as it is around them, with the help of modern science, with the earnestness of the men of the thirteenth and fourteenth centuries, they will, as I said, found a new and noble school in England."[118] In short, the quest for truth to nature, sustained by science, would purify artistic practices and produce art resolutely of the moment and thus truly modern. Thus Ruskin encouraged the young artists he supported to participate in the increasingly visible and culturally significant practice of scientific observation.

Ruskin's pedagogical aims and inductive commitment also informed his *Elements of Drawing* (1857), written in response to numerous requests for advice. Aimed at the erudite amateur, *Elements of Drawing* provides a series of guidelines and exercises of value for my argument here. In his chapter "Sketching from Nature," Ruskin attends to three key principles: subordination, individuality, and incomprehensibility. Subordination refers to the artist's need to recognize that all forms in nature are governed by underlying laws or principles. Ruskin condemned the bad artist who "breaks his mountain side into rugged fragments, wholly unconscious of the lines of force with which the real rocks have risen."[119] As Ann Bermingham explains, Ruskin's idea of truth was "closely tied to a mechanical view of things, to a belief that the appearances of objects, landscapes, and views were determined by the physical and biological laws governing their formation and perception, and that representational truth consisted in the revelation of these laws."[120] Individuality, for Ruskin, must always be in balance with these laws of nature. The artist recognized the telling detail through "the earnest and complete study of facts" in nature. The last of Ruskin's three principles, incomprehensibility, refers to a sense of mystery associated with nature, that is, the recognition of what cannot be seen. Only Turner, Ruskin added, truly had the ability to depict this character in landscape, achieved after considerable study of nature.[121]

How to see was an important theme in *Elements of Drawing*, and in this text Ruskin developed his concept of the "innocence of the eye." In his first exercise, he states, "Everything that you can see in the world presents itself to your eyes only as an arrangement of patches of different colours variously shaded." For readers who did not comprehend his point, Ruskin added this note:

> *The perception of solid Form is entirely a matter of experience. We* see *nothing but flat colours; and it is only by a series of experiments that we find out that a stain of black or grey indicates the dark side of a solid substance, or that a faint hue indicates that the object in which it appears is far away. The whole technical power of painting depends on our recovery of what may be called the* innocence of the eye*; that is to say, of a sort of childish perception of these flat stains of colour, merely as such, without consciousness of what they signify,—as a blind man would see them if suddenly gifted with sight.*[122]

Whewell had made a similar point in *The Philosophy of the Inductive Sciences* when he asked rhetorically, "Does anyone fancy that he sees a solid cube?" Whewell distinguished between information derived optically—the perception of surfaces—and that produced by mental interpretation—the existence of solid bodies in space.

While Ruskin's concept of the "innocence of the eye" was located in a footnote, it is highly significant for his model of artistic practice and the ways in which he sought to constrain the subjectivity of sight. Pragmatically, it offered artists lacking Turner's memory or imagination a means by which to gain proficiency. Theoretically, it elucidated how Ruskin construed the act of perception; he believed it possible for the artist to approach nature without a priori knowledge, as a strict inductionist might, and then to conduct a series of visual experiments in order to discern and then assign meaning to what is observed. Ruskin acknowledged that this was quite difficult and proposed various stratagems, such as viewing an object with one eye, to render the familiar unfamiliar. Such dodges suppressed the perceiving self.

It is worthwhile, at this juncture, to situate Ruskin's theories with respect to Klonk's analysis of phenomenalism and its inherent instability. It is clear that Ruskin conceived a continuum, from those artists who should focus on developing their skills in rigorous observation, to those, namely, Turner, who possessed sufficient experience to be guided by imagination and who could use nature to form associations and express emotion.[123] He resolved the potential tension between artist as recorder and as autonomous perceiving subject by holding that both roles were possible, depending on the mind and experience of the artist. Above all, both were subject to nature as the source of all truth.

William Michael Rossetti, in an essay on Pre-Raphaelitism first published in the *Spectator* in 1851, offered a similar solution. A work of art, he explained, was composed of two halves: nature and artist. While the artist possessed an "individual mind," what was perceived was "absolute data of fact." In his conclusion, he reiterated the Pre-Raphaelite commitment to working directly from nature: The Pre-Raphaelites "recur to the one primary school—nature, as interpreted by their own eyes and feelings." While the verb *to interpret* might suggest that they could give their faculties free rein, Rossetti imposed the limitation that the Pre-Raphaelites should build upon "the acquired knowledge of the day."[124] Truth to nature, in short, was not incompatible with the recognition of individual perception. The task was how best to undertake those acts of perception. Ruskin's proposition of the innocent eye suggested that no preparation was necessary, whereas Rossetti's reference to acquired knowledge, data, and facts suggests the need for rigorous study. Ultimately, both approaches took nature as the measure against which success would be judged.

Ruskin's vision of a new and noble school of art born of science is exemplified in Millais's portrait of the critic, executed between 1853 and 1854 (fig. 8). Ruskin predicted that the "picture of a torrent among rocks . . . will make a revolution in landscape painting."[125] The painting, representing Ruskin standing astride rocks in a stream rushing through a woodland, was executed largely on site. Millais's and Ruskin's letters to friends and family are filled with the challenges that painter and sitter faced by choosing to paint from nature at Glenfinlas in the Trossachs of Scotland, with its temperamental weather conditions (fig. 9).[126] The stream, rocks, and surrounding foliage were executed outdoors in the late summer and fall of 1853 and late spring of 1854. The figure, originally outlined in the summer of 1853, was completed in the studio.

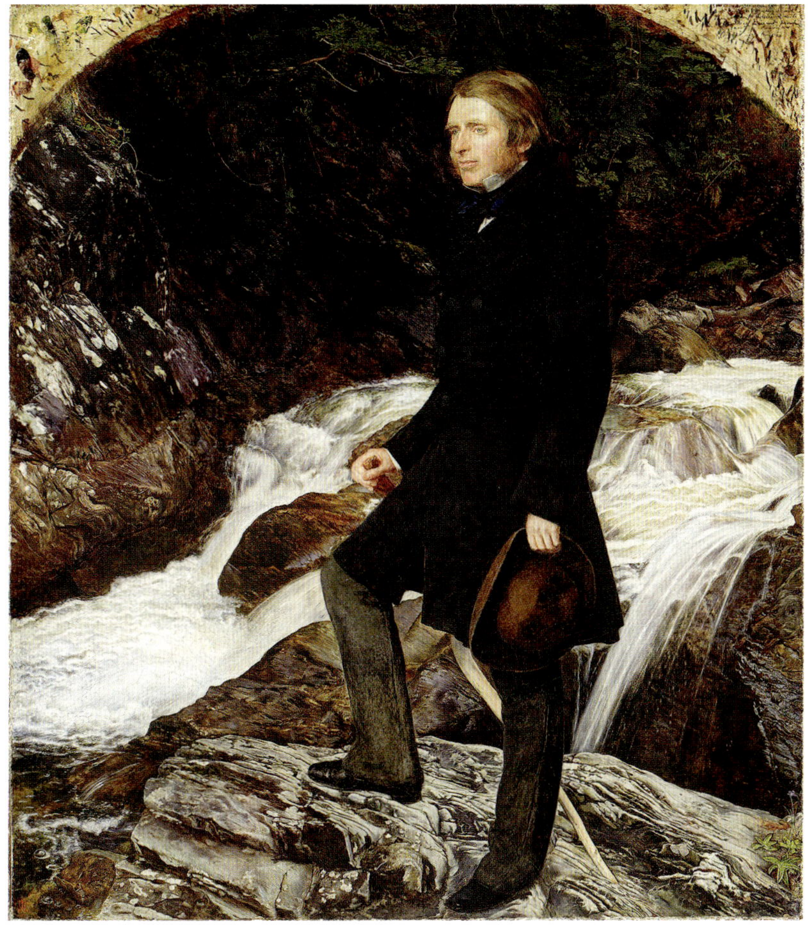

FIG. 8
John Everett Millais, *John Ruskin*, 1853–54. Oil on canvas, 71.3 × 60.8 cm. © Ashmolean Museum, University of Oxford.

The months of scrupulously observing nature are evident in Millais's meticulous depiction of various components of the scene; for example, the foreground rock is so specific that it can be identified as gneiss because of the pattern of striation and fracturing.[127] This

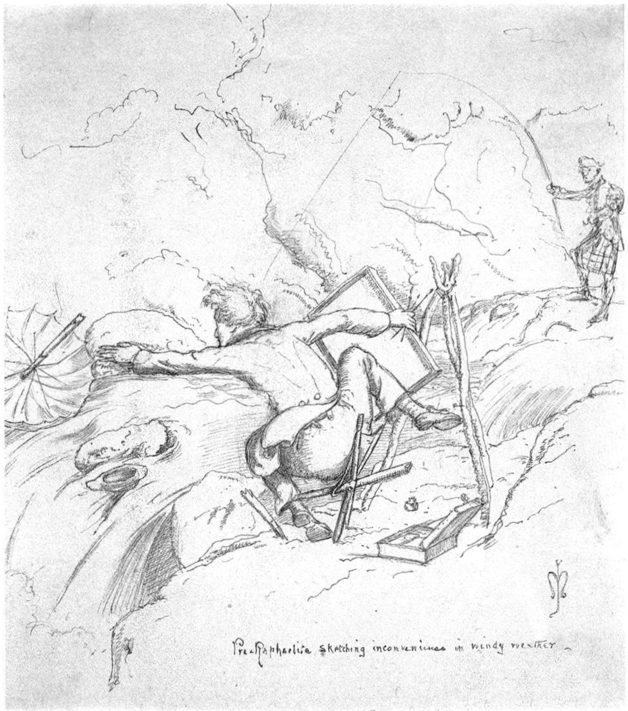



FIG. 9
John Everett Millais, *Pre-Raphaelite Sketching Inconvenience in Windy Weather*, 1853. Pen and ink on paper, 8 × 7½ in. Victoria & Albert Museum, London.

level of attentiveness exhausted the artist and earned him Ruskin's praise, as recorded in the critic's letters to his father: "Millais has put all his strength into the ash bough at the top of the picture—and the dark rocks and creepers below this. Nothing can be finer"; "Millais has done some glorious bits of rock in the picture these last two or three days with all their lichens gleaming like frosted silver—most heavenly."[128] Yet, although the painting may have epitomized all that Ruskin hoped for in terms of a new direction in landscape painting, the personal difficulties between the two men caused by the dissolution of Ruskin's marriage and Millais's desire to marry Effie Ruskin meant that Millais never took up the position of acolyte that Ruskin envisioned for him.

## Pre-Raphaelitism and Science: Geology and Chemistry

Ruskin was joined in the cause to establish a favorable reputation for the Pre-Raphaelites by the critic F. G. Stephens. Art criticism had not yet emerged as a profession, and Stephens, like many critics of his day, originally trained in art. He did exhibit publicly in the first half of the 1850s but increasingly turned to writing, contributing to the Pre-Raphaelite journal the *Germ* and other periodicals. In 1861, he was hired as the critic for the *Athenaeum*, a post he held until 1901. As the art historian Dianne Sachko Macleod observes, while Stephens helped to advance the cause of Pre-Raphaelitism by voicing the artists' goals and giving greater visibility to their productions, he occasionally ventured negative opinions of their work, as in the case of Hunt's *Triumph of the Innocents* (1883–84), leading to decades-long antagonism between the two men.[129] But in the 1850s, there was a great deal of sympathy between Stephens and his fellow members of the Brotherhood.

In 1850, Stephens, writing under the pseudonym of John Seward, offered a critically important exegesis on the aims of the Brotherhood in the context of an article on early Italian art for the *Germ*. Writing for an audience already predisposed toward Pre-Raphaelitism, Stephens, even more emphatically than Ruskin, marked a break between the art of the past and that of the "modern school," with which he associated Pre-Raphaelitism. Modern artists, according to Stephens, produced "pure transcripts and faithful studies from nature, instead of conventionalities and feeble reminiscences from the Old Masters"; their artworks "undoubtedly surpass all others in the simple attention to nature in detail as well as in generalities." Here are the key terms of the emerging discourse of Pre-Raphaelitism: transcript, faithful study, and attention to detail—phrases analogous to those deployed to describe photogenic drawing and the calotype and also privileged in inductive science.

Clearly informed by Ruskin's *Modern Painters*, Stephens dismissed the idea that nature possesses blemishes that should be corrected by the artist in pursuit of the ideal; instead, he urged "closer communion with nature, and . . . a more exact adherence to all her details, (for nature has no peculiarities or excentricities [*sic*]) in whatsoever direction her study may conduct."[130] He also perceived, as Ruskin had done, a great compatibility between modern art and modern science; both shared a similar foundation—the search for truth—that Stephens promoted as the new moral guide for humanity:

> *The sciences have become almost exact within the present century. Geology and chemistry are almost re-instituted. The first has been nearly created; the second expanded so widely that it now searches and measures the creation. And how has this been done but by bringing greater knowledge to bear upon a wider range of experiment; by being precise in the search after truth? If this adherence to fact, to experiment and not theory,—to begin at the beginning and not fly to the end,—has added so much to the knowledge of man in science; why may it not greatly assist the moral purposes of the Arts? It cannot be well to degrade a lesson by falsehood. Truth in every particular ought to be the aim of the artist. Admit no untruth; let the priest's garment be clean.*[131]

Even more than Ruskin, Stephens turned to science for cultural authority. Moreover, like contemporaneous philosophers of induction such as Whewell, Stephens believed that the rigorous study of nature was an act of faith in the deity:

> *The modern artist does not retire to monasteries, or practise discipline; but he may show his participation in the same high feeling by a firm attachment to truth in every point of representation, which is the most just method. For how can good be sought by evil means, or by falsehood, or by slight in any degree? By a determination to represent the thing and the whole of the thing, by training himself to the deepest observation of its fact and detail, enabling himself to reproduce, as far as possible, nature herself, the painter will best evince his share of faith.*[132]

What is particularly noteworthy in Stephens's joining of the practices of modern art and modern science is his specific mention of geology and chemistry, two disciplines that had undergone radical change by the early nineteenth century.[133] A few key episodes from the history of geology and chemistry illuminate why Stephens would specifically refer to these fields and how scientific developments informed the production and reception of works of art by Pre-Raphaelite artists.

In 1842, W. T. Brande, who succeeded Davy as professor of chemistry at the Royal Institution, explained in his *Dictionary of Science, Literature, and Art* that geology had been transformed over the previous twenty-five years, shifting from "a very limited collection of facts, strung upon amusing but insufficient theories and hypotheses," to a rigorously practiced inductive science.[134] Recent scholarship, including that of Charles

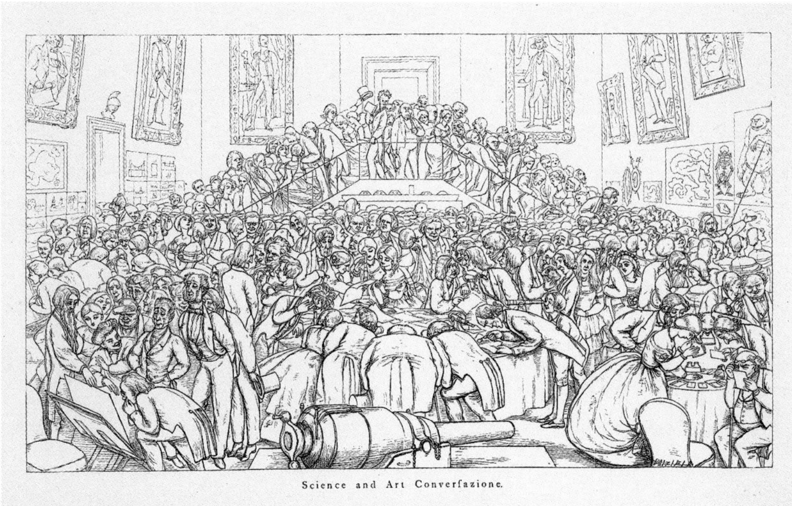


FIG. 10
Richard Doyle, *A Science and Art Conversazione*. In Doyle, *Bird's Eye Views of Society* (London: Smith & Elder, 1864), facing p. 45. The Huntington Library, San Marino, California.

Lyell, William Buckland, John Phillips (who developed a framework for geological dating), Robert Bakewell, and Gideon Mantell, had established sound causes for the appearance of the earth, with its stratified and unstratified layers demonstrating change over time and the impact of powerful forces of fire and water in combination with changes of temperature. Brande also noted that the Geological Society was an important arena for new research. Twelve years earlier, in his *Reflections on the Decline of Science in England*, Babbage had given the Geological Society high marks, citing its "freshness" and "vigour" and praising the instructive oral discussions that followed each paper.[135]

At the time of Babbage's treatise, Lyell had issued his response to James Hutton's theory, published in *Theory of the Earth* (1795), that geological change could be explained by slow, natural, observable processes, as opposed to climactic culminating events. Lyell built upon Hutton's concept of vast geological time and the constancy of geological forces to develop the theory of uniformitarianism in his *Principles of Geology, Being an Attempt to Explain the Former Changes in the Earth's Surface* (first published between 1830 and 1833).[136] *Principles of Geology* was a decisive influence on Charles Darwin, as the younger scientist readily acknowledged.[137]

The gradual popularization of geology is easily measured by the Geological Society's establishment of the Museum of Practical Geology, the origins of which can be traced to the society's collecting activities. In 1841, after the purchase of a building at Craig's Court, Whitehall, the society put its specimens on display at the Museum of Economic Geology. The collections and public interest quickly outgrew the premises, and in the mid-1840s the organization began to work with the government's Office of Works to develop a purpose-built structure, located between Piccadilly and Jermyn Street and designed by James Pennethorne, which Prince Albert opened in 1851.[138]

The public visibility of such collections, along with lectures and writings on geology, meant that visitors to art exhibitions, drawn from the same educated middle classes that attended scientific displays, would have brought considerable knowledge to the act of examining landscape paintings (fig. 10). Ruskin carved out a role for himself as a guide to these acts of looking in his *Academy Notes*, begun in 1855, in which he discussed the merits and defects of what he considered the most important works exhibited at the Royal Academy each year. In his first year of this venture, Ruskin championed the work of John William Inchbold and his *Moorland: Tennyson* (possibly the painting *The Moorland*

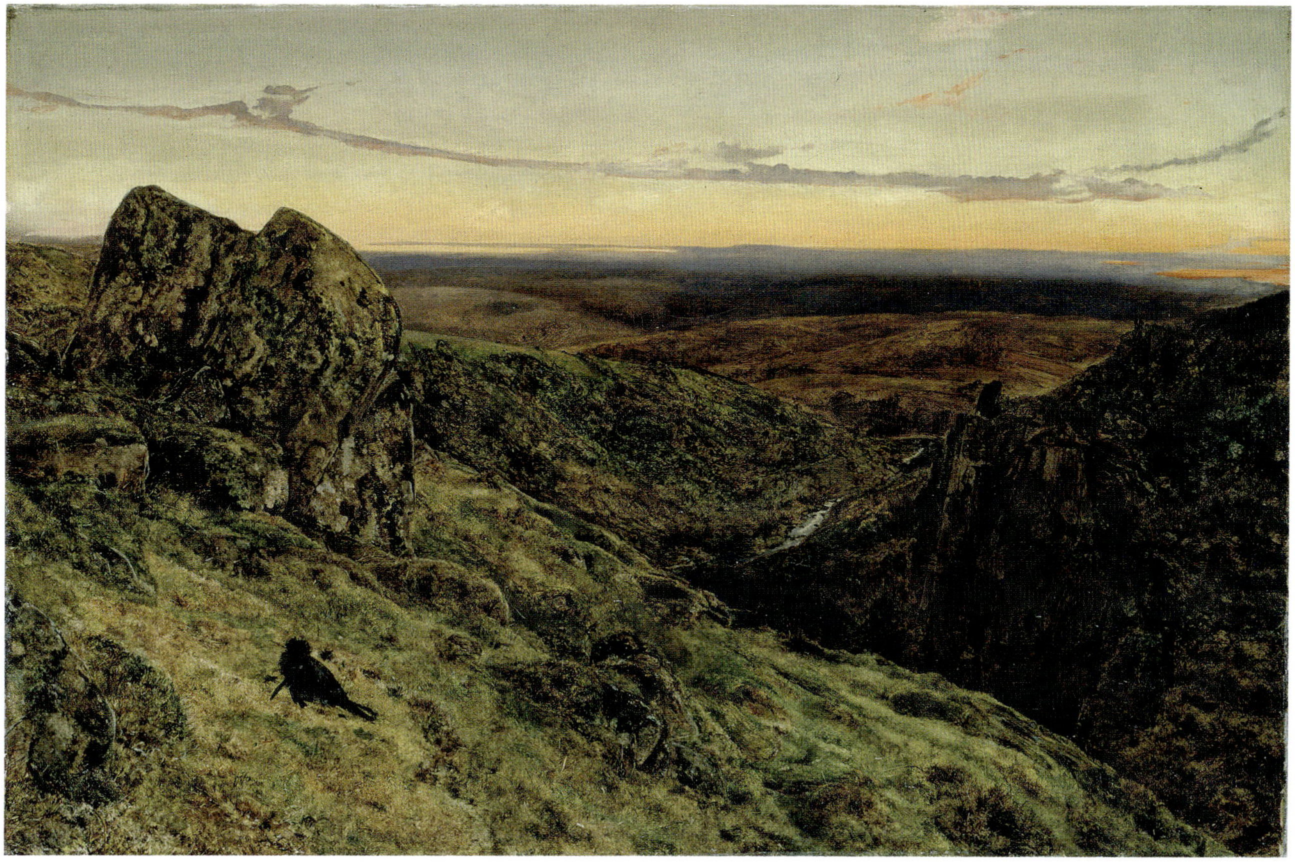

FIG. 11
John William Inchbold (1830–1888), *The Moorland (Dewar-stone, Dartmoor)*, 1854. Oil on canvas, support 356 × 533 mm. Bequeathed by Sir J. Russell Reynolds, Bt. 1896. © Tate, London 2015.

*[Dewar-stone, Dartmoor]*, 1854) as "the only thoroughly good landscape in the rooms of the Academy" (fig. 11). Inchbold, the son of a Leeds newspaperman, had come to London to study lithography, and by the 1850s had begun to exhibit landscape oils in the Pre-Raphaelite manner. Ruskin described Inchbold's painting in superlative terms, suggesting that Inchbold had superseded Millais in terms of truth to nature: "It is more exquisite in its finish of lichenous rock painting than any work I have ever seen. Its colour, throughout, is as forcible as it is subtle and refined; and although it appears as yet to display little power of invention, the appreciation of truth in it is so intense, that a single inch of it is well worth all the rest of the landscapes in the room."[139] While Ruskin noted that Inchbold had failed to invent, as Turner might have done, he did not fault Inchbold

because the artist had instead supplied a detailed representation of nature. In addition, Ruskin's criticism explains how these new paintings should be viewed—inch by inch, with an appreciation for finish—rather than by stepping back and embracing the breadth of the scene.

When Inchbold undertook painting expeditions to the Alps in 1856 and 1858, Ruskin, who knew the region well, took it upon himself to guide and instruct the young artist. The lure of the mountains also tugged at the painter John Brett. Like Inchbold, Brett was not a proper member of the Pre-Raphaelite Brotherhood but was deeply attracted to their ideal of direct, unmediated vision. Brett's diary entries from around age twenty reveal his interest in fields outside art, including French, Latin, music, mechanics, and astronomy, and he owned a telescope to which he devoted considerable time and effort.[140] His diary reveals that in May 1852 he read Ruskin's pamphlet on Pre-Raphaelitism, finding it "gloriously written & containing much earnest sound healthy thruth [*sic*], but much also that is false & senseless—wrong," a streak of defiance that would reemerge later, once he made Ruskin's acquaintance. He visited the Royal Academy exhibition of 1852 and saw the Pre-Raphaelite canvases there. Then, in May 1853, while attending the Royal Academy schools, he declared his intention of "going on fast towards preraphaelitism. . . . Millais & Hunt are truly fine fellows I greatly admire them & honor them— Have resolved in future to go through a severe course of training & close childlike study of nature in short to follow in their steps." Brett's reference to childlike study seems to anticipate Ruskin's innocent eye; moreover, he has clearly absorbed the rhetoric of labor associated with Pre-Raphaelitism. His appreciation of the natural world was also in keeping with Ruskin's natural theology; he wrote exultantly about "the dew of Gods [*sic*] morning that perfectly pure translucent pearl of heaven rests on all blades of grass and surely on every fibre of mans [*sic*] heart too, yes it is Gods [*sic*] world."[141] In September 1853, Brett met Hunt, and by May 1854 his familiarity with Pre-Raphaelitism was such that he participated in a public discussion of the movement.[142]

In 1856, Brett ventured to the Swiss Alps, where he met Inchbold and, in response to Inchbold's work, resolved to alter his approach: "then saw that I had never painted in my life, but only fooled and slopped; & thenceforth attempted in a reasonable way to paint all I could see." Brett's emphasis on sight as the key to knowledge is very much in keeping with Pre-Raphaelitism at this moment. The result of his efforts was the painting *The Glacier of Rosenlaui*, which both Hunt and Rossetti admired, the latter so much so that he borrowed the work in order to show it to Ruskin (fig. 12).[143] The small landscape portrays what Brett could see of the glacier as it swept through the valley between the Wetterhorn and Dossenhorn.

The painting presents the glacier dispassionately, and the composition calls attention to the differences between the types of stones littering the glacier's path.[144] The foreground rocks are precisely delineated specimens—the pink tones of the center boulder indicate granite, whereas the combination of nougat and layer-cake effects in the neighboring boulder describes gneiss. The thin, broken layers of the boulder farthest to the right illustrate metamorphic marble. These three geological types, as Allen Staley points out, were those that Ruskin highlighted in *Of Mountain Beauty* (1856).[145] Through its illusion of materiality, the painting makes visible what previously had been considered invisible—the forces of nature that shaped the appearance of the earth's crust.[146]

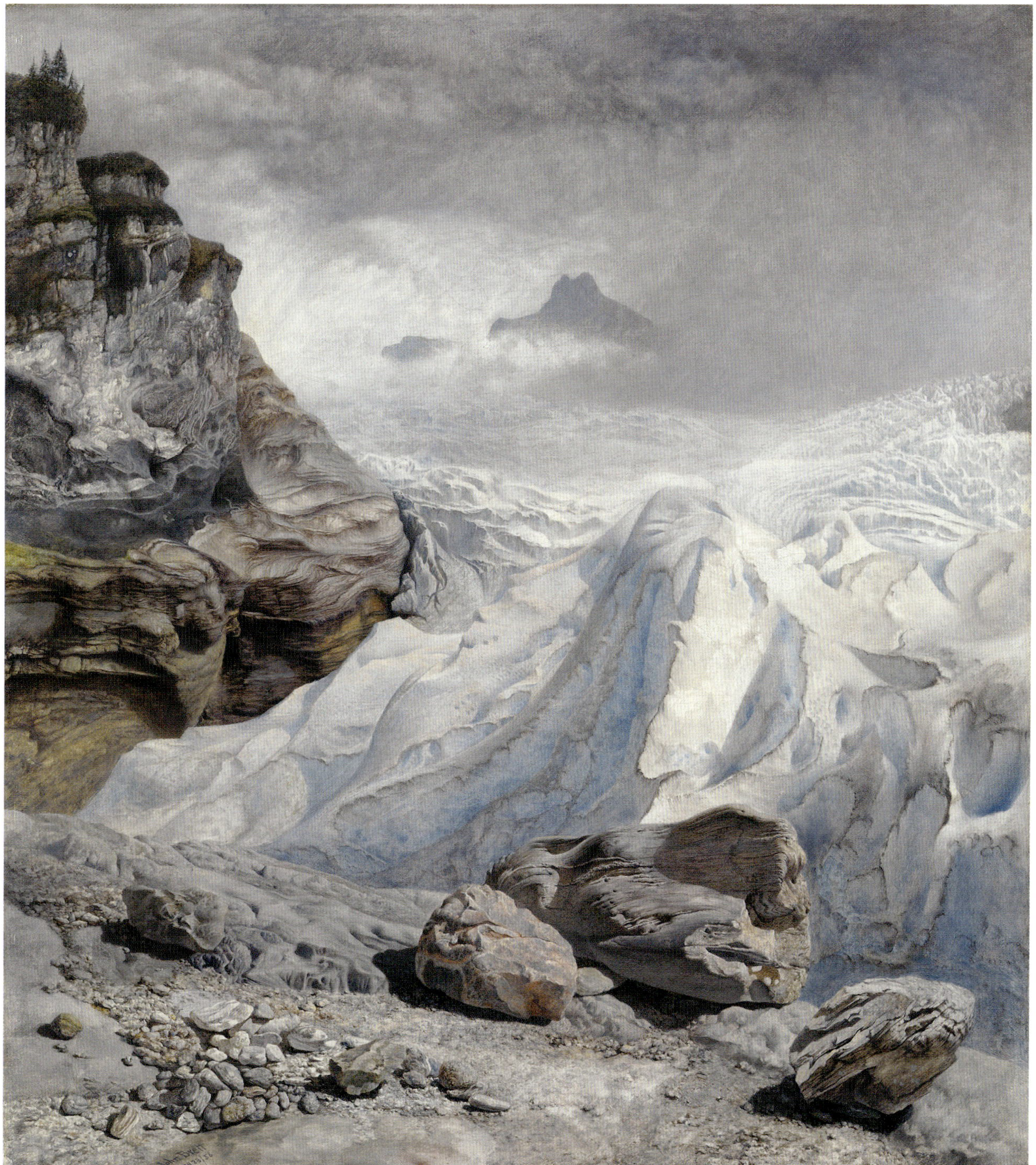

FIG. 12
John Brett (1831–1902), *The Glacier of Rosenlaui*, 1856. Oil on canvas, support 445 × 419 mm. Purchased 1946. © Tate, London 2015.

The disparate congregation of rock that figures in the foreground of Brett's painting is indicative of topographical features that had spurred investigation by geologists, and the glaciers of Switzerland were the subject of much recent attention. Brande, in his popular dictionary, explained to readers that the dominant theory concerning the movement of glaciers was that of Horace-Benedict de Saussure, who attributed their motion to gravity and their inherent weight. But Brande also alluded to Louis Agassiz's theory that the expansion of ice produced both movement and the formation of erratic material.[147] Lyell, in his *Principles of Geology*, also endorsed the view that erratic material was the result of the "transporting power of ice" and that glaciers might behave like icebergs in terms of breaking apart and moving.[148] James Forbes, a physicist and professor of natural history at Edinburgh University who traveled extensively in the Alps and met Ruskin there in 1844, developed from his observations the viscous theory of glaciers. He recounted, in an essay for the Royal Society, the surprising resistance he encountered to his theory: "I had not expected to meet with so determined an opposition to the proposition that the stupendous aggregation of freezing water and thawing ice, called a glacier, subjected to the pressure of thousands of vertical feet of its own substance, might not under these circumstances possess a degree of yielding, moulding, self-adapting power, sufficient to admit of slight changes of figure in long periods of time."[149] In the course of developing this theory, Forbes carefully delineated various visual features of glaciers that he felt had not yet received sufficient attention, including "the internal veined or ribboned structure presented by all glaciers."[150]

The debates over the origins and movements of rock were part of a larger controversy regarding creation—did the biblical book of Genesis offer a viable account of the origins and age of the earth, and was this account compatible with more recent findings in the material sciences, or were the latter sufficiently persuasive on their own terms? Brett's painting adds weight to the material scientists' point of view. The artist emphasizes the dirt bands embedded in the blanketlike glacier, evidence of the aggregation of material as the ice moved. Brett delineates the exposed layers of rock, as fine as sheets of puff pastry, in the outcropping at left, as well as the undulating curves carved by the glacier hugging the rock at the bottom of the cliff. A long crack in the ice betrays the presence of the rock beneath as the glacier attempts to pass over it. The glacier's path thus explains the crags of the valley walls. The small rocks and pebbles in the immediate foreground suggest the grinding and polishing force of the glacier.

John Brett's signature and date, illusionistically inscribed on the rock in the left foreground, are the exception to the artist's attention to the forces of nature distributed throughout the painting. The signature and date function as a witnessing mark: "I was there," "I saw this scene." Indeed, the painting's composition confirms the sensation that it results the artist's direct observations, as opposed to from an already established formula for structuring landscape paintings. While the mountain peak in the far distance creates a vanishing point for implicit orthogonals, the space does not unfold in a coherent fashion. This has less to do with the undulations of the landscape as the glacier makes its way through the valley than with the heightened focus that Brett brings to nearly each aspect of the middle ground and background. The jarring nature of the painting's pictorial space is not immediately perceptible; the composition invites the viewer to enter in the far left corner, amid the piled pebbles, move up the boulders to the glacier peak beyond, following the blue of the ice crevasse, and then over to the outcropping at left, topped with minute pine trees, before plunging into the hazy, atmospheric far distance. The attention demanded by each of these quadrants of the painting, picked out in extreme detail, makes it difficult, if not impossible, to perceive the painting as a unified whole—to hold the entire composition in a single glance. The tiny scale of the pine trees at the upper left, for example, suggests that the outcropping should be at a far remove from the foreground rocks, but because the cliffs on which they stand present nearly the same level of geological specificity as the foreground boulders, the sense of space is awkwardly collapsed. In many ways, Brett was attempting to replicate human vision—the ability of the human eye to rapidly scan and refocus—and the ways in which the foreground space seems to extend into that of the viewer reinforce this impression. The painting is reminiscent of stereoscopic views in which forms in the foreground are often dramatically three-dimensional and seem to project out against a flatter background space. Brett appears to be attentive to Ruskin's notion of the "innocence of the eye" and the need to overturn previous conventions of linear perspective in order to convey a fresh, authentic vision.

Brett's evident mastery of the complexities of the natural world proved to be both the path to his future success and his undoing, in Ruskin's opinion. In 1858, Brett again ventured to the Alps to take up the painting *The Val d'Aosta* (fig. 13). Ruskin apparently proposed the subject and also monitored the artist's work over the summer. The result demonstrates close study of the local topography; Brett's previous attention to geological formations is now matched by a fascination with botany and enhanced by a new interest in broad views.[151] Ruskin, when he commented on the work upon its exhibition at the Royal Academy, praised the visual knowledge conveyed by the painting:

> *And here, accordingly, for the first time in history, we have, by help of art, the power of visiting a place, reasoning about it, and knowing it, just as if we were there, except only that we cannot stir from our place nor look behind us. For the rest, standing before this picture is just as good as standing on that spot in Val d'Aosta, so far as gaining of knowledge is concerned; and perhaps in some degree pleasanter, for it would be very hot on that rock to-day, and there would probably be a disagreeable smell of juniper plants growing on the slopes above.*[152]

Yet Ruskin's entire review indicates that he found Brett incapable of producing more than a transcript.[153] The artist failed to exploit the full capacity of the scene to convey the power and vastness of the universe. Moreover, he failed to elicit emotion in the scene or to impress his individualistic perception upon it. He had fallen short of Ruskin's ideal as represented by Turner.[154] Brett's work was caught up in a paradox inherent in the truth-to-nature project, one that persisted for decades: Was truth best defined by faithfulness to observed detail, or faithfulness to the artist's personal response to the scene? This paradox can be rephrased in terms of an irresolvable dialectic between fact and the mysterious or incomprehensible. As the literary historian Lindsay Smith notes, "What seems to occur is that Ruskin wishes to botanise, to use an inductive method of sorts, while retaining a belief in the transcendental, which means that his botanising is, in a sense, always working by processes contrary to those he acknowledges."[155]

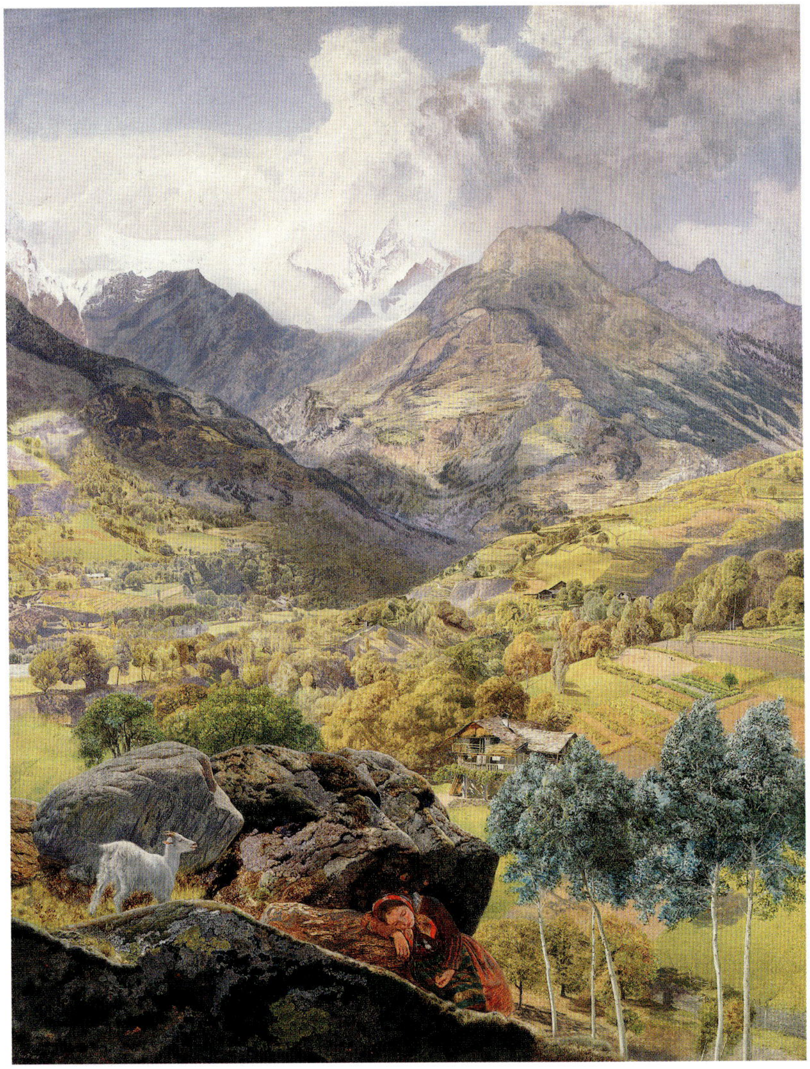

FIG. 13
John Brett, *The Val d'Aosta*, 1858. Oil on canvas, 87.6 × 68 cm. Private collection.

But despite such paradoxes and the risk of severe criticism, artists continued to adopt the Pre-Raphaelite methodology. In the 1850s, the painter William Dyce was drawn into the Pre-Raphaelite orbit, largely through his scientific interests. Both Dyce and his father were members of the Royal Society of Edinburgh, a natural history organization in which scientists such as David Brewster participated. In 1828, the young artist won a prize for an essay on electromagnetism. Dyce, who was older than the members of the Brotherhood, was largely occupied with the project of painting frescos for the Houses of Parliament throughout the 1840s and 1850s. He returned to his interest in science near the end of his career with the painting *Pegwell Bay, Kent—A Recollection of October 5th 1858* (1858–60?), which was executed in the studio after a watercolor study (fig. 14).[156] Dyce's image demonstrates how nature was a vital source of knowledge for many different disciplines. In the foreground, members of Dyce's family gather shells, illustrating the mid-Victorian craze for conchology. Indeed, numerous writers in this period, among them Margaret Gatty, Philip Henry Gosse, Charles Kingsley, George Henry Lewes, and Rev. John George Wood, recommended beachcombing as an appropriate family activity because of its didactic potential. The cover of Wood's book, *The Common Objects of the Sea-Shore*, like Dyce's painting, foregrounds fashionably dressed women collecting specimens against the backdrop of the seaside cliffs (fig. 15). In Dyce's scene, the artist himself is present in the far right middle ground; the portfolio under his arm is a reminder of the practice of sketching out of doors. This figure directs visual attention to the cliffs, a source of geological knowledge owing to their discernible erosion, as well as of paleontological information, as evidenced by the fossils embedded in the chalk. Arcing in the sky above is Donati's comet, first noticed by the Florentine astronomer for whom it was named, signifying astronomical knowledge. The date in Dyce's title—5 October 1858—refers to a high point in

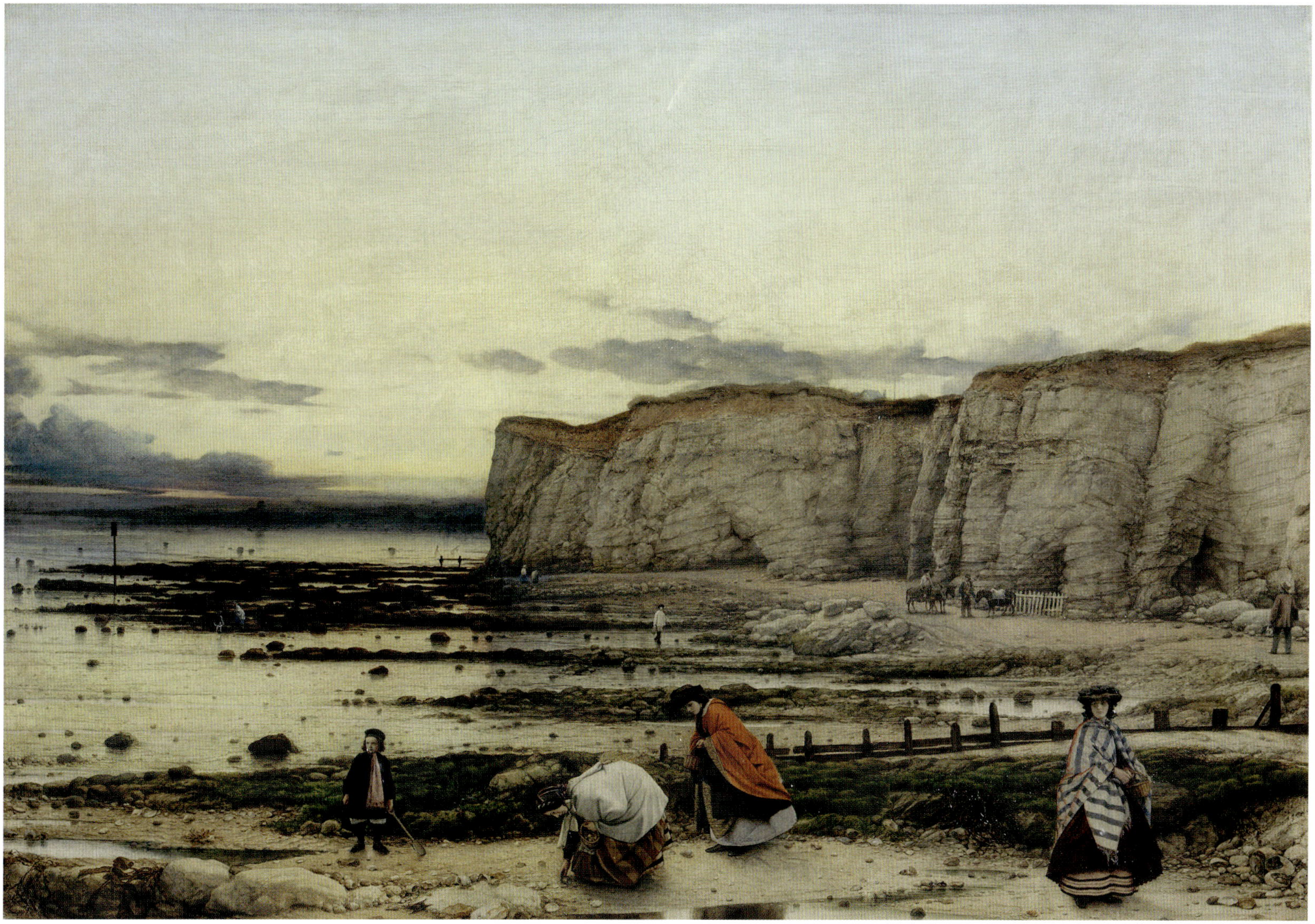

FIG. 14
William Dyce (1806–1864), *Pegwell Bay, Kent—A Recollection of October 5th 1858*, 1858–60? Oil on canvas, support 635 × 889 mm. Purchased 1894. © Tate, London 2015.

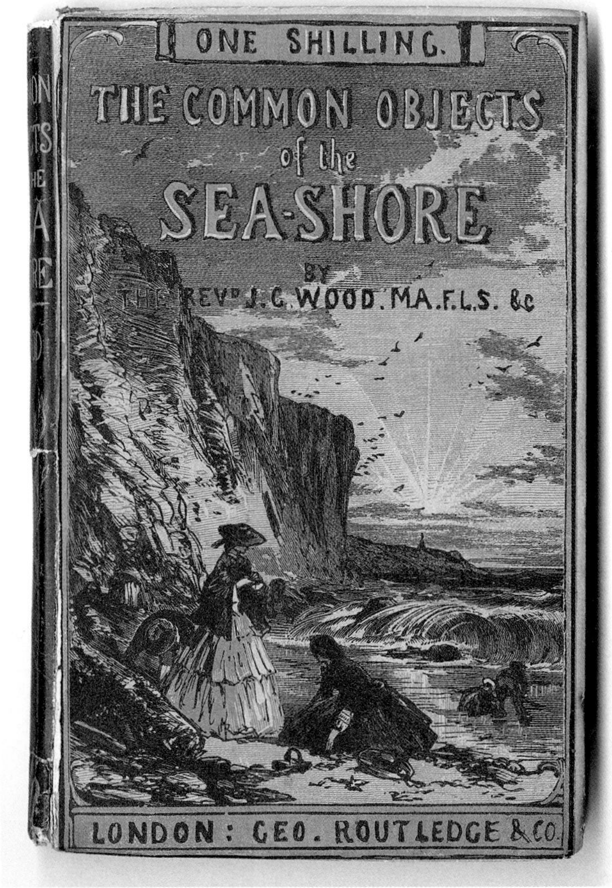


FIG. 15
Rev. J. G. Wood (1827–1889), *The Common Objects of the Sea-Shore* (1857), with illustrations by G. B. Sowerby. Library Special Collections, University of California, Los Angeles.

the comet's visibility, when it reportedly could be seen from sunset until at least 9:00 P.M., competing with the brightest stars in the sky and capturing the public's attention.

Marcia Pointon has argued that the painting is an allegory about time, both a particular point in time and such ongoing processes as the cycles of a comet and the erosion of cliff walls that register the age of the earth.[157] Dyce alludes to the two disciplines—astronomy and geology—that had most forcefully challenged the biblical narrative of creation set forth in Genesis and produced some of the most heated debates and profound anxieties in Victorian culture. As Malcolm Warner has noted, Dyce's choice of a site redolent with Christian associations—Pegwell Bay was near Richborough, where Saint Augustine first landed in England—underscores the competing explanations of creation.[158] Humanity is represented as just one element in the natural world, not the commanding influence.[159]

But did viewers at the time engage in such symbolic readings? The nonhierarchical composition and even distribution of pictorial elements and nodes of activity across the picture plane did not encourage viewers to privilege any single motif over another. The evidence suggests that the subject in and of itself, without associative readings, was sufficient for viewers. The reviewer for the *Times*, for example, argued that the detailed passages added up to establish the quotidian modernity of the scene—in dress and the taste for scientific topics—and nothing more.[160]

The geological focus in these paintings by Inchbold, Brett, and Dyce amply demonstrates Stephens's point that the pursuit of knowledge and truth in modern geology could inform art. And chemistry, like geology, was deeply dependent upon inductive experiment at midcentury. Chemistry, both theoretical and applied, encompassed a broad range of inquiries, from the analysis of elements and the behavior of gases to atomic theory. Brande, in an overview of the field in his *Dictionary of Science, Literature, and Art*, drew particular attention to the work of Joseph Priestley, who discovered oxygen, as well as to Humphry Davy, who advanced the study of electricity. Brande also explicated how Davy and William Hyde Wollaston had advanced atomic theory.[161] Wollaston, who also worked in metallurgy, built upon the findings of John Dalton, who pursued Isaac Newton's conviction that all matter was made up of invisible particles. The question of invisible particles was pertinent not only to the study of gases, which were central to Dalton's work, but also to investigations of light, color, optics, and human vision. These subjects were of great topical interest for artists.

David Brewster's *Treatise on Optics* (1831) was highly influential among artists, as it advanced the systematization of color begun by Newton.[162] Through a series of experiments, Brewster established that red, yellow, and blue constituted the primary colors, and he also explained the phenomenon of complementary contrasts. His interest in the reflection and absorption of light led him to study the behavior of the eye. Upon reviewing current theories of vision and supporting evidence, Brewster reverted to the earlier notion that the retina was the "seat of vision" and that images of objects could be conveyed via the retina to the brain, a trope repeated in art criticism of the period.[163]

The chemist George Field built upon the physical and chemical experiments of inductive philosophers like Brewster and translated their findings into practical applications for artists in a reference manual that went through several editions beginning in 1835.[164] Field developed color charts, as well as a scale by which to measure contrasting colors. Understanding how each color could affect the perception of adjacent tones was of particular concern to Field, who noted that when nature is studied "microscopically," there are no "uniform tints" but instead "a variety of hues and shades." Field explained how colors could be blended on the palette, or applied in pure tones on the canvas and then combined to produce secondary or tertiary colors. The latter approach, most akin to that of the Pre-Raphaelites, required, according to Field, "a better knowledge of colours,—it is also more favourable to the brilliancy, purity, and durability of colours."[165]

William Holman Hunt owned an abridged version of Field's text *Rudiments of the Painter's Art, or A Grammar of Colouring*, which he acquired from Charles Roberson & Co., where most of the Pre-Raphaelites obtained their painting materials.[166] Hunt clearly consulted it, as evidenced by a letter of February 1858 to a Mrs. Blackburn in which Hunt cited Field as a source of his advice on making yellow, blue ochre, and orange vermillion.[167] Hunt and his colleagues understood color as a critical area of distinction between themselves and the preceding generation of British painters. As early as 1843–44, Millais, writing to Hunt, recommended the benefits of using pure rather than blended colors, the latter being the preferred method of the earlier generation: "For what reason is the sky in a daylight picture made as black as night? And then about colour, why should the gradation go from the principal white, through yellow to pink and red, and so on to stronger colours? Why all this subservience to early examples, when the turn of violet comes, why does the courage of the modern imitator fail? . . . If you notice, a clean purple is scarcely given these days, and pure green is as much ignored."[168] Indeed, green and purple were among the colors that Hunt purchased from Roberson and Field.[169] The Pre-Raphaelites preferred transparent and high chroma pigments, such as madder.[170]

The painters enhanced the brilliancy of the colors they selected by applying them in thin layers or glazes over a white ground that increased the refraction of light. To decrease the likelihood that paint would be absorbed into the ground and lose its glossiness, they sometimes applied additional ground to the prepared canvas and then worked wet into wet—applying thin glazes of color directly onto wet ground—in order to achieve even more luminosity.[171] In addition to thinly layering colors, they also juxtaposed colors or mixed in contrasting pigments, as Hunt explained to Mrs. Blackburn: "I never *glaze* according to the general meaning for completing work. The effect of one uniform map of transparent color on either the prepared canvas or under work preparation is too vague in surface for my eye. I am obliged to work opaque or semi-opaque hues of the opposite tints into it to give solidity or atmosphere before I can leave it."[172]

Ruskin, like F. G. Stephens, championed Hunt in particular for his innovative approach to color and shadow. Famously, Ruskin recalled in 1883 that Hunt's *Our English Coasts* of 1852 (fig. 16) "showed us, for the first time in the history of art, the absolutely faithful balances of colour and shade by which actual sunshine might be transposed into a key in which the harmonies possible in material pigments should yet produce the same impressions upon the mind which were caused by the light itself," praise that recalls his exultation of Brett's verisimilitude in *Val d'Aosta*.[173] In other words, Hunt's depiction of strayed sheep along the cliff tops at Covehurst Bay at Fairlight Glen near Hastings could produce the same result upon the mind as had the original scene.

This sense of immediacy was achieved not only by the color but also by the composition. The cropping of the sheep at the lower right, for example, creates the sensation that the viewer is standing at the edge of the cliff, looking out on the scene beyond as the sheep mill about. A sense of physical intimacy—of a scene patiently observed by the painter—is suggested by the gaze of the sheep in the foreground toward the viewer. The sensation of proximity is reinforced by the meticulous rendering of textures, ranging from the semiglossy leaves of the brambles to the clotted matting of sheep's wool. Such attention to the details of nature, evenly distributed throughout the foreground of the canvas, suggests that the composition, overall, offers the full complexity of nature without hierarchy.

The sensation of immediacy is also conveyed by Hunt's depiction of light and color. Touches of highlight on the dark green upward-curving bramble sprigs in the foreground suggest the reflection of light, while the lighter, more yellowy-green touches on the downward-hanging leaves imply the passage of light through the delicate cells of the leaves. The sheep's wool, like the rocks along the cliff line, alters in coloration in accordance with the play of light. In the middle ground, trees cast shadows over the sunny meadow, coursed with walking paths traced out in darker lines along the cliff's edge. The sun's light, entering the scene from the right, skims the surface of the cliff's top, leaving the descent toward the sea in shadow. Beyond, as the sea stretches out to the horizon, the light plays over the surface of the reflective water, its movement and depth suggested by alternating passages of darker blue and lighter blue-green.

In his overview of Hunt's career, published in 1860, F. G. Stephens explicitly linked Hunt's approach to color to the work of Davy and Brewster. "He was probably the first figure-painter," Stephens wrote, "who gave the true colour to sun-shadows, made them partake of the tint of the object on which they were cast, and deepened such shadows to pure blue when he found them to be so."[174] It is very likely that Hunt was aware of Brewster's work

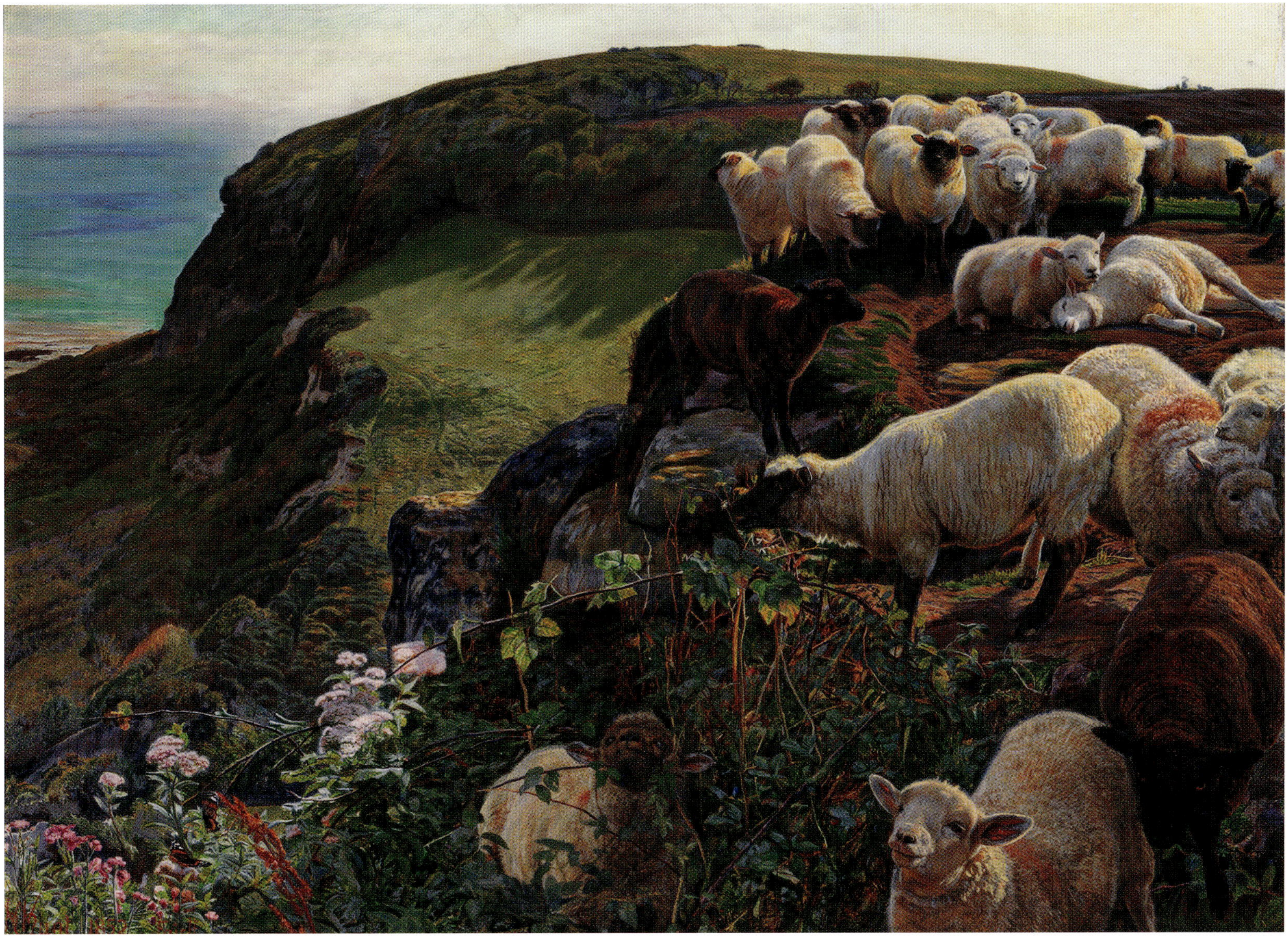

FIG. 16
William Holman Hunt (1827–1910), *Our English Coasts, 1852 ("Strayed Sheep")*, 1852. Oil on canvas, support 432 × 584 mm. Presented by the Art Fund, 1946. 

through D. R. Hay's *The Laws of Harmonious Colouring*, which engaged with Brewster's theories and had been given to the painter by his father.[175] In the third edition, of 1836, Hay, a Scottish housepainter, included a discussion of the scientific laws on which he had based his advice. Dissatisfied with Newton's theories of color, Hay reported his delight in hearing a lecture by Brewster that definitively established, through the use of a prism, the existence of only three primary colors, and proved that secondary colors were composed of combinations of these three primary colors. Referring also to Field, Hay described how the perception of color intensity could be enhanced through contrast and harmony.[176] Hunt's interest in color theory is also demonstrated by a watercolor he made of a kaleidoscope, which Brewster had invented to analyze color contrasts and complementarity.[177]

Hunt's intensely bright and saturated color choices, enhanced by his use of white ground, astonished contemporary viewers, given that prevailing practices dictated using a brown ground and working from dark to light and from fat to lean, with white highlights added in the final stage. Hunt, in his autobiography, recalled that the Pre-Raphaelite break with accepted practice was fueled by the search for truth and the practice of painting directly from nature:

> *When we Pre-Raphaelites were charged with exaggeration in our key of colour, and were told that our pictures had all the hues of the rainbow, we replied that the brown shadows of old professors did not give the impression of open-air effect which we had been surprised to discover while searching for the truth before Nature herself. We registered prismatic hues because we found that each terrestrial feature mirrored the blue sky and the tints of its neighbouring creations; and we maintained that while a part of our picture by itself might appear over-coloured, it was consistent in the impression it gave of truth.*[178]

Here, Hunt explains that by using colors as found in nature, the Pre-Raphaelites enhanced the truth-value of their pictures. He could have added that by drawing on contemporaneous theories of color derived from close study of the behavior of the light spectrum, he was rooting his practice in the cultural authority of science.

### Tensions, Resistance, and Challenges

While the midcentury archive of text and image offers up a preponderance of evidence to support my argument concerning the intersection of art and science at the point of truth to nature, I want also to indicate the fissures, weak spots, and disjunctures that developed over the course of the formation and consolidation of these practices and discourses. The pursuit of truth to nature was highly debated, its outcome never a foregone conclusion. I therefore want to conclude this chapter with a series of brief episodes, drawn from both photography and painting, that point to tensions, skepticism, and paradoxes produced by the conjoined enterprise of modern science and modern art.

I want to begin by probing further Ruskin's notion of the innocent eye, a metaphor for the ways in which he hoped young artists would broach nature. Obviously, even in itself, this is a suspect notion, for seeing without assigning meaning would require scrupulous acts of mental discipline. I am particularly interested in those arguments that push against Ruskin's aim of untutored vision or its photographic corollaries, the photogenic drawing and calotype, modes of art making that implied dispassionate transcription.

In 1839, Talbot proposed to the naturalist William Hooker, then chair of botany at the University of Glasgow, a book "on the plants of Britain, or any other plants, with photographic plates." Hooker raised objections to the photographic examples that Talbot sent him, however, because of their failure to communicate necessary information. For example, he wrote the photographer, "Your beautiful Campanula hederacea was very pretty as to general effect:—but it did not express the swelling of the flower, nor the calyx, nor the veins of the leaves distinctly."[179]

When Hooker eventually produced a study of ferns, he turned to the botanical illustrator Walter Fitch for drawings, reproduced as color lithographs, to illustrate his text (fig. 17). Fitch was Hooker's first choice for most of his publications, and he built up a considerable body of knowledge concerning plant material and a corresponding reputation. According to Hooker, informed vision, as opposed to the "innocent eye," ensured truth to nature.[180]

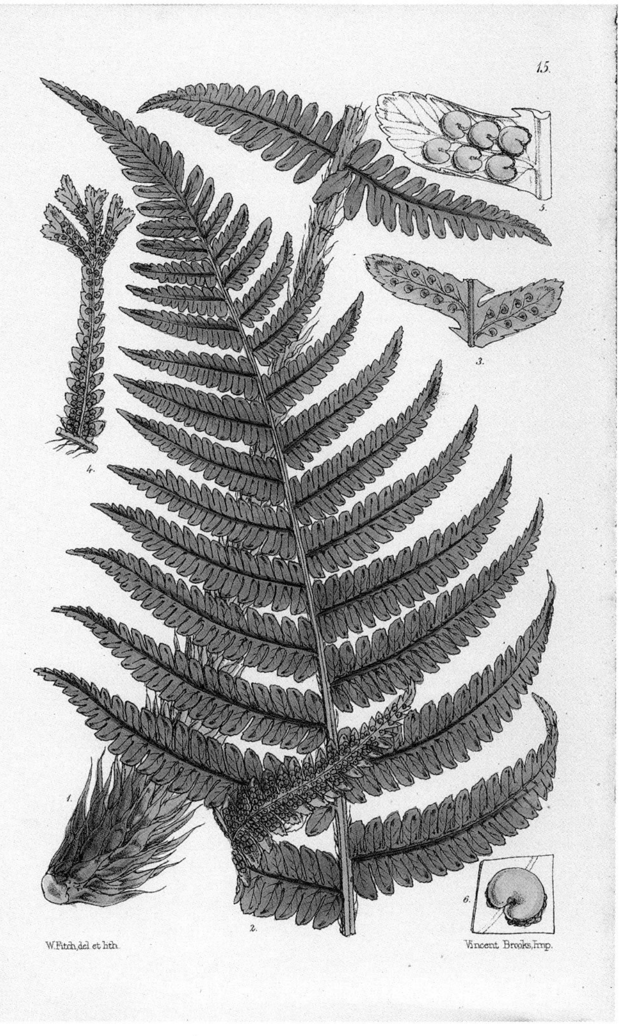

FIG. 17
Walter Fitch (1817–1892), "Male Buckler Fern." In William Jackson Hooker, *The British Ferns* (London: L. Reeve, 1861), plate 15. Gernsheim Collection, Harry Ransom Center, The University of Texas at Austin.

When Thomas Moore, curator of the Botanic Garden of the Apothecaries, and John Lindley, secretary of the Royal Horticultural Society, collaborated on a book titled *The Ferns of Great Britain* (1855), they chose to use a process dubbed "nature printing" rather than photography or lithography (fig. 18). As a commercially viable process, nature printing had been developed by Alois Auer, who conceived the method by which dried specimens were impressed into lead plates in an intaglio manner.[181] When the plate is printed, particularly dense aspects of the plants, such as stems, are slightly raised above the planar surface. Thus the image is endowed with a physical and material presence that two-dimensional representations lack. In addition, unlike the calotype or cyanotype, nature printing allowed for the replication of texture as well as the easier application of color. The resulting book is a large folio,

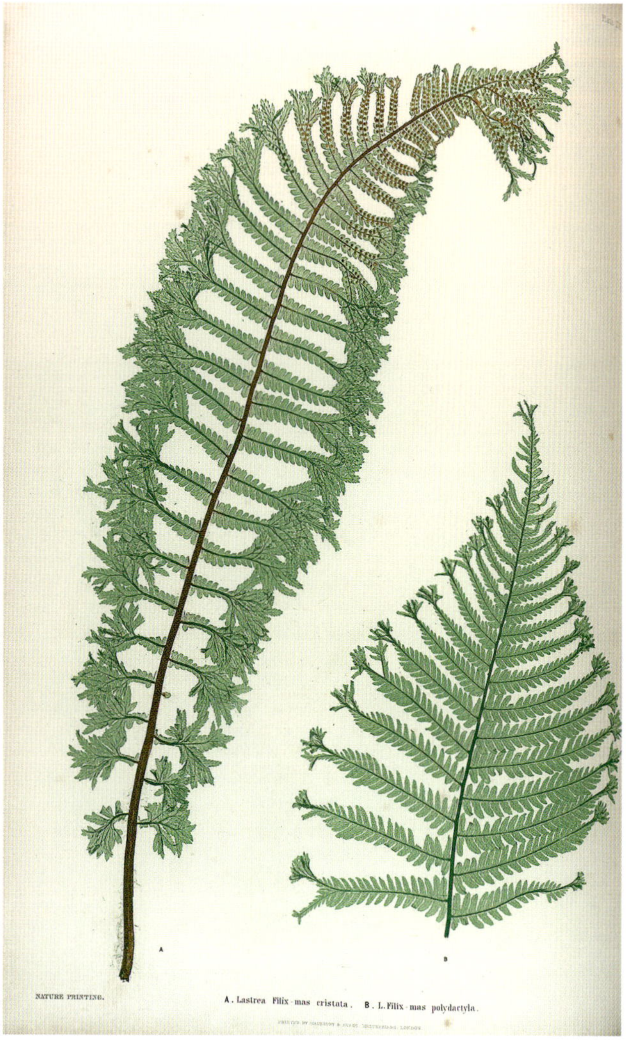


FIG. 18
"Plate XIV, Lastrea Filix-Mas." In John Lindley and Thomas Moore, eds., *The Ferns of Great Britain and Ireland, Nature Printed by Henry Bradbury* (London: Bradbury and Evans, 1855). Gernsheim Collection, Harry Ransom Center, The University of Texas at Austin.

with each fern shown in its actual size and identified by its Latin name, a format derived from Enlightenment herbariums.

There are notable similarities between nature printing and Talbot's photogenic drawing. Both are indexical to the object, and, as Carol Armstrong observes, the authorial function has been taken away from a human agent and given to nature itself. Yet the producers of nature prints claimed that their medium was superior.[182] Henry Bradbury, the publisher of *The Ferns of Great Britain*, insisted that readers had to be convinced that the prints were not the thing itself. Moreover, the medium could "represent faithfully those minute peculiarities by which natural objects are often best distinguished," eliminating the need for any "human hand to execute" the reproduction of nature.[183] Here, nature printing has encroached on the same field of expertise claimed for photogenic drawing—the faithful, detailed representation of nature.

The botanist Lindley argued that nature printing was superior to both daguerreotypes and calotypes, which failed to depict "facts with the necessary accuracy." Lindley added that although nature printing could represent only the surface of a plant, "and not the whole even of that," it nonetheless provided the best means of acquiring visual knowledge. The internal organization of the plant, he insisted, was "not practically employed in the first-sight recognition of a Fern"; the "practised eye" relied upon external features for identification. For Lindley, incompleteness was not incompatible with truth to nature, and photography did not have a unique purchase on truth. Moreover, the practiced eye could overcome such limitations because of repeated experience. Lindley concluded that nature printing depicted nature "with unerring truth" by providing a visual record

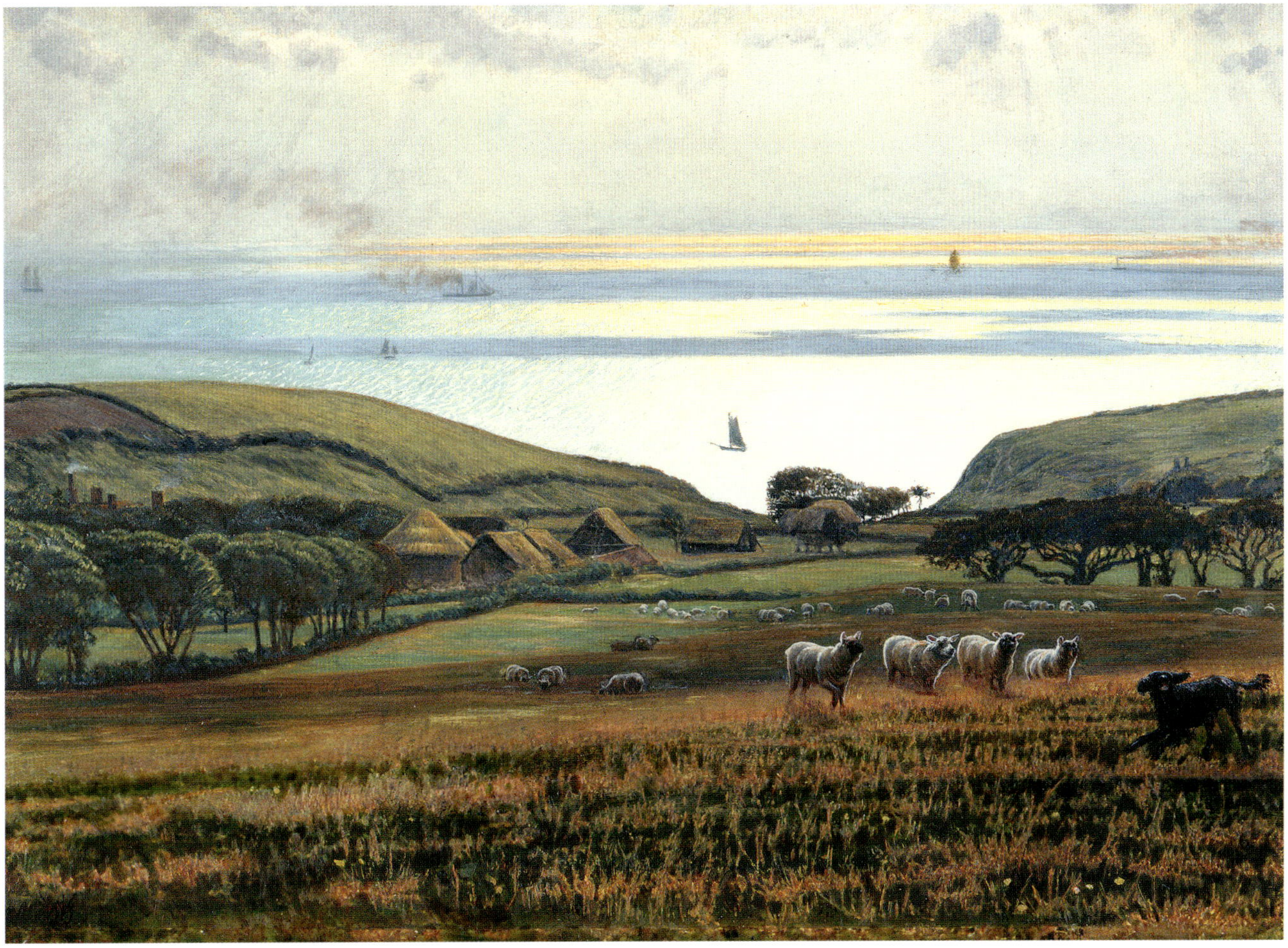

FIG. 19
William Holman Hunt, *Fairlight Downs, Sunlight on the Sea*, 1852–58. Oil on panel, 21.6 × 31.1 cm. Private collection.

of the fern's "general manner of growth, the ramification of its stipes, and the form of its leaflets."[184]

Yet, while Lindley denied that truth was the unique province of photography, his reference to the medium indicates its cultural valence as authoritative. This conviction informed the critic Philip Gilbert Hamerton's assessment of William Holman Hunt's painting *Fairlight Downs, Sunlight on the Sea*, which Hamerton declared "as true [as] and truer than the photograph" (fig. 19).[185] The painting was begun when Hunt was painting *Our English Coasts* at Hastings. In the foreground, at right, a black dog appears poised to dash forward, toward the left side of the canvas. The gaze of the dog, echoed by that of the four sheep just behind it, suggests a point of visual interest beyond the frame of the canvas. An unseen figure has presumably launched the stick that nearly touches the ground at the far left side of the canvas. Between these two nodes of activity lies a shallow valley sloping to the sea. At the water's edge, near the lip of the valley, stands a clump of trees silhouetted against the brightly lit reflective surface of the water. A series of implied lines in the composition—a hedgerow and contour of trees at left, echoed by a cluster of buildings and the crest of the hill farther to the left, and the more sharply rising cliff at right—creates another field of visual interest at the center of the canvas, where a sailboat stands out against the nearly white sea. The alternating passages of white and blue that mark the sea's surface register both waves and the clouds dotting the blue sky above. When the painting was exhibited at Gambart's French Gallery in 1858, the critic for the *Athenaeum* lauded the work: "There is such an intensity of conception about this picture that where other pictures seem smeared, stained, or scratched, this seems sculptured out of slabs of jewel colour, and then soaked and washed in air and sunlight." The critic underscored the thrill of viewing the painting by advising readers that "this is no vulgar transcript of nature, this is a fine sea-poem,—a poet's hand rendering what a poet's eye has seen."[186] While Ruskin had hoped that artists, once they had acquired sufficient experience, would explore the territory of poetry as Turner had done, he supposed that they would do so through mystery and incomprehensibility. Hunt, according to this critic (probably Stephens), achieved poetry through truth to nature, collapsing the distinctions that Ruskin had attempted to maintain.

Hamerton discussed the veracity of Hunt's painting in his treatise on painting out of doors, comparing it to one of Gustave Le Gray's marine photographs, which had first been exhibited publicly in Great Britain at the Norwich Photographic Society in 1856. Hamerton found Le Gray's combination photograph (produced from multiple negatives) wanting. Although the critic admitted that the effect of light on the water was accurate, he argued that the clouds and water along the edge of the picture were far too dark. Hunt, by contrast, deployed a rich variety of colors that could not yet be achieved in photography, and also conveyed such atmospheric effects as "hot haze in the blinding distance miles away, and . . . a sultriness in the accumulated clouds." Hamerton explained that "all these other facts Hunt could get into his picture because painting is a great intellectual art." Photography, however, was a mechanical process, "an art-science; narrow in range, emphatic in assertion, telling one truth for ten falsehoods, but telling always distinctly the one truth that it is able to perceive."[187]

Human vision, unlike the purely mechanical camera eye, could achieve the most comprehensive truthfulness, according to Hamerton. While photography and Pre-Raphaelite painting shared similar intellectual origins in inductive science, their practitioners competed to claim superior veracity and to garner cultural approbation for their achievements.

This sense of competition was fueled by the relatively rapid spread of photography, despite Talbot's constraining patent and license structure. At the same time, Talbot moved photography away from its inductive origins, obscuring its ontological status. The recurring association of the "picturesque" with Talbot's enterprise, a label applied by critics and even by Talbot himself, troubles the account I have given thus far of photography's disciplined truth to nature. In 1841, Talbot wrote to Herschel, "I must now really transport my apparatus to some locality where picturesque objects are to be met with, such as a Cathedral or seaport Town, for my own neighborhood is not particularly suited to the Artist, and offers no great variety of subjects."[188]

This statement suggests that Talbot regarded the subjects of his photographs as key to their persuasive power and the more widespread adoption of the medium. Talbot pursued a picturesque locality for *Sun Pictures in Scotland* (1845), which conforms to the notion of a picturesque tour in the repertoire and conventions of the images, which include Sir Walter Scott's home of Abbottsford, the ruined abbey of Melrose, and Loch Katrine (fig. 20). In a view of Loch Katrine, the visual trope of roughened clumps of trees and craggy hills framing bright patches of water evokes the picturesque as depicted in Gilpin's *Observations on the River Wye*, first published in 1782 (fig. 21).

But such recourse to the picturesque complicates Talbot's claim of objectivity, as well as his assertion that nature alone produced the photograph. In the "Notice to the Reader" on the errata leaf for *Sun Pictures in Scotland*, he states, "the plates of the present work are impressed by the agency of Light alone, without any aid whatever from the artist's pencil. They are the sun-pictures themselves, and not, as some persons imagined, engravings in imitation." The picturesque paradoxically claimed naturalness—being inherent in nature—while concomitantly relying upon a replicable system that required written explanation. Gilpin's definition of the picturesque, with its recourse to painting, reveals a central paradox or circularity in the concept. It was a system for viewing landscape that derived from painting, and a system for representing landscape in paintings and drawings as well as in gardens. It was a means of selecting and framing the scene; but, as Gilpin was careful to insist, it could not be reduced to "the *rules of painting*." One needed to absorb the precepts of the picturesque fully in order to identify it in the external world and then to represent that scene of picturesque nature in the chosen medium. In both life and art, it required an educated eye: "the province of the picturesque eye is to *survey nature*. . . . The picturesque eye, it is true, finds its *chief* objects in nature, but it delights also in the images of art, if they are marked with the characteristics, which it requires."[189]

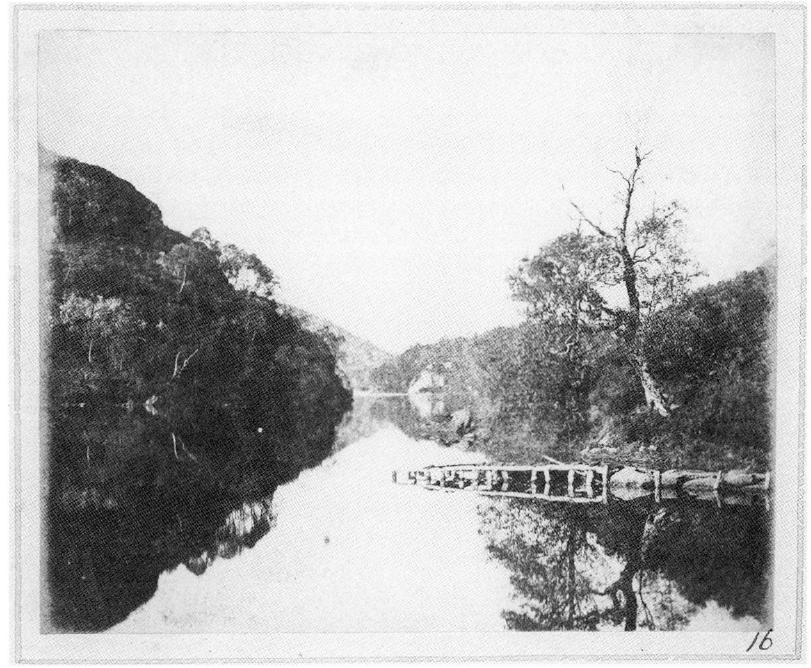


FIG. 20
William Henry Fox Talbot, *Loch Katrine*. Salted paper print from a paper negative. In Talbot, *Sun Pictures in Scotland* (London, 1845). Gernsheim Collection, Harry Ransom Center, The University of Texas at Austin.

So what of the camera's eye—the one that was supposedly unmediated and that communicated only what

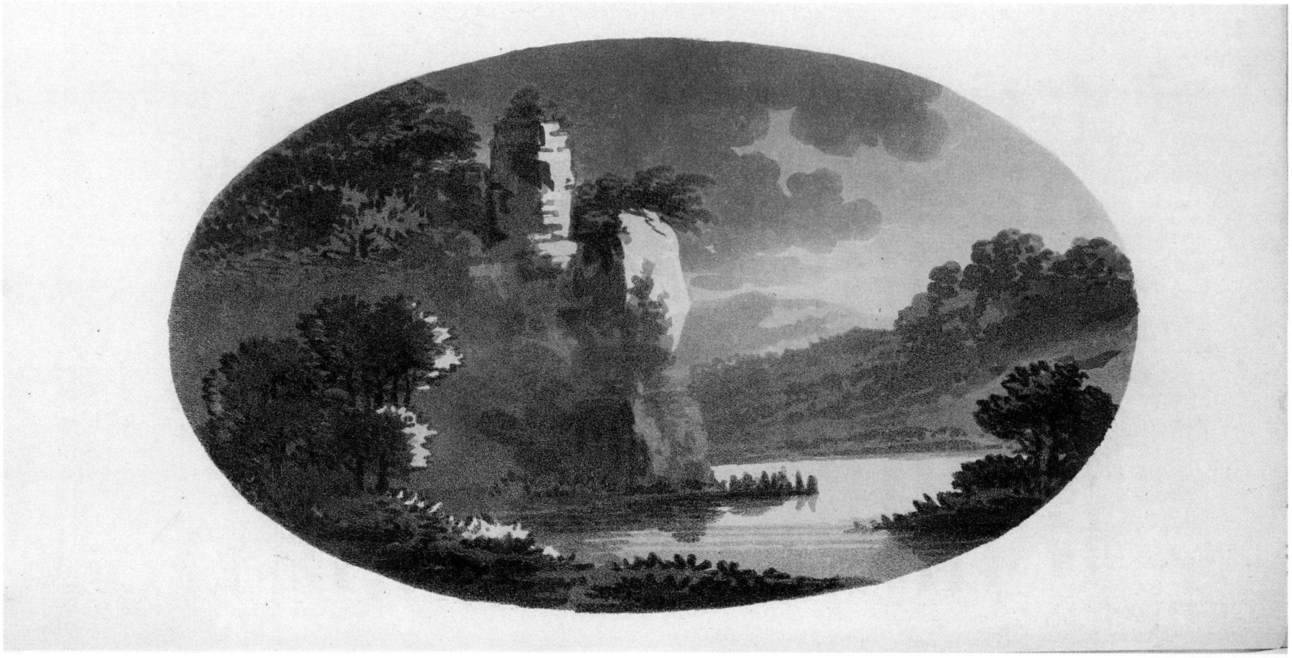

FIG. 21
William Gilpin (1724–1804), *New Weir*. In Gilpin, *Observations on the River Wye*, 3rd ed. (London: Printed for R. Blamire, 1792). The Getty Research Institute, Los Angeles (92-B25037).

lay before it? In fact, photographers quickly absorbed the aesthetic of the picturesque. This was not a surprising development, for the picturesque had become widely popular by the turn of the eighteenth century and had become closely tied to touring. "Nature was always already a picture in waiting," perceived by "a figure who looks with discernment while simultaneously being constituted as *discerning* by 'the habit of viewing,'" as the historian of photography Geoffrey Batchen argues.[190] The presence of this figure disassembles the argument that nature produced the photograph. Or did it, in fact, if the picturesque was inherent in the scene? Did the photograph instead reify the picturesque, and confirm, by virtue of its documentary, truth-telling mode, that the picturesque existed, waiting to be captured by the image? If the photograph did confirm the picturesque, did photography thus deny its unique status by mining the same vein as painting, drawing, and garden design?

The lineage of photographically illustrated books of landscape scenery confirms the direction that Talbot pursued in his views of Scotland. A few examples of this genre will suffice to demonstrate this point. In 1859, the photographer Philip H. Delamotte edited a collection of images by himself and colleagues titled *The Sunbeam: A Book of Photographs from Nature*. Although the book lacked the polemical tone of Talbot's earlier *Pencil of*

*Nature*, the reference to sun in the title situated the book in the rhetoric of the photograph as the work of nature. J. D. Llewelyn's opening print, *The Woods at Penllergare*, makes recourse to pictorial convention in the path that enters the scene in the lower right-hand corner and wends its way into the middle ground (fig. 22). There it meets a railroad track. This definitively unpicturesque element, however, receives no comment either pictorially or textually. The compositional arrangement does not overtly call attention to the structure; it is treated equally with the rustic bridge that leads from the path to the track. Indeed, the upward thrust of the trees pulls visual interest away from the rails that nestle at the foot of the trees. Moreover, the accompanying poem, by Mary Russell Mitford, lacks any reference to a railroad track, and recounts how the path leads through a wood to a valley of meadowland with a farm in its midst, a place of "peace, / A deep repose, a silent harmony, / Of nature and of man." The text is thus essential for a picturesque reading of the image.

The family tree of resemblances extends from *The Sunbeam* to *Sunshine in the Country: A Book of Rural Poetry, Embellished with Photographs from Nature* of 1861. As the subtitle suggests, the emphasis in this book has shifted from the photograph to the text. Since the photographs typically act as illustrations to the poems, or are interspersed with the text, they create a context and extend the possibilities of the words at the same time that they concretize the verbal imagery (fig. 23).

In these books, the discourse of truth to nature and of images produced solely by nature has disappeared in favor of *ut pictura poesis*, "as is painting so is poetry." Thus, despite photography's origins in the inductive method, certain practitioners began to eschew its scientific genealogy. Moreover, many photographers

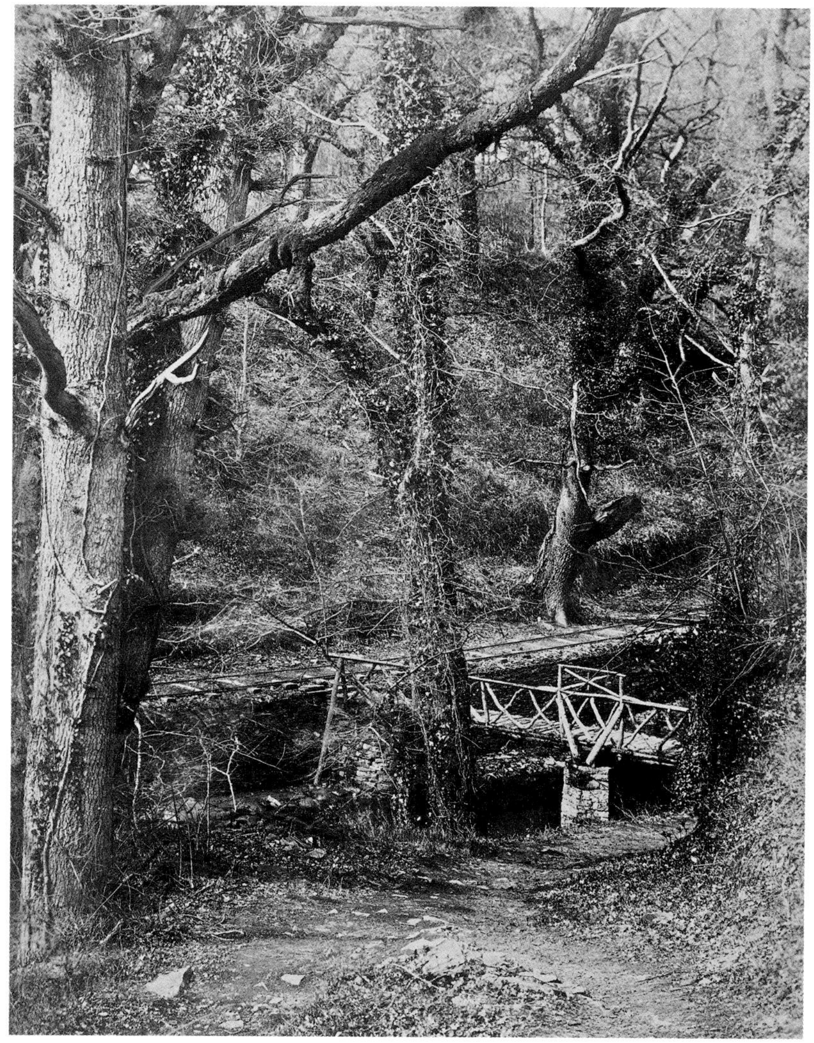

FIG. 22
John Dillwyn Llewelyn (1810–1877), *The Woods at Penllergare*. Albumen silver print, 20.8 × 15.9 cm. In Philip Henry Delamotte, ed., *The Sunbeam: A Book of Photographs from Nature* (London: Chapman and Hall, 1859). The J. Paul Getty Museum, Los Angeles.

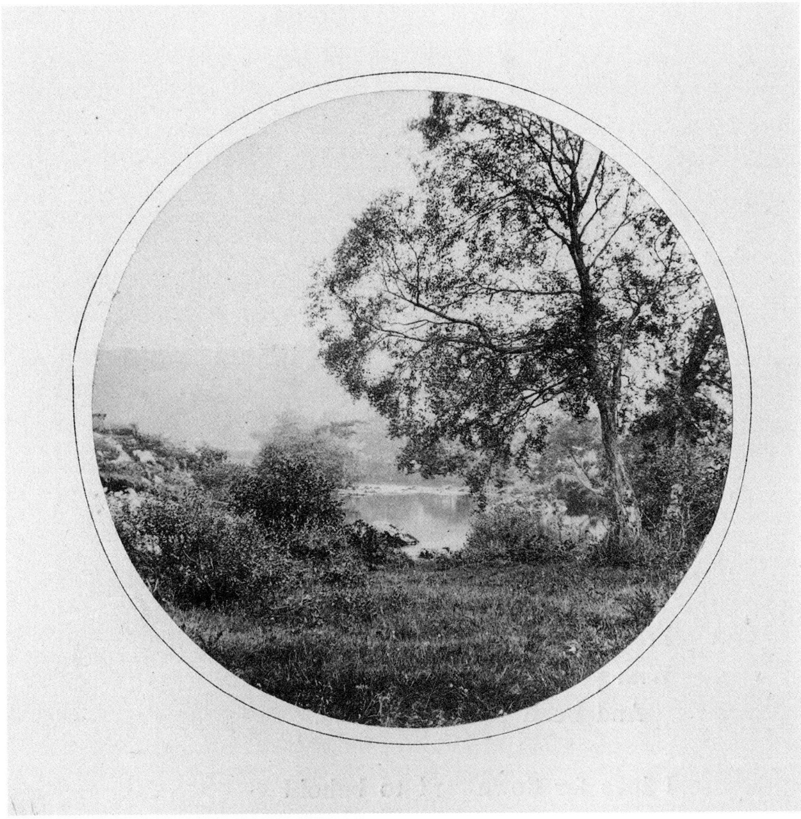

FIG. 23
William Morris Grundy (active 1860s), "A Landscape." In Grundy, *Sunshine in the Country: A Book of Rural Poetry, Embellished with Photographs from Nature* (London: Richard Griffin and Co., 1861). Gernsheim Collection, Harry Ransom Center, The University of Texas at Austin.

began to reject its most telling characteristic: the detail. In 1853, William Newton presented a paper to the Photographic Society of London in which he distinguished scientific photography from fine art photography; the former was "perfect in detail," while the latter should "endeavour at producing a broad and general effect," which might be achieved by throwing the subject out of focus. Furthermore, Newton championed what the photographer—a subjective individual—could contribute to the process rather than disavowing his or her presence: "Consequently, I conceive that when a tolerably faithful and picturesque effect can be obtained by chemical or other process, applied to the negative, the operator is at full liberty to use his own discretion."[191] Thus the photograph, as a publicly exhibited and commercially published form, had become absorbed into existing picture-making systems—specifically, the picturesque—that ran alongside scientific developments.

But this shift did not take place without resistance; in the *Journal of the Photographic Society of London*, for example, Brewster argued that "there is no poetry in the pencil of the Sun. The photographer cannot separate what is beautiful from what is common."[192] And neither was this shift complete; later photographers, such as P. H. Emerson, still accorded the telling detail significant value. The photograph, in sum, was unstable, its connotations highly dependent on the context in which the image was deployed and discursively situated.

Ironically, to reform painting, Ruskin proposed the equanimity associated with photography—"rejecting nothing, selecting nothing, and scorning nothing." Following this aesthetic path, however, was a considerable challenge, as Ford Madox Brown's landscape paintings of the 1850s reveal. Brown never joined the Pre-Raphaelite Brotherhood, but he was quite close to both its members and its principles. Born in Calais, Brown studied with the Flemish history painter Egide Charles Gustave, Baron Wappers. In 1846, he moved to England, after submitting cartoons for the competition to decorate the Houses of Parliament. Around 1848, he met Rossetti, who studied with him briefly, and through this connection Brown entered the orbit of the Pre-Raphaelite Brotherhood. While he aspired to history painting, in the mid-1850s he set his sights on producing a series of small views that he hoped would find a ready market and thus improve his meager finances. He turned to the River Brent for subject matter.

*The Brent at Hendon* (1854–55) refuses to conform to common typologies of landscape. Its nonhierarchical and densely organized composition rejects the classic landscapes of Poussin, organized like stage sets, and the picturesque of Gilpin. Like Brown's earlier painting *An English Autumn Afternoon*, *The Brent at Hendon* is oval in format and difficult to apprehend in terms of spatial recession and figure-ground relationship. Alison Smith has pointed out that the oval format approximates the human eye, and Allen Staley argues for an affinity with binocular vision.[193] The act of viewing demanded by this diminutive painting, which measures roughly eight by ten inches, also mimics the type of looking entreated by objects that Victorians used to encase nature, such as the terrarium and the bell jar.[194]

Brown recorded in his diary his response to the scene he witnessed on 1 September 1854 and the challenges it presented: "a mere brooklet running in most dainty sinuosity under overshadowing oaks and all manner of leafgrass. Many beauties & hard to chuse amongst for I had determined to make a little picture of it. However nature that at first sight appears so lovely is on consideration almost always incomplete, moreover there is no painting intertangled foliage without lossing half its beauties. If imitated exactly it can only be done as seen from one eye & quite flat & confused therefore."[195]

Brown could not decide what to focus upon and, moreover, found it difficult to capture the whole because of its complexity. His response was to attempt to short-circuit his visual perception of three dimensions by closing one eye, as Ruskin had recommended, but the result was a flat field that thwarted his effort to paint the scene. It was as if Brown was attempting to mimic the experiments in perception that fueled Brewster and others.[196] The embowered scene contained more than he could represent; its complexity eluded him. He experienced considerable frustration when he attempted to resume painting after a rainstorm had passed and he "found the weight of water quite displaced the different branches from their normal possition making confusion."[197] Faced with rapidly changing conditions, the very stuff that inspired Turner, Brown struggled.

While painting the Hendon landscape, Brown embarked on several other landscape studies in which he grappled to depict in meticulous detail the mutability of nature, although he had already acknowledged to himself "altoghether these little landscapes take up too much time to be proffitable." He attempted to depict a turnip field only to encounter "sun shine when I did not want it, cold & wind when it went . . . [and] the turnips too difficult to do anything with of a serious kind," adding, "I dont know if it would be possible to paint them well, they change from day to day." In early October he despaired, with respect to a cornfield he had selected as a subject, that "very small trees in the distance are very difficult objects to paint or else I am not suited to this sort of work for I can make nothing of the small screen of trees though I have pottered over [them] sufficient time to have painted a large landskape." Pre-Raphaelite ideals of strenuous self-discipline and rigorous study of nature were at odds with the reality of the capitalist marketplace. Brown finally sold *The Brent at Hendon* and *Carrying Corn* of 1854–55 (fig. 24) to the dealer D. T. White on 27 June 1855 for £10 and £12, respectively.[198] While £20 a year was

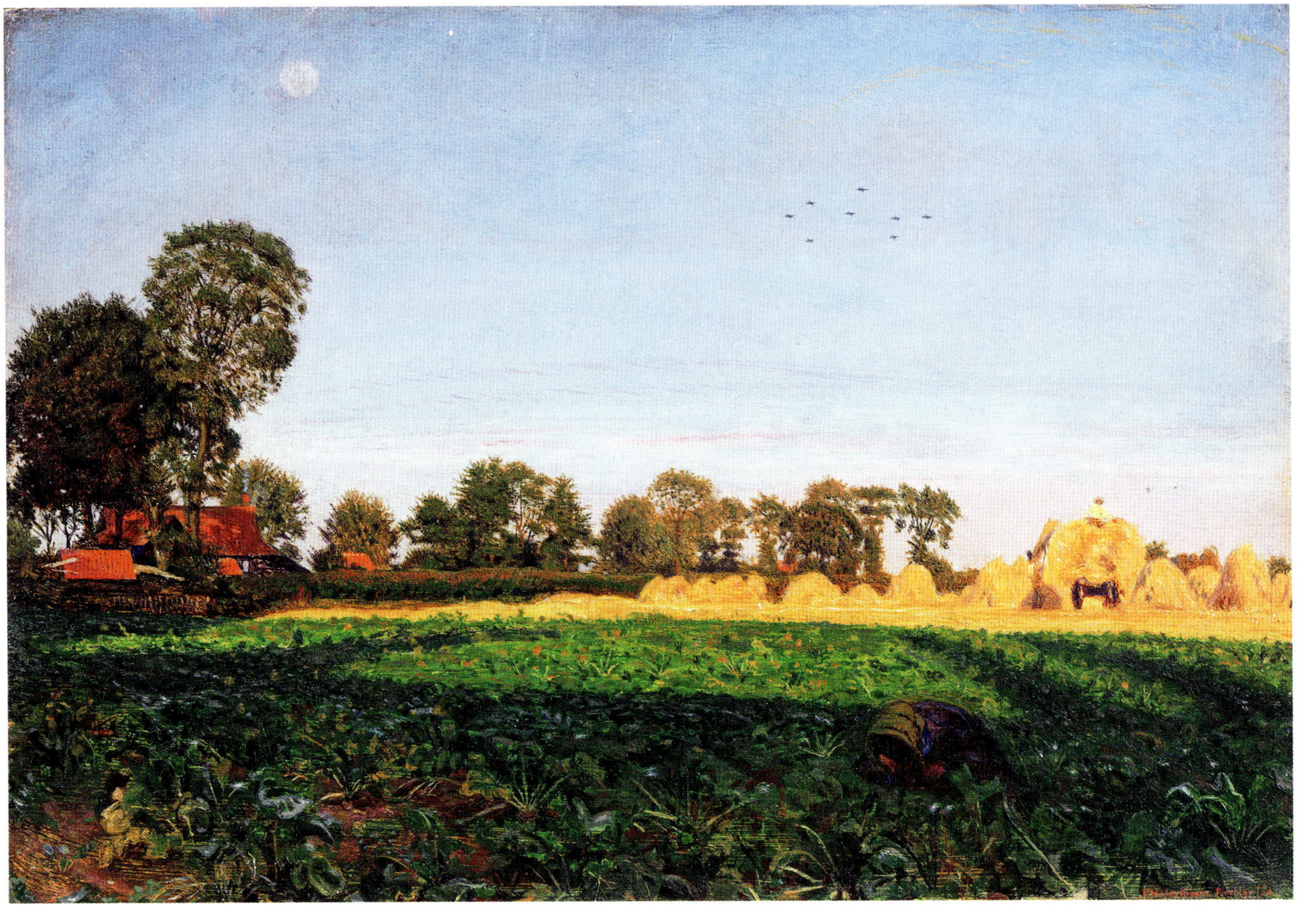

FIG. 24
Ford Madox Brown, *Carrying Corn*, 1854–55. Oil on mahogany, support 197 × 276 mm. Purchased 1934. © Tate, London 2015.

regarded as a middle-class income, it was equivalent to a very low annual wage for a clerk; £150 a year was the typical threshold for income tax.[199]

A year or so later, Brown sought out the subject of *The Hayfield* (1855–56), which shows both agricultural laborers and the artist at the end of their workday (fig. 25). The laborers harness the horses and begin to gather together, while the artist reclines against a heap of hay with a palette in his hand and the tools of his trade as an outdoor painter—umbrella and stool—nearby. He views the scene with one hand raised to his head, perhaps reenacting Brown's moment of contemplation that sparked this painting. In his diary entry of 21 July 1855, Brown recalled that he "looked out for landscapes this evening" and again experienced anguish when faced with the ever-changing rural scene:

*How despairing it is to view the loveliness of nature towards sunset & know the impossibility of imitating it, at least in a satisfactory manner as one could do would it only remain so long enough. Then one feels the want of a life study such as Turner devoted to landskape & even then what a botch is any attempt to render it. What wonderful effects I have seen this eveng in the hay fields, the warmth of the uncut grass, the greeny greyness of the unmade hay in furrows or tufts, with lovely violet shadows & long shades of the trees thrown athwart all & melting away one tint into another imperceptibly, & one moment more & cloud passes & all the magic is gone. Begin tomorrow morning all is changed, the hay & the reepers are gone most likely, the sun too, or if not it is in quite the opposite quarter & all that was lovelyest is all that is tamest now, alas! it is better to be a poet—still better a mere lover of nature, one who never dreams of possession.*[200]

This passage reveals that Brown never doubted his perception; he was clearly entranced by the scenes he witnessed. Contemporary theories of vision that described images painted on the retina and promised that even "the minutest eye of the minutest animal, which itself requires a microscope to make it visible, contains in the invisible image which is painted on its retina, a representation of the external world," as Brewster put it, would have confirmed to Brown his ability to represent such a mutable scene.[201] But he found his own study of nature insufficient to the task he set for himself. Brown yearned for the practiced eye of Turner and the mental warehouse that could inform his act of representation.

The trope of the closely observed detail that cost so much of Brown's labor also occasioned criticism of such Pre-Raphaelite works as Charles Allston Collins's painting *May, in the Regent's Park* of 1851 (fig. 26). The scene, probably painted from the window of a building facing the park, offers a view of *rus in urbe*. A fenced road courses horizontally through the center of the image; sheep graze in the field behind the fence; and a carefully planned lake, curved according to the principles of eighteenth-century garden designer Capability Brown, sparkles in the background, while the foreground is dominated by a large flower bed, planted in the gardenesque style. A man in a top hat and a child stroll along the fence at right, confirming the modernity of the scene. The critic for the *Athenaeum*, probably F. G. Stephens, faulted the artist for exaggerating his attention to detail, insisting that "these objects, at the distance imagined in the picture, and reduced to its scale, could by no means be seen with such *hortus siccus* minuteness." This review reveals that despite the legitimacy that scientific and artistic discourses granted the detail as a synecdoche for direct observation and fact, it did not automatically signal truth to nature. The critic even warned that detail could become a "mannerism," an affectation that undermined artistic merit.[202]

Put another way, once painters claimed that their work was truthful to nature, they were inviting critics to measure it against nature, and this comparison was not always favorable to the painter. In the case of Dyce's *Pegwell Bay*, for example, the *Athenaeum*'s critic was likewise critical of the painter's execution, while admitting that a higher bar was being set because of the painter's seriousness of purpose. In particular, the critic found fault with Dyce's handling of light: "in fact, the eye facing such a light would hardly perceive the immense amount of detail shown in the lines of white chalk before us. The figures keep the tone and tint of open, subdued

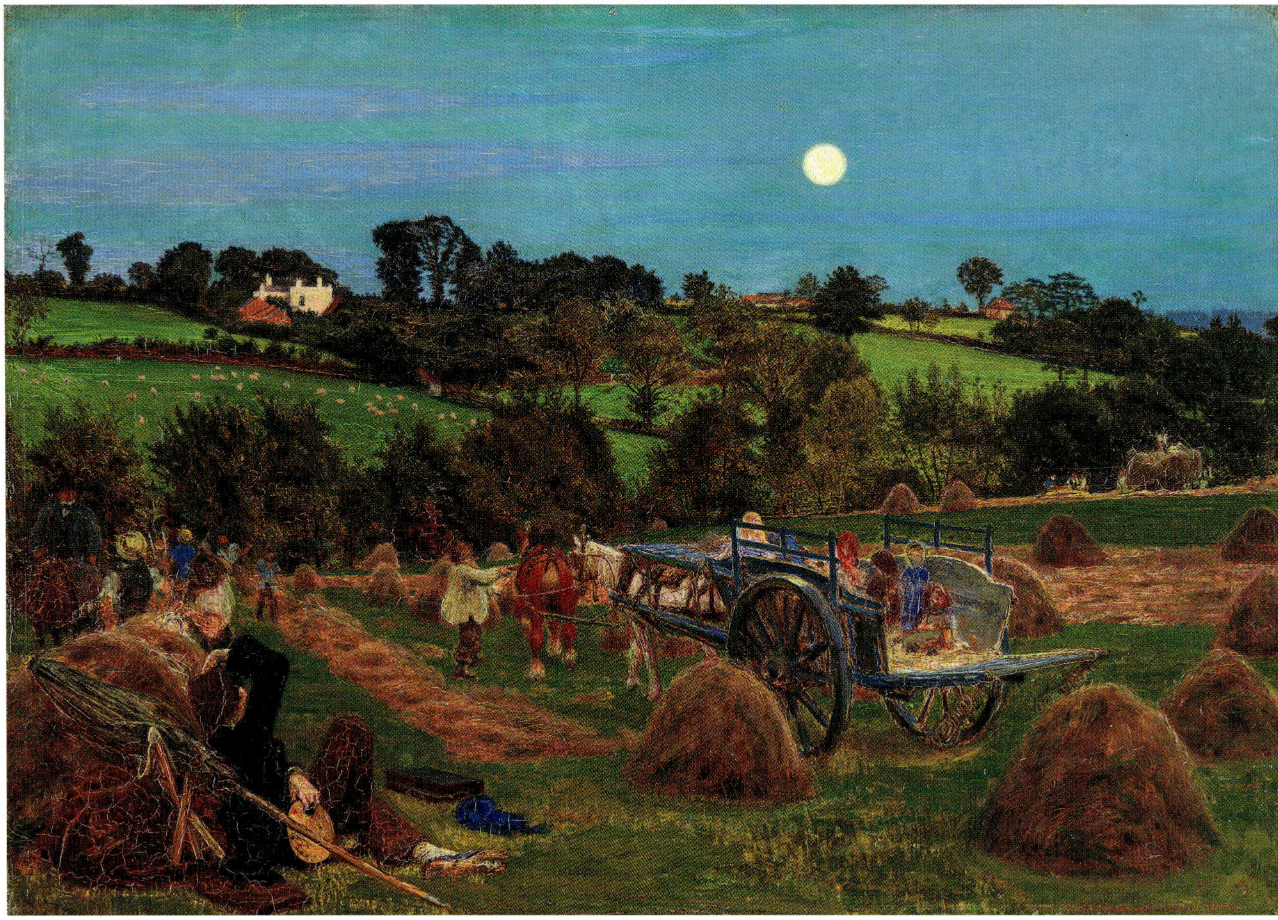

FIG. 25
Ford Madox Brown, *The Hayfield*, 1855–56. Oil on mahogany, support 241 × 333 mm. Purchased 1974. © Tate, London 2015.

daylight; whereas, facing the sunset, the reds, as in the lady's dress who stands in front, would be nearly purple; and her figure tell as a whole, opaque, solid and darkly, instead of being bright, thin and transparent, as it is shown to be."[203]

This question of inaccuracy and exaggeration troubled the reception of Hunt's late watercolors, in particular *Sunset at Chimalditi*, which was exhibited with *Festa at Fiesole* at the Old Watercolour Society in 1870.[204] The watercolor encompasses the hilltop view of the valley and sea below; a classical temple is visible in the valley floor just beyond the lip of the hillside. The artist's presumed viewing spot is indicated only by the cropped foliage in the foreground. The halation around the plants

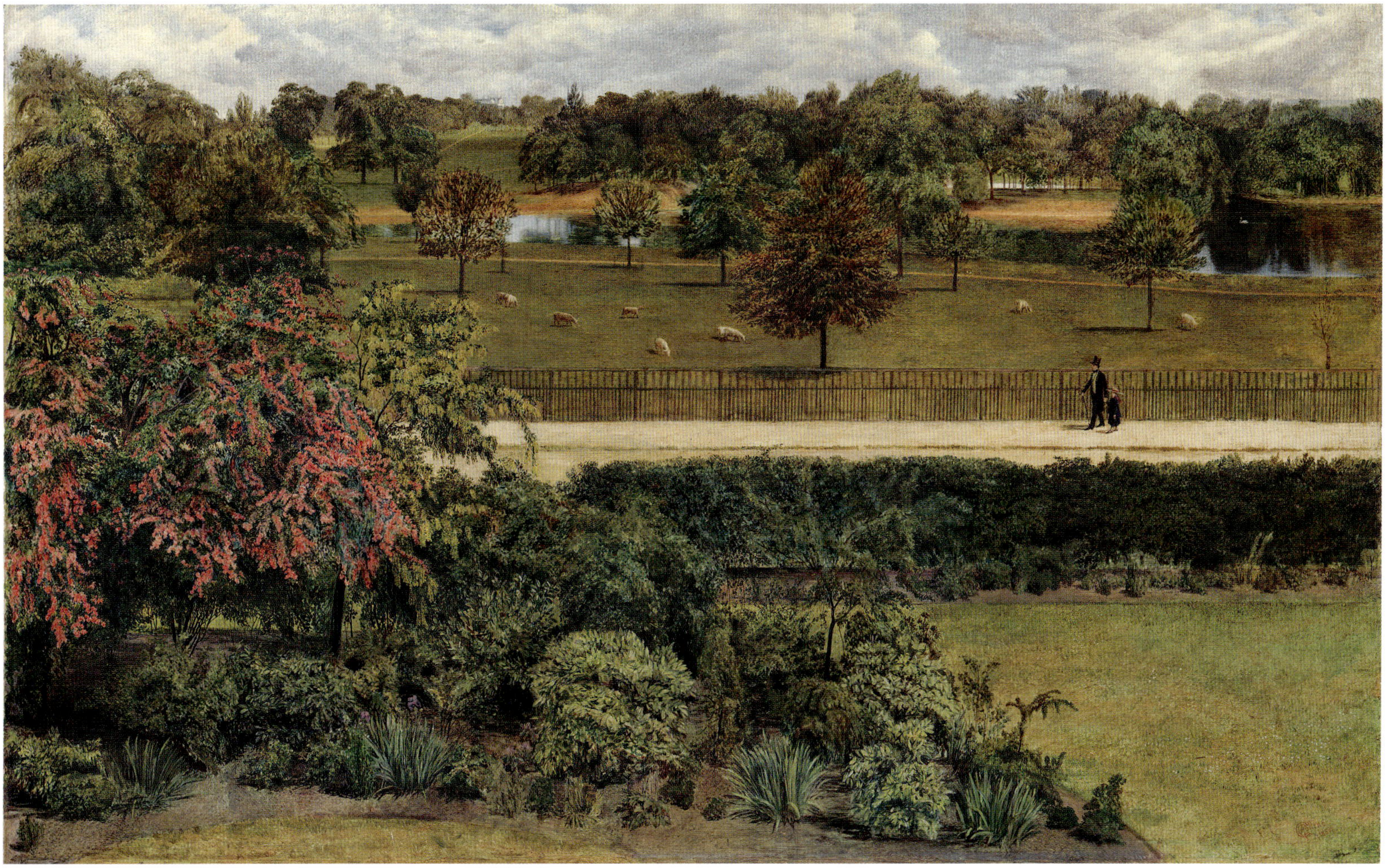

FIG. 26
Charles Allston Collins (1828–1873), *May, in the Regent's Park*, 1851. Oil on mahogany, support 445 × 692 mm. Purchased 1980. © Tate, London 2015.

registers the source of light as the setting sun, directly behind them and in front of the artist. The passage of the setting sun is also indicated by the wine-red stain on the ocean's surface, the pinkish tinge cast over the green fields of the valley, and the colorful undertones to the dark, thin clouds hovering over the sun. An even stronger halation is registered around the sun itself, which is surrounded by a circular whitish haze.

F. G. Stephens, in his *Athenaeum* review, reasserted the discourses of verisimilitude and inductive experimentation established thirty years earlier by explaining to his readers:

> *The aim, as we understand it, of the painter has been to render by means of painting two of those effects of light on the retina which, because of their extreme delicacy and marvelous difficulty, have not been attempted before; unless by Turner, with circumstances less favourable than those under which Mr. Hunt has worked, the appearance of intense radiance upon the eye in looking at a landscape has not yet hitherto been successfully attempted. . . . No. 58 is styled* Sunset at Chimalditi. . . . *The sun of the picture, i.e. of our eyes, forms a halo of such power that where the blood-red portion interposes to the green landscape behind, that part seems to quiver and fluctuate in its fierceness, where the upper part of the luminary appears with the sky as its background, if we may say so, it blazes in fiery light. . . . The drawings are to be looked at as pictorial experiments by one of the most powerful hands guided by keenest observation, and rendering with surpassing vigour and science two almost unapproachable effects of sunlight.*[205]

The implication is that perception occurs in the optical apparatus, the eye, a position Brewster had staked out earlier in debates regarding vision. But by 1870 Brewster's formula was several decades old and was being displaced by arguments developed by physiologists such as Hermann von Helmholtz, who held that "visual perception is not produced directly in each retina, but only in the brain itself by the means of impressions transmitted to it from both eyes." Helmholtz added, "All that we apprehend of the external world is brought to our consciousness by means of certain changes which are produced in our organs of sense by external impressions, and transmitted to the brain by the nerves."[206]

Thus it is perhaps no surprise that the *Art Journal*, in its review of Hunt's watercolors, offered an alternative reading to that supplied by Stephens. The critic argued that the scene did not replicate the optical perception of external nature but instead the mental phenomena induced by the scene. In criticizing *Sunset at Chimalditi*, he returned to the dichotomy first articulated by Ruskin in his essay "Pre-Raphaelitism," in his comparison of the methods of Turner and Millais.

> *But we do not regard the effect depicted as true to outward nature, it is rather the result of a frenzied vision; the eye when dazzled by the effects of light sees colours which do not exist outwardly and physically, but only inwardly and within the mind's consciousness. It is an interesting question, not easy of solution, how far a painter is justified in throwing into his picture a poetic cast of thought, a frenzy of colour, which correspond not to nature but to his individual idiosyncrasies. Turner's practice must be taken as an argument that cuts both ways, and we cannot but consider that Holman Hunt falls into an extravagance which even in Turner is not excused. The "sunset" before us wants the moderation of Art and the modesty of Nature.*[207]

While this critic judges Hunt's work a failure for being overly individualistic, he implicitly concurs with Stephens in evaluating the painting against external nature. Like an inductive experimentalist, the *Art Journal*'s critic wishes for effects that can be experienced and replicated by others. Yet, although the critic questioned

Hunt's methodology, Ruskin, who purchased the work, claimed that it was "so *exactly* true that everybody disbelieves its being true at all."[208] In other words, the truth-value of the work was so great that it produced its opposite, doubt.

Hunt himself, later in life, recast himself as an artist interested in capturing the mental perception of nature, as opposed to transcribing unmediated optical data: he recalled that when he and his fellow Pre-Raphaelites were charged with being realists, he retorted that art could never simply be "a representation, elaborate or unelaborate, of a fact in nature"; it "must be the reflex of a living image in his own mind, and not the icy double of the facts themselves."[209] These questions of perception would continue to plague the Pre-Raphaelites long after their Brotherhood had dissolved.

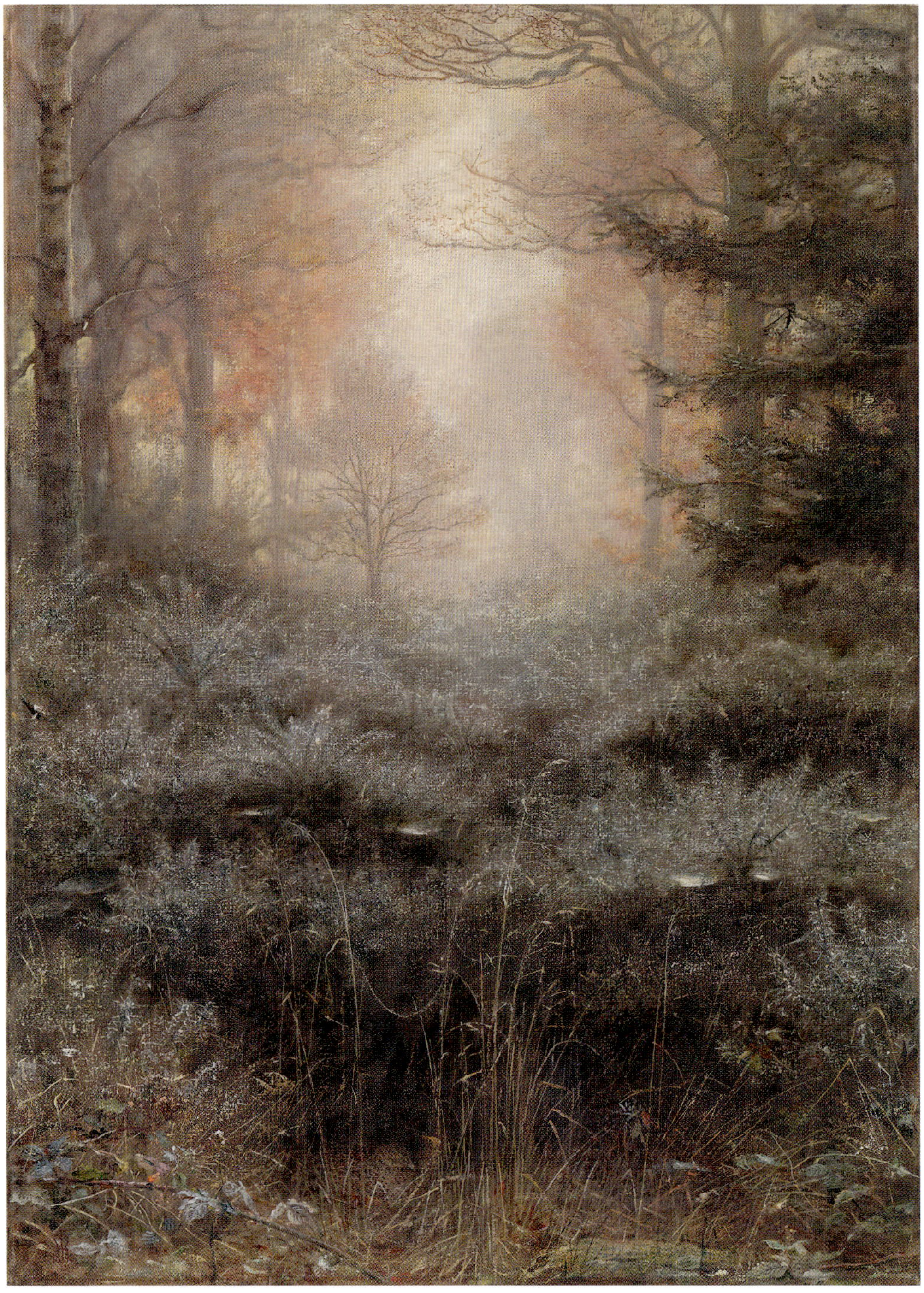

FIG. 27
John Everett Millais, *Dew-Drenched Furze*, 1889–90. Oil on canvas, support 1732 × 1230 × 27 mm. Presented by Geoffroy Millais in memory of his late father, Sir Ralph Millais, Bt. 2009. © Tate, London 2015.

{ CHAPTER TWO }

# John Everett Millais and John Brett
## *The Rise of Imagination and the Crisis of Pre-Raphaelitism*

A dazzling net of highlights, rendered by whitened flecks of paint, is cast over the greens and browns of the foreground shrubs and trees. The cool tones of the moisture-laden air contrast with the warm browns and russet tones of the fall foliage visible in the middle ground. Curtains of trees frame a distant haze. A partridge nestles under the gorse in the immediate foreground, tucked beneath a few tendrils of grass that peek above the brambles. Cupped cobwebs among the gorse branches catch the dew. The quiet mood and atmospheric effects convey a sensibility much different from that in Millais's portrait of Ruskin, in which the writer authoritatively claims his place in the Scottish stream, astride rocks that scientists decoded to understand nature's processes and laws. The critic's stance and the deft, precise handling of paint impart a sense of mastery. Millais's brushwork is adroit again in *Dew-Drenched Furze* (1889–90) (fig. 27), but to a different end: to obscure the scene, to hang a melancholy veil over nature.

As this chapter will show, Millais eventually rejected the inductive approach of his portrait of Ruskin, in which the whole is made up of scrupulously observed and rendered details, all of them given equal weight. Instead, he began to focus on the overall effect or sensation of a painting and its potential emotional resonance. *Dew-Drenched Furze* was inspired by Alfred Tennyson's poem *In Memoriam A. H. H.* (1849), which grappled with Christian theology and new scientific theories of evolution. At midcentury, Tennyson and Millais had discussed, according to Tennyson's son, "the limits of realism in painting." Tennyson had suggested to the painter that it was possible that a painting could be "*too* minutely drawn," and that this level of detail could detract from the main subject.[1]

Millais's doubts about realism contrast vividly with the certitude of John Brett. In 1899, Brett published his essay "Realism in Painting," in which he acknowledged that realism, understood as truth to nature, was difficult to achieve but could nonetheless be attained, he insisted, by rigorous training and practice. The Pre-Raphaelites, according to Brett, had established "modern realism" and drew their inspiration from "modern science." In describing realism, he wrote, "where certain facts underlie the sentiment that your picture is intended to convey, you have only to represent the facts faithfully in order to ensure the sentiment following as a matter of course." He disparaged artists who "hate science or the mechanism of nature, as many weak persons do, . . . [and] care only to be the passive recipients of a sensation or a sentiment."[2]

How did Brett and Millais, both former protégés of Ruskin, come to avow such divergent views of realism? Brett's sketchbooks, paintings, and published writings reveal an ongoing pursuit of accurate observation as well as supreme confidence in his ability to paint his precise vision convincingly. Millais increasingly adopted a manner of painting that evoked the Old Masters and sought out subjects that carried poetic and emotive associations. These connotations had the potential to override the truth-to-nature paradigm associated with Pre-Raphaelite paintings, to which Brett continued fiercely to adhere. These two positions, and the processes that led to them, are indicative of the debates and changes within the twinned discourses of art and science that unfolded over the second half of the nineteenth century.

In this chapter, I trace the fate of the Pre-Raphaelite pursuit of truth to nature as embodied in the work of Millais and Brett. The second half of the nineteenth century would appear to be fertile ground for the truth-to-nature paradigm in art, given the gathering strength of scientific naturalism as represented by such figures as Thomas Henry Huxley and John Tyndall. But, in fact, the relationship between art and inductive science became increasingly contested during this period, which also witnessed the rise of psychology, which emerged as a battleground for scientific naturalism. The new discipline stimulated new theories of perception and aesthetic judgment that became critically important within artistic circles. Nonetheless, the eventual exhaustion of the Pre-Raphaelite model did not mean the end of the quest for truth; rather, it demonstrated the need for new paradigms of realism, discussed in succeeding chapters.

## Art and Science: The Royal Academy Banquets

To map out the relationship between scientific naturalism and art over the course of the second half of the nineteenth century, an essential context for understanding the production and reception of canvases and published statements by Millais and Brett, I want to draw upon a series of public statements made by leading scientists—specifically, their speeches, or toasts, delivered at the annual Royal Academy banquet. This body of texts has not yet been fully plumbed for what it reveals about the positions staked out by scientists in the cultural field.[3]

Late in 1867, the Royal Academy moved to Burlington House, near Piccadilly Circus, which had been purchased by the British government in 1854; it joined the Royal Society, as well as the Linnean and Chemical Societies, which had located there in 1857. From 1867 to 1873, all four organizations shared space until the completion of new wings that accommodated the Royal Society, Linnean Society, and Astronomical Society, on the west side, and the Geological Society and the Chemical Society, on the east. This physical proximity was intended to foster greater collaboration between the organizations, and the Royal Academy symbolized this intent by inviting members of these learned societies to its annual banquet, held in honor of the opening of its spring exhibition of contemporary art. In turn, leaders of these societies offered toasts in honor of the Royal Academy. Over time, these speeches increased in length and substance.

At the annual banquet in May 1867, which occurred before the move to Burlington House, the president of the Royal Academy, Francis Grant, offered a toast to the Royal Society in which he looked forward to the shared premises, which he predicted would foster "a more intimate and tangible union . . . between science and art."[4] One might remark that this union had already been achieved by the Pre-Raphaelites, but it is worth remembering that they were a small group of artists—although the critical reception of their work was considerable—and their reforms had not been undertaken through the auspices of the Royal Academy. While Millais had become a full member of the academy in 1863 and was widely seen as one of its leading lights, John Brett did not become an associate member until 1881. William Holman Hunt made well known his objections to joining the academy, and Rossetti never joined its ranks. Moreover, the cohesiveness of the Pre-Raphaelite Brotherhood had loosened considerably by the 1860s. The letters among the artists make clear that while a sensibility of personal friendship persisted, the members of the Pre-Raphaelite circle did not meet as frequently as they had in the late 1840s. The Brotherhood had, for all intents and purposes, dismantled by the early 1850s. In July 1852, Thomas Woolner departed for Australia, and in early 1854 Hunt left for the Holy Land to pursue his religious subjects.

The account of Grant's speech of 1867 in the London *Times* indicates that a certain degree of skepticism greeted his utopian vision of a future union between artists and scientists. Grant's hope that "members of the Royal Academy [would acquire] a more extensive knowledge and a deeper interest in the marvels of science" was met with "a laugh," and his prediction that "men of science in all probability will acquire an increased interest in the mysteries of art" received the mixed response of "a laugh" and cries of "Hear, Hear."[5]

Given this inauspicious beginning to a proposed union between art and science, the speeches at Royal Academy banquets in subsequent years are particularly significant. Speakers on behalf of the Royal Society consistently argued that art and science were commensurate or compatible enterprises, although the terms of correspondence changed over time. This theme could be explained away by politeness—after all, members of the Royal Society were guests at a banquet that was also attended by important political and social figures, among them the Prince of Wales (later Edward VII). But in fact the speeches were of a piece with the larger agenda of the speakers, who included John Tyndall and Thomas Huxley, chief spokespersons for scientific naturalism at its height in the 1850s and '60s. Tyndall and Huxley were eager to claim cultural authority for their enterprise, using public lectures and publications, and the Royal Academy banquet was an ideal platform.[6]

In 1871, the Royal Society was represented at the Royal Academy banquet by Huxley; the president of the Royal Society, Edward Sabine, declined to attend, as he found such speeches superfluous.[7] Huxley, a biologist, was a popular public interpreter and demonstrator of science and well recognized for his support of Darwin. He had been elected to the Royal Society in 1851, following his participation in the important survey expedition of the HMS *Rattlesnake*.[8] In 1852, he earned the society's gold medal for his essay "On the Anatomy and the Affinities of the Family of the Medusae." That same year, Huxley gave his first public lecture, "On Animal Individuality," at the Royal Institution; Huxley's subsequent demand and reputation as an orator indicate that he had mastered addressing amateurs and professionals alike.[9] He finally secured a permanent position in 1854 at the Government School of Mines, which had begun three years earlier in conjunction with the Museum for Practical Geology and had merged in 1853 with the Royal College of Chemistry. The curriculum included natural history, and this became the focus of Huxley's research; simultaneously, he pushed for the introduction of science into higher education.[10] Huxley advocated powerfully that his generation look to science, and its rational and increasingly secular methodologies, to solve problems and interpret experience.[11]

Perhaps the best known of Huxley's speeches is the one he delivered in Oxford in 1860 at the meeting of the British Association for the Advancement of Science in defense of Charles Darwin's *Origin of Species*, which had been published the preceding year. Bishop Wilberforce was called upon to respond to Darwin's theory of natural selection, which he argued was not sufficiently grounded in the practices of inductive science and therefore should be regarded as a hypothesis and not a definitive conclusion. Huxley defended Darwin's theory, stating that it was, the *Athenaeum* reported, "an explanation of facts; and his book was full of new facts, all bearing on his theory. . . . [Huxley] maintained that it was the best explanation of the origin of species which had yet been offered."[12]

As Bernard Lightman explains, Huxley, along with his close friend Tyndall and other scientific naturalists of their generation, deliberately sought to challenge the gentlemen of science. They operated outside the orbit of Oxbridge, argued for a new level of professionalization, and, perhaps most significantly, strove to secularize science.[13] Huxley, according to Lightman, "wanted science to be associated with expertise, laboratory research, and naturalism, and he wanted to break its connection with the Anglican clergy and natural theology."[14] To expand the sphere and influence of science, Huxley and other scientific naturalists strove to make their research accessible and even popular.[15] In addition, Huxley joined forces with literary realists in order to foster, as Lightman describes it, a "culture shaped by science."[16] It is thus not surprising that Huxley would take the opportunity to speak to a prestigious gathering of artists.

In his academy speech of 1871, Huxley declared that artists and scientists shared a common purpose, namely, "to seize the idea which lies hidden under the shifting phenomena of nature, and to bind it in such fetters that it may increase the pleasure and profit of endless generations of men. We both seek truth and we both seek beauty."[17] Huxley's argument for correspondence rested upon the foundation of inductive philosophy: observing nature closely to discern its laws and thereby establishing verifiable truth.

The following year, the newly elected president of the Royal Society, George Airy, declined to speak at the academy's annual banquet and John Tyndall was deputized to speak in his place. Like Huxley, Tyndall premised his speech upon a close relationship between art and science. Both artists and scientists, he explained, bring skills in visual observation, a well-prepared mind, and "the warm glow of the emotions" to the study of nature. But, he observed, science had robbed nature of "hopes and pleasures" as well as "grandeur and beauty," which fortunately could be found in art.[18]

Tyndall, like Huxley, was a leading voice in Victorian science. He first worked as a surveyor, then as a railway engineer, before taking a position teaching mathematics at Queenswood College. In 1848, he decided to pursue formal studies at the University of Marburg. When he returned to England, he quickly rose to the attention of fellow scientists through his advances in physics, and in 1852 he became a fellow of the Royal Society. The following year he became a professor of natural philosophy at the Royal Institution, and his lectures and popular writings earned him a considerable reputation. As his biographer Ursula DeYoung records, audiences of more than a thousand people attended Tyndall's evening discourses at the Royal Institution, "numbers only matched by Huxley."[19]

Tyndall's first lecture at the Royal Institution, "On the Influence of Material Aggregation upon the Manifestations of Force," concerning diamagnetism, reflected his first major research effort. He then focused his efforts on Alpine glaciers, challenging prevailing theories of viscosity proposed by Forbes and proposing instead that ice moved because of a process of melting and freezing. He first articulated this theory in a paper co-authored with Huxley, published in the *Transactions of the Royal Society*, and then expanded upon it in his highly popular *Glaciers in the Alps* (1860) and *The Forms of Water in Clouds and Rivers, Ice, and Glaciers* (1872). Tyndall's challenge to Forbes erupted into a major controversy that eventually included John Ruskin, who continued to publish on geology well into the 1880s and was a supporter of Forbes.[20]

In the late 1850s, Tyndall initiated his investigations into the behavior of heat, gas, and vapors, which would occupy him for more than a decade. In the 1870s, he was concerned with the function of bacteria and processes of contagion. At the same time that he was advancing his research agenda, Tyndall was rethinking inductive methodology to allow for a greater role for hypothesis, thus revisiting the debates of the 1840s.[21] Through a series of lectures, subsequently published as essays, most notably "Scientific Limit of the Imagination" (1868) and "Scientific Use of the Imagination" (1870), Tyndall critiqued strict materialism, pointing out that the experimentalist relied heavily on "guesses" and "conjectures" to push beyond the boundaries of confirmed knowledge. "Thus," he explained, "the vocation of the true experimentalist may be defined as the continued exercise of spiritual insight, and its incessant correction and realization. His experiments constitute a body, of which his purified intuitions are, as it were, the soul."[22] Tyndall's studies of light in particular had forced him to recognize that there were phenomena that eluded the present capacities of human senses (and instruments) to observe and analyze. Thus the scientist needed the "power of imagination . . . and by this power we can lighten the darkness which surrounds the world of the senses. . . . Bounded and conditioned by coöperant Reason, imagination becomes the mightiest instrument of the physical discoverer."[23]

Ruskin felt explicitly challenged by Tyndall's notion of the scientific imagination. Although he repeatedly described imagination as the artist's greatest power, Ruskin argued that it should be grounded in material conditions.[24] Tyndall and his circle, he charged, failed to observe accurately, record faithfully, or draw the appropriate conclusions from visual evidence.[25] In the face of ongoing changes in inductive methodology, Ruskin maintained his faith in naïve vision as the foundation of true art. "You are, in drawing, to try only to represent the appearance of things, never what you know the things to be," he announced in *The Eagle's Nest* (1872).[26] As the literary historian Jonathan Smith puts it, "Tyndall's failure, in Ruskin's eyes, is not that he lacks an emotional, imaginative response to nature but that this response is too strong."[27]

Moreover, Tyndall was associated with the dismantling of natural theology. In his 1874 presidential address to the British Association for the Advancement of Science, he famously declared that "the impregnable position of science may be described in a few words. We claim, and we shall wrest from theology, the entire domain of cosmological theory."[28] In Ruskin's view, Tyndall unjustly jettisoned the wisdom of Oxbridge professors and unnecessarily dispensed with natural theology. In time, Ruskin would despair that a gulf had opened up between art and science. True artists, according to Ruskin, rely only on their own senses and "imperatively and scornfully refuse, either the force, or the information, which are beyond the scope of the flesh and the senses of humanity."[29] Scientists, he implied, no longer held to this course of action, and they had destroyed the possibility of cooperation between the disciplines.

The degree to which Tyndall's notion of the scientific use of the imagination caught hold can be measured by the Royal Academy banquet speech of 1877 by the botanist Joseph Hooker, who assumed the presidency of the Royal Society in 1873. Hooker, unlike Ruskin, still believed that art and science could have a productive relationship. Both owed their advances, he proclaimed, to "accurate observation," a claim that would have pleased Ruskin. But then Hooker warned that observation alone was not sufficient; the artist and scientist alike needed imagination: "it is by the exercise of the imagination that the student of science sees a discovery and grasps it, before even the steps by which he attained to it have been formulated in his mind." Similarly, Hooker asked his audience, "is there a man in this room that has attained eminence in art who has not, by a similar exercise in imagination, seen his picture and its effects on the walls of this Academy before even the palette has pressed his thumb or his pencil touched the canvas?" The audience answered with cheers.[30] Thus, while Ruskin viewed Tyndall's position as a wedge that drove art and science apart, others in the art world saw imagination as a shared point of reference. Indeed, Martin Willis, in his study of literature from 1870 to 1920, argues that this period was characterized by an ongoing negotiation between "the real and the imagined" that forged "a new epistemology of vision that held the actual and the imagined in fragile suspension."[31]

In 1883, Huxley again represented the interests of science at the Royal Academy banquet, when he proposed that science was the prevailing cultural influence. Indeed, he recognized that many in the audience might consider science a "monster rising out of the sea of modern thought with the purpose of devouring the Andromeda of Art." Should anyone seek to take on the role of Perseus to rescue the poor maiden, Huxley warned them off, "because the creature is hard of head, strong of jaw, and for some time past has shown a great capacity for going over and through whatever comes in his way." Huxley then proposed a better metaphor, depicting art and science as two sides of the coin of nature, "the one expressing the eternal order of things in terms of feeling, the other in terms of thought."[32]

When Tyndall addressed the Royal Academy in 1888, introduced by President Frederic Leighton as both an "old friend" and a "priest within the temple of Science," he reiterated Huxley's formulation, assigning art to the province of emotion and science to the sphere of the intellect.[33] This represented a pronounced shift in the proposed interaction between science and art. Rather than sharing a common methodology of visual observation and inductive inquiry, or even imaginative investigation, art and science had been cleaved into separate, albeit connected, realms of expertise.

The notion that the province of art was affect, as opposed to cold rationalism, was the result of arguments among philosophers and psychologists concerning perception and aesthetic judgment. Huxley's and Tyndall's effort to divide knowledge between feeling/emotion/art and thought/intellect/science also corresponds closely to the critical reception of the work of Millais and Brett, with Millais more often aligned with feeling and Brett attempting to associate himself with intellect and science.

### Millais and Brett in the 1860s: Between Pre-Raphaelitism and Psychology

Perhaps it is no surprise, given the strained personal relationship between Millais and Ruskin, that Millais began to question his former mentor's aesthetic judgment. In 1858, he critiqued Ruskin's preference for John Brett's *Val d'Aosta* over his own work: "There is a wretched work like a photograph of some place in Switzerland, evidently painted under his guidance, for he seems to have lauded it up sky-high; and that is *just where it is* in the miniature room! He does not understand my work, which is now too broad for him to appreciate, and I think his eye is only fit to judge the portraits of insects."[34]

Two years earlier, Millais marked a new direction in his practice with his painting *Autumn Leaves* (1855–56), his first aesthetic painting, which arguably was painted in response to his conversation with Tennyson regarding the limits of realism (fig. 28). To paint the work, Millais drew on his recollection of gathering and burning leaves while visiting Tennyson and also worked directly from the motif, assembling the landscape scene in his garden in Perth and then arranging the sitters in the studio. The resulting painting weaves together landscape and figures, each reinforcing the mood of the other. The staring gazes of the girls at left and center convey a dreamlike sensation echoed by the gathering gloom. This sensibility accords with the poetry on which the artist drew for inspiration, including James Thomson's *The Seasons*, Dante Gabriel Rossetti's "The Fall of the Leaf" (also known as "Autumn Song"), William Allingham's "Autumnal Sonnet," and Tennyson's *The Princess* ("In looking on the happy Autumn-fields, / And thinking of the days that are no more").[35]

The painting, with its autumnal setting and poetic references, hints at mortality. At the close of day, four girls, who clearly represent different stages of youth, from childhood to young womanhood, gather around a pile of smoldering leaves. The apple held by the youngest child, at right, signals the Old Testament narrative of the expulsion from paradise. F. G. Stephens, in his review in the *Crayon*, praised the painting and associated it with biblical passages describing God's recognition of sin through the metaphor of fire.[36]

Millais responded enthusiastically to Stephens's reading, explaining that he had considered including a passage from the Psalms when the work was exhibited at the Royal Academy but had decided against it out of "fear that it would be considered an affectation and obscure." He then clarified his intent, revealing a gradual abandonment of his former practice of scrutinizing nature's details closely. He now privileged feeling—the overall sensations conveyed by the painting—above all else. But Millais's interest in feeling was very different from that of his eighteenth-century predecessors. He was not turning to a well-developed typology or system of landscape, such as the sublime or picturesque, to achieve his end. He was still keenly engaged with a particularized scene of nature, but one that he hoped might trigger literary associations, whether the Bible or poetry: "I have

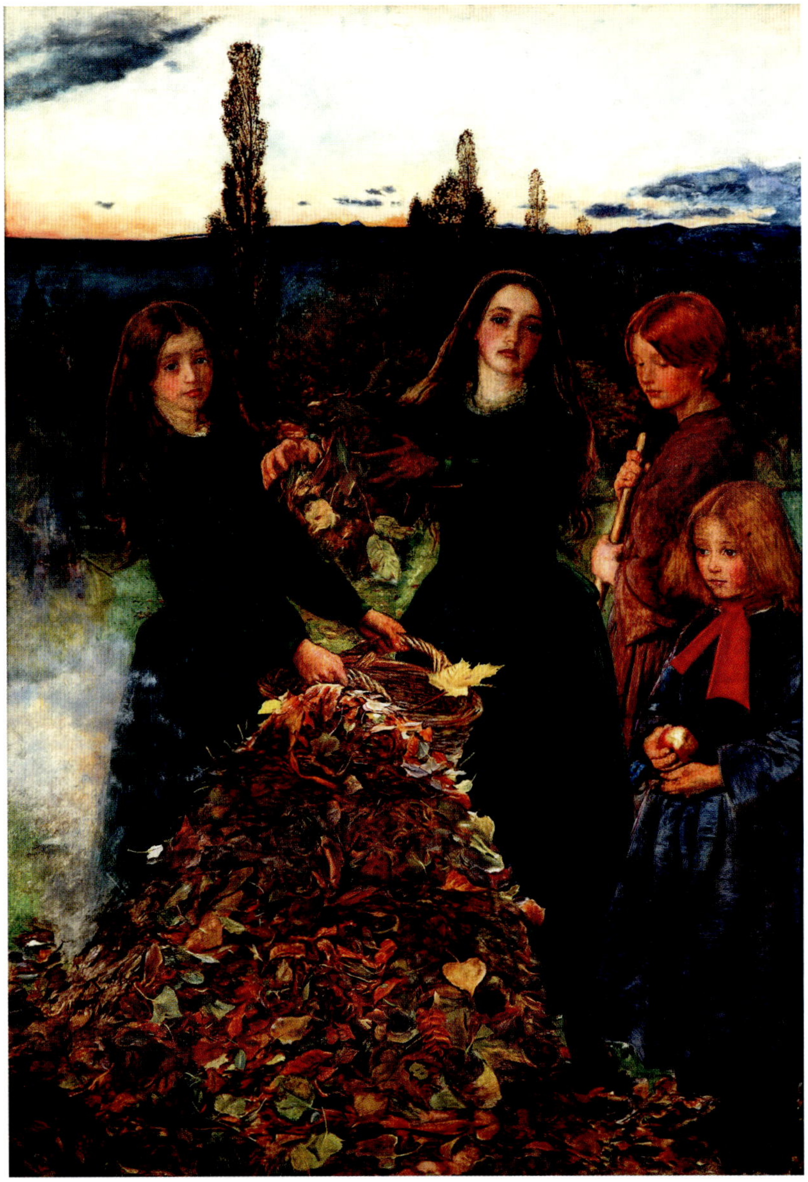

FIG. 28
John Everett Millais, *Autumn Leaves*, 1855–56. Oil on canvas, 104.3 × 74 cm. Manchester Art Gallery.

always felt insulted when people have regarded the picture as a simple little domestic episode, chosen for effect, and colour, as I intended the picture to awaken by its solemnity the deepest religious reflection. I chose the subject of burning leaves as most calculated to produce this feeling, and the picture was thought of, and begun with that object *solely* in view. These kind of pictures are really more difficult to paint than any other, as they are not to be achieved by faithful attention to Nature, such effects are so transient and occur so rarely that the rendering becomes a matter of feeling and recollection."[37] Millais's reference to feeling and recollection also indicates that he was not attempting Ruskin's ideal of unmediated vision but was bringing memory and experience to bear. His was a practiced, not an innocent, eye.

Millais's wife offered another explanation for the painting that should dissuade us from unpacking its typological symbolism in search of a specific narrative, encouraging us instead to focus on the overall sensation; she recorded that her husband aspired to "paint a picture full of beauty and without subject."[38] The surrounding landscape reinforces this sensibility; the gloom of twilight and the gradual withdrawal of light obscure forms in the painting, particularly in the background, where buildings and trees form dark silhouettes against the red-tinged golden sky.

The art historian David Peters Corbett argues that the very surface of the painting, rendered in far looser brushstrokes than in Millais's previous work, indicates a shift from the burden of the painting's argument being made by content, or illusionism, to technique, or material means of expression.[39] Millais's attention to surface qualities and materiality, and the ways in which he invites viewers to partake in the pleasure of painterly brushwork that is both self-sufficient as painterly gesture and illusionistic, should again dissuade us from "decoding" the work.

Ruskin noticed the painting in his *Academy Notes*, describing it as "by much the most poetical work the painter has yet conceived; and also, as far as I know, the first instance of a perfectly painted twilight," considerable praise considering that Ruskin had earlier claimed that transient effects eluded Millais's skills. Ruskin's dual interpretation—truth to nature and poetic association—is noteworthy in that it anticipates readings of Millais's later landscape paintings. It gains further significance when compared to Ruskin's adjacent review of William Holman Hunt's *The Scapegoat*, a religious painting depicting a suffering sacrificial animal caught in the salt-encrusted edges of the Dead Sea. While Ruskin praised Hunt's choice of subject and diligence in executing the scene on site, he warned his readers that the artist had succumbed to "a too great intensity of feeling" that prevented him from attending to the details of accurately representing the landscape. Ruskin hoped that Hunt would "paint a few pictures with less feeling in them, and more handling," in the future.[40] The expression of feeling continued to trouble Ruskin, and he cast himself as the arbiter of which artists had the capacity to undertake it properly.

Further support for this reading of *Autumn Leaves* as poetic and redolent of feeling can be found in another body of scientific literature, the nascent field of psychology. Psychology shared points of connection with the discourse of scientific naturalism and induction, particularly with regard to studies of physiology and Tyndall's proposition for the scientific use of imagination. Yet psychology was also emerging as a distinct discipline in the second half of the nineteenth century. As Bernard Lightman explains, psychology was a "relatively open field" from about 1850 to about 1880; its formation took place across a wide range of publications and within at least four different lines of inquiry. The first, Lightman notes, was "the discourse of the soul," which emerged from natural theology. The second, more important for the discussion here, was largely located within philosophy and was built upon theories of association and directly affected aesthetics.[41] Associationism, which derived from the work of seventeenth- and eighteenth-century philosophers such as John Locke, David Hume, and David Hartley, refers to the processes by which the mind linked ideas.[42] Associationism was largely displaced by, or absorbed into, the rise of physiological studies in the 1870s. Of particular significance for my inquiry are studies that examined the relationship between vision and cognition.[43] This third line of inquiry, physiology, was the closest to scientific naturalism, as both were concerned with evolution and the relationship between organism and environment. Near the close of the century, the discourse of medicine, diagnosing and correcting psychological problems, developed.[44]

John Stuart Mill laid an important foundation for psychology and theories of perception in his midcentury writings. In 1842, for example, in an essay in the *Westminster Review* on Samuel Bailey's revision of George Berkeley's theory of vision, Mill explained that visual knowledge is composed of optical information, or "sensations," derived from the mechanism of the eye, and of subsequent associations, or "inferences," that the mind, drawing on experience, attaches to that optical information.[45] The following year, in his *System of Logic*, Mill explored further the topic of sensation in the context of what he called psychology, or "mental philosophy." Sensations, he explained, all derived from some bodily trigger. Once experienced, a state of consciousness or thought could be recalled without the original trigger. This, Mill explained, constitutes an idea, and by the laws of association it can be brought to mind again either by "our impressions, or by other ideas."[46]

The philosopher Alexander Bain, who was closely affiliated with Mill and the *Westminster Review*, expanded on this mind-body problem, beginning with his *Senses and the Intellect*, published in 1855. Bain, even more than Mill, developed a materialist model to support associationism. Bain recognized the mind as possessing three capacities: (1) feeling (sensation and emotion)—which he also referred to as consciousness, (2) action (spurred on by feeling), and (3) thought or intelligence.[47] Bain's categories anticipate the Royal Academy speeches by Huxley and Tyndall, in which they associated art with feeling/emotion and science with intellect/thought. Indeed, Bain knew Huxley; and Tyndall, in his essay on the scientific use of the imagination, recalled Bain's later treatise, *Logic* (1870), which inspired his new paradigm of imaginative induction. Bain devoted a chapter of *Logic* to psychology, "the science of the mind," wherein he reiterated his tripartite division of the mind into "Feeling, Volition, and Intellect."[48]

In *The Senses and the Intellect*, Bain located the five senses, including sight, under the category of feeling and, more specifically, within the field of sensation. Analyzing sight entailed understanding the physiological operations of the eye as well as links between sight and feeling. Following this chain of logic, Bain updated Burke's theory of the sublime, now locating the precise "sensations" that caused that feeling: "As regards the emotional effects of the visual sensations of objects, I have already remarked on the influence of light and colour, and also on the effect of the single element of distance. When this element is combined with great visual expanse, we have then the sensation of *largeness*, and of a wide prospect. This feeling is very powerful. . . . It is a massive and keen emotion allied with the sense of power or great expenditure of force, yet without the feeling of painful exhausting effort. . . . This is the simplest form of the feeling that we term the *sublime*."[49]

Such arguments fed easily into Bain's larger exegesis regarding "constructive association"—that is, the mind's ability "by means of association . . . to form combinations or aggregates different from any that have been presented to it in the course of experience."[50] Here, Bain began to explore new territory beyond the boundaries of empirical, inductionist methodology by allowing for the significance of ideas accrued outside strictly observational experience.[51] Like Tyndall after him, Bain paid particular attention to the role of imagination, a power that he attributed to painters, poets, musicians, and inventors. According to Bain, the primary task of art was to produce emotions or feelings through the representation of objects that will have a profound effect on the viewer. He insisted that "the artist's standard is *feeling*, his end is refined pleasure; he goes to nature and selects what chimes in with his feelings of artistic effect, and passes by the rest."[52]

Bain was clearly aware that he was challenging the prevailing, Ruskinian dictum that the artist should "go to Nature . . . having no other thoughts but how best to penetrate her meaning . . . rejecting nothing, selecting nothing." Indeed, Bain explicitly stated that he was proposing an alternative model for artistic production: "conceiving as I do that the first object of an artist is to gratify the feelings of taste, or the proper aesthetic emotions, I cannot assent to the current maxim that nature is his standard, or truth his chief end." The scientist, he claimed, should observe nature and rigorously record all its details; but the artist should use only those aspects of nature that appeal to his taste or will. Furthermore, he dismissed artists who acted "as if the exhibition of truth were his prime endeavour and his highest honour." This aim, Bain countered, was a mistaken attempt to "subject imagination to the conditions of truth and reality" that had led to an unnecessary "transference . . . whereby the definition of science has been made the definition of art." He continued, "now I have every desire to do justice to the merits of the truth-seeking artist," but he nonetheless maintained that "there is, and always will be, a distinction between the degree of truth attainable by an artist, and the degree of truth attained by a man of science or a man of business." He concluded his entire study with the statement that "we ought not to look to an artist to guide us to truth; it is enough for him that he do not mis-guide us."[53] Bain thus answers for us the question of why the Royal Academy speeches changed over time from promising a utopian union of art and science to imagining a partnership between the two in which each occupied a separate arena—art assigned to emotion and feeling and science to thought and intellect. Indeed, Tyndall, a leading advocate of this dualism, was a keen admirer of Bain.

Bain's work was also important for helping artists and critics to unlock the problem of realism and the nature of truth.[54] Nature, according to Bain, was no longer the final arbiter of truth because the external world could be known only through the mind, through sensations. But Bain was not promoting pure individualism. Bain's measure of success was the artist's ability to provoke emotional responses in the viewer, presumably responses that could be experienced collectively and repeatedly.

George Henry Lewes, whose intellectual career spanned literary writing, science, and philosophy, explicated further the topic of sensation in his two-volume study *The Physiology of Common Life* (1859–60). While Lewes's writings have been associated with literary realism, they were arguably equally important for the visual arts.[55] Lewes and Millais knew each other; an undated letter from the Mary Millais album at the Morgan Library, for example, describes a pending visit of Lewes and George Eliot to Millais's studio, and visitors to the home of Lewes and Eliot in the 1870s recall meeting Millais there.[56]

In his *Physiology of Common Life*, Lewes tackled the problem of perception. But unlike Mill and Bain, as the literary historian Peter Dale explains, Lewes was engaged less in how the recipient interacted with the work of art—associationism—than with the laws of human nature and biology on which reception rested. Lewes offered a hierarchical classification of sensations that was deeply influenced by the writings of the physiologist Hermann von Helmholtz, whose three-part *Handbuch der Physiologischen Optik* was published between 1856 and 1867.[57]

Helmholtz, a physician, studied the behavior of the eye closely; in 1851, he developed the ophthalmoscope to study the physical composition of vision. He was also interested in art; in 1848–49, he served as professor of anatomy at the Akademie der Künste in Berlin. Thereafter, he occupied a series of university positions in physiology and anatomy, finally settling at the University of Berlin as professor of physics in 1871. Through his studies of vision, Helmholtz arrived at the conclusion that visual perception was largely learned rather than innate.[58] As the historian of science Timothy Lenoir explains, "the act of seeing itself, for Helmholtz, is a purely psychological phenomenon. . . . Moreover, the visual image for Helmholtz is a sign . . . a purely symbolic representation constructed by the mind in order to facilitate our physical interaction with things."[59] Helmholtz built his arguments upon Bain's earlier conclusions; for example, the British author had insisted, in his analysis of sight and imagery, that "the mind supplies from the past what the eye does not distinctly see at the time, so that the picture actually realized is not the bare optical impression of the moment, but a much fuller picture which that impression suffices to suggest."[60]

These findings informed Lewes's materialist understanding of sensations, which he described as "the reaction of Sensibility, stimulated by an impression—the activity of a nerve-centre awakened by the Neurility of a nerve." He further distinguished two categories of sensations: "*systematic sensations*," or those "not localized in any special organ," and "*sense-sensations*," which derived from our five senses. In Lewes's hierarchical classification, sight was no longer given primacy; touch was the most important sense, followed by taste, smell, hearing, and, finally, sight. One of the reasons he demoted sight as a source of accurate knowledge was that he recognized, thanks to Helmholtz, that "the formation of an image on the retina is the precursor of a visual sensation; but this image is not transmitted to the brain." In other words, there is no direct conduit from object to perception via the eye; the optic nerve mediates this process. He explained further that "that which is in each case transmitted is the excited sensation." Thus, he concluded, "we can no longer suppose that we see the objects themselves. Our visual sensations are simply excited states of our sentient organism [i.e., the brain or mind]." In short, vision was "*subjective*," to use Lewes's terminology.[61]

In other words, the process of sight cannot be severed from the process of mental perception. Moreover, if mental perception is highly individualized, then one cannot claim an objective universal truth rooted in the visual. In the 1830s and 1840s, belief in a universal truth, based on rigorous optical observation, had led to a set of agreed-upon and shared visual practices for signifying truth to nature, such as (and most significantly) the particularized detail. But Lewes had now offered a different proposition, one that shifted the burden of perception from the eye to the mind. How would this be translated into visual practices? And how would works of art be evaluated without the common denominator of truth to nature? What were the new terms by which the art critic should judge art? There were no immediate answers to these questions, and consensus did not emerge quickly.

Lewes's theory of perception was highly sympathetic to Millais's ambition to craft in *Autumn Leaves* a painting "calculated" to incite by its "solemnity the deepest religious feeling." Moreover, Lewes gave scientific authority to Millais's intention to work from "feeling and recollection" rather than pursue only the faithful study of nature.

Around the time that Millais was at work on *Autumn Leaves*, John Brett was engaged in his first Alpine campaign, which resulted in *The Glacier of Rosenlaui*. This was followed several years later by his second campaign, which yielded *The Val d'Aosta*. During this time, Ruskin expended considerable energy in directing the younger painter, although Brett did not always follow his advice.[62] Ruskin, for example, criticized the painter for attempting figural subjects and for "making big studies and supposing they are pictures."[63]

In the 1860s, Brett began to focus on marine and coastal views, as in his Royal Academy submission of 1864, *Massa, Bay of Naples* (fig. 29). In order to render the complexity of his chosen subjects accurately, he began sketching from aboard ship, eventually purchasing his own yacht. While in Italy in 1863, he navigated the Sorrento peninsula, en route to Naples, via the SS *Scotia*. His sketchbooks from this trip reveal that he studied the coastal landscape from a variety of angles made possible by his nautical perch, and likewise observed the water intently, trying to determine how best to capture its actions and colors in paint. His sketchbook notations include measurements and language that reveal his scientific predilections: "Convex elements (scarce) reflects the eastern clouds—very coppery purple and in the midst of each flake is a dash of pure blue—the eastern sky high up. It is only in the nearest flakes that the blue dark appears. Western sky is of a tender blue slightly greenish at bottom leaden with heavy flakes of dk grey forming low down a flat layer 3 degrees or 4 degrees high behind which appears here and there lace work of silvery gold very finely divided into little scales & whisps—2 or three dots of crimson breaking out over the edge of the hills."[64]

This attention to detail is manifested in *Massa, Bay of Naples*, which depicts the coastline from shipside. Despite the different vistas that Brett compiled in his sketchbook, in his final painting he adopted a single, fixed point of view, as is common in his canvases, from which to organize the pictorial space so that the scene unfolds like a stage set, with foreground giving way to middle ground and then background. The ripples in the foreground water suggest the slow forward motion of the artist's boat. The rising sun has cast the middle ground, where a town nestles in a hillside valley, in bright light, whereas the foreground still lies in shadow. Hazy morning mist clings to the tree-lined hillsides and is particularly discernible in the sunlit passages, hovering like an aura over the landscape. The mist is absent in the shadowed areas, where the air would have been cooler. The surface of the water reflects the sky above, and in the far distance, recognizably cumulus clouds gather.

The critical reception of the painting was divided, even within the space of a single notice in the *Times*, probably written by Tom Taylor, who had become the art critic for the newspaper in 1857, an informal post he held until his death in 1880.[65] Taylor easily associated *Massa, Bay of Naples* with Ruskin's inductive methodology. The painting was "the most consummate expression of that school of landscape-painting which upholds the dogma that all the facts in a scene should be painted to the utmost of the painter's power, as they appear during long and close study; that the painter's mood cannot profitably contribute anything to his picture."

But this choice was not popularly received, Taylor observed. I want to quote Taylor at length here, because his review offers an elusive piece of evidence—the response of the crowd. "It is curious," he wrote, "to note the criticisms of the crowd on a picture painted, and so well painted, on this theory. Few of the commentators on it, in our hearing, admitted the mere truth of its colour; still fewer seemed to be arrested by the intense realism of the painting." Taylor then imagined Brett's response: "Mr. Brett will no doubt say, and say fairly, that such criticism is worthless,—that he appeals to more careful and better-informed judges." From this imagined exchange, the critic constructs a lesson that indicates the degree to which Ruskin's inductive methodology had failed to secure permanent and universal acceptance, and helps to explain why Millais was propelled to search for an alternative:

> *Is it not the fact that the obvious display of labour in such work breeds involuntary and instinctive*

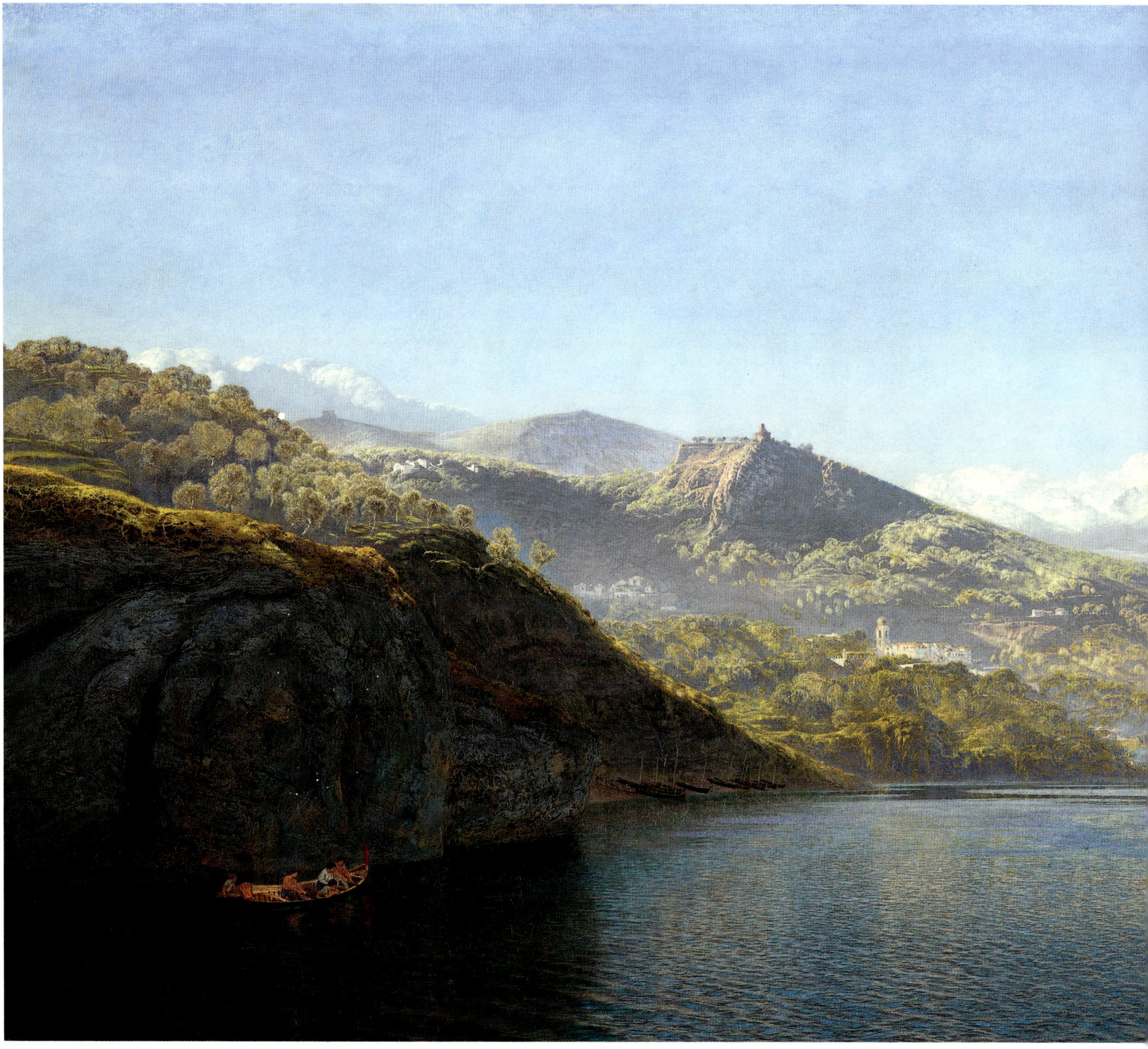

FIG. 29
John Brett, *Massa, Bay of Naples*, 1864. Oil on canvas, 34 ¾ × 51 ¼ in. Indianapolis Museum of Art. James E. Roberts Fund, 64.14.

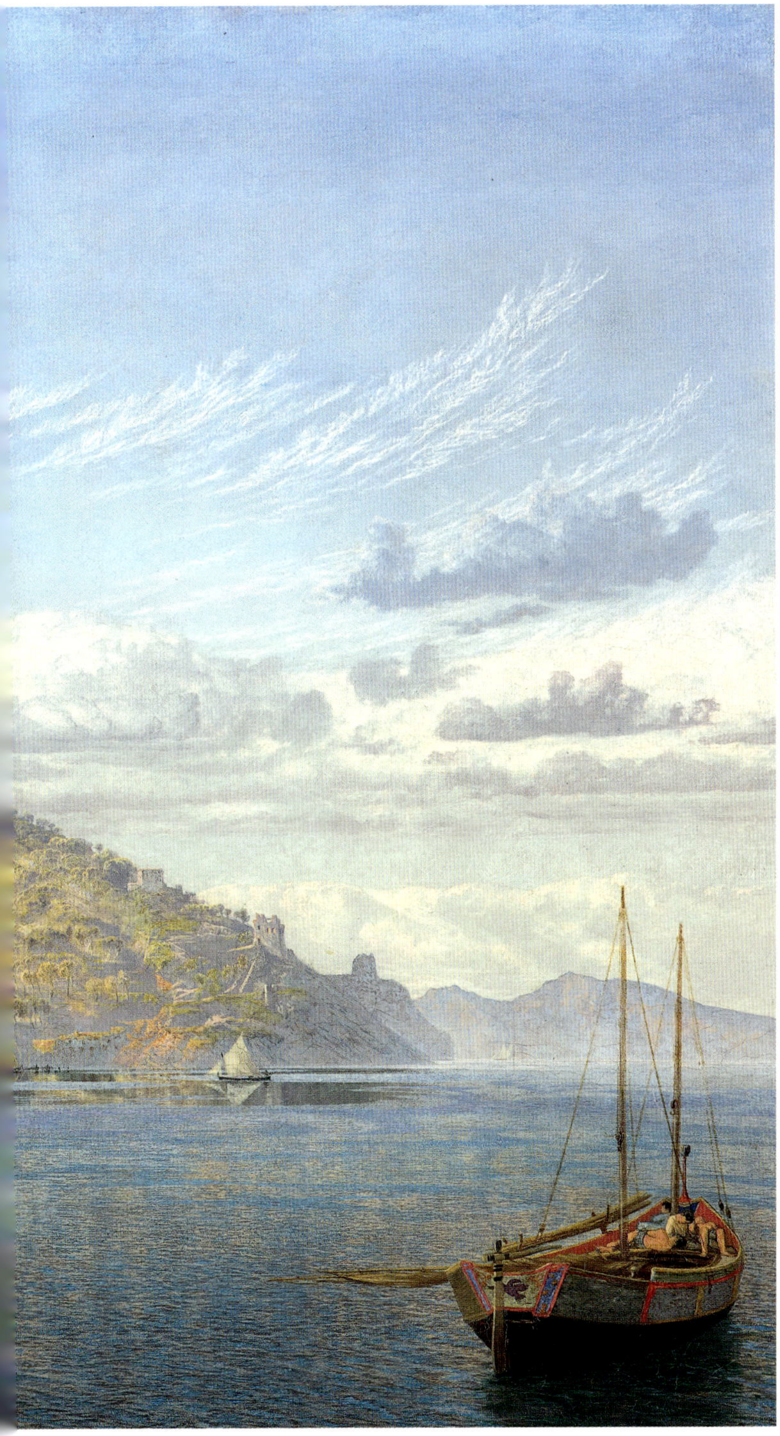

*distaste? There may be scenes of nature grand enough, significant enough, pregnant enough with interest or association to admit of this dwelling on every fact, this elaboration of every feature. But such scenes are rare, and, applied to all but these, this resolute forcing of details upon the attention is wearisome. . . . We contest Mr. Brett's theory at an earlier point, and assert that nature* cannot *be truthfully painted on his principle. She is ever varying in every detail, and the utmost that we can preserve is a general impression of her changing features, to say nothing of that limitation of the painter's means and materials—for the expression of light and shade, above all—which must perforce drive the painter to compromises, substitutions, and artifices. With the fullest recognition of the conscientiousness of such work as Mr. Brett's, and the strongest sense of the value of that close and patient study of which it is the extreme expression, we should rather maintain that all of landscape-painting destined to take permanent high rank, must bear strongly the impress of the painter's moods, tastes, and, indeed, character, conveyed by some "manner," some short-hand of the painter's own invention, the swifter, broader, more compendious and inscrutable the better. This is not, indeed, the mode of landscape painting now in vogue with us.*[66]

Taylor's vision of ideal landscape painting is quite close to J. M. W. Turner's, although the artist is unnamed in the text. For Taylor, Brett's painting practice is objectionable not just because it is not like Turner's, full of mystery and the personal insights of the painter, but also because optical truth to nature is unachievable owing to nature's mutability and the artist's limited repertoire. Like Bain, Taylor wants the artist to select a scene that could provoke associations in the mind of the viewer and to set aside the impossible task of mimicking nature.

This mixture of measured admiration and deep reservation in the critical reception of Brett's *Massa, Bay of Naples* permeates the essay "The Present Position of Landscape Painting in England," published in 1865 in *Cornhill Magazine*, then edited by Lewes.[67] Nature, according to the anonymous author, held great value for the modern age—for the romantic poets it was the "source of intellectual enjoyment and moral instruction," and for scientists, a source of endless discovery and progress. In this context, landscape painting had achieved a place of prominence in the modern age. The quality of modern landscape art should be judged, the author explained, by how much truth it conveyed; the ideal painting was "a faithful transcript of the truth in nature."[68]

But, as the essay unfolded, the author's argument became inconsistent. For example, while repeating key components of the Pre-Raphaelite discourses, the author at the same time criticized landscape painters for merely copying what they saw in nature, and for filling their scenes with details without exercising either "unity" or a "controlling intellect." The artist should judiciously choose what to copy from nature with the aim of arriving at a "characteristic scene" or, even better, the sensibility of a "painted poem." Landscapes should be "the vehicles of thought and feeling," and artists should work from poetical inspiration, associations triggered in the mind, and intuitions. The detail should be subordinate to the dominant concept of the painting—a rule, the author added, that the Pre-Raphaelites had failed to follow.[69]

This essay underscores the confused state of landscape painting in the mid-1860s. Should artists continue to adhere to the Pre-Raphaelite paradigm, holding on to the modern emphasis on observable fact, or should they pursue the expression of sentiment and feeling? In sum, over the second half of the nineteenth century, as scientists and philosophers reinvented induction in order to encourage the use of imagination, investigated the behavior of forces invisible to the human eye, such as electricity and gases, and dismantled the feasibility of unmediated perception of the external world, art critics and artists debated the nature of landscape painting for the modern age.

In 1870, the critic Philip Gilbert Hamerton offered his response to the question of what should constitute modern landscape painting in an essay titled "Can Science Help Art?" As a younger man studying art, Hamerton had considered himself a follower of Ruskin (he had met the older critic in 1853), and he regarded nature as the "sole authority" in matters of art. But he later came to repudiate his earlier work for obsessively dissecting a scene into parts. He also chastised himself for failing to recognize that "art is the product of the human mind," not of the perceiving eye.[70]

By the time he wrote "Can Science Help Art?," Hamerton was carving out a space for himself as a leading art critic. In 1866, he had succeeded Francis Palgrave as the art critic for the *Saturday Review*. In 1870, he founded his own specialized art journal, the *Portfolio*, with the publisher Richard Seeley, and served as its editor until his death. The journal set a new standard for luxury art publishing, featuring even higher-quality illustrations than the *Art Journal* and *Magazine of Art*.

That the title of Hamerton's essay is put in the form of a question indicates a shift from the position staked out by Stephens twenty years earlier, when he enthusiastically proposed that the new premises of science—"this adherence to fact, to experiment and not theory"—be applied to art. Hamerton's position was far more qualified. On the one hand, Hamerton cautioned, science could ruin an artist if the "artistic spirit" was not sufficiently strong. On the other hand, he acknowledged that science could assist an artist, functioning as an aide-mémoire, helping the artist to observe nature, understand what he sees, and commit it to memory.[71] The last point is worth underscoring—while Hamerton was a strong advocate of working on site, he was equally forceful about the benefits of working from memory.

Hamerton had already developed a rationale for working from memory, as opposed to directly from the motif, in his essay "Painting from Memoranda," published as part of his *Thoughts About Art* collection in 1862. Acknowledging that painting directly from nature was considered progressive, Hamerton nonetheless made a case for working up a final painting from sketches in the studio. He argued that the artist's imagination was likely to be stronger when the actual motif no longer exerted any pull and the artist was no longer preoccupied with careful observation. And, Hamerton added, wasn't it often miserable to paint outdoors? Nonetheless, the artist must spend time in nature to gather "memoranda"—notations and sketches of such elements as trees, sky, and the behavior of local color—that could be reused for later compositions or help to educate or train the artist. When sketching (or photographing) outdoors, Hamerton urged the artist to be observant and truthful. The notion of truth also applied to working from memory: Hamerton encouraged artists to delay rendering sketches from nature until they returned to the studio in order to work purely from memory and strengthen their skills of recall. For Hamerton, the final painting was an act of transposition—a series of reinterpretations of the scene observed in nature—not transcription.[72]

Hamerton's engagement with memory and imagination brought his work to the attention of the physician William Carpenter, a pioneer of psychology who was deeply interested in the workings of the unconscious mind and also believed in the possibility of mental training.[73] Carpenter had developed the theory of "unconscious cerebration," which, as he explained in the fifth edition of *Principles of Human Physiology* (1855), referred to the habitual or automatic processes by which the mind linked previous associations with new objects through identifying and comparing.[74] In his study *Principles of Mental Physiology, with Their Applications to the Training and Discipline of the Mind, and the Study of Its Morbid Conditions* (1874), Carpenter explored processes of mental activity more fully. In a chapter titled "Of Ideation and Ideo-Motor Action," he explained how ideas formed at the intersection of sensation and perception. Ideas were the result of first external and then internal prompts: "the sensational stimulus, acting on the Cerebrum, excite[s] the changes which give rise to the Ideational form of consciousness." To prove his point that the same object or occurrence could provoke different responses in different individuals, Carpenter turned to Hamerton's observations on artists. Hamerton had demonstrated that the artist was guided more by "his *mental* than by his *sensorial* view of certain objects." Thus "the same face or landscape shall be depicted by different Artists, each expressing in his peculiar 'manner' that representation of the object which his Mind had formed."[75] Different artists would depict the same subject differently because they perceived it differently.

This view aligned with Carpenter's argument that "preconceived *notions*, or of the *feelings* by which the Mind is habitually pervaded," modify the "forms of what is presented to . . . observation." That is, the perceiver's

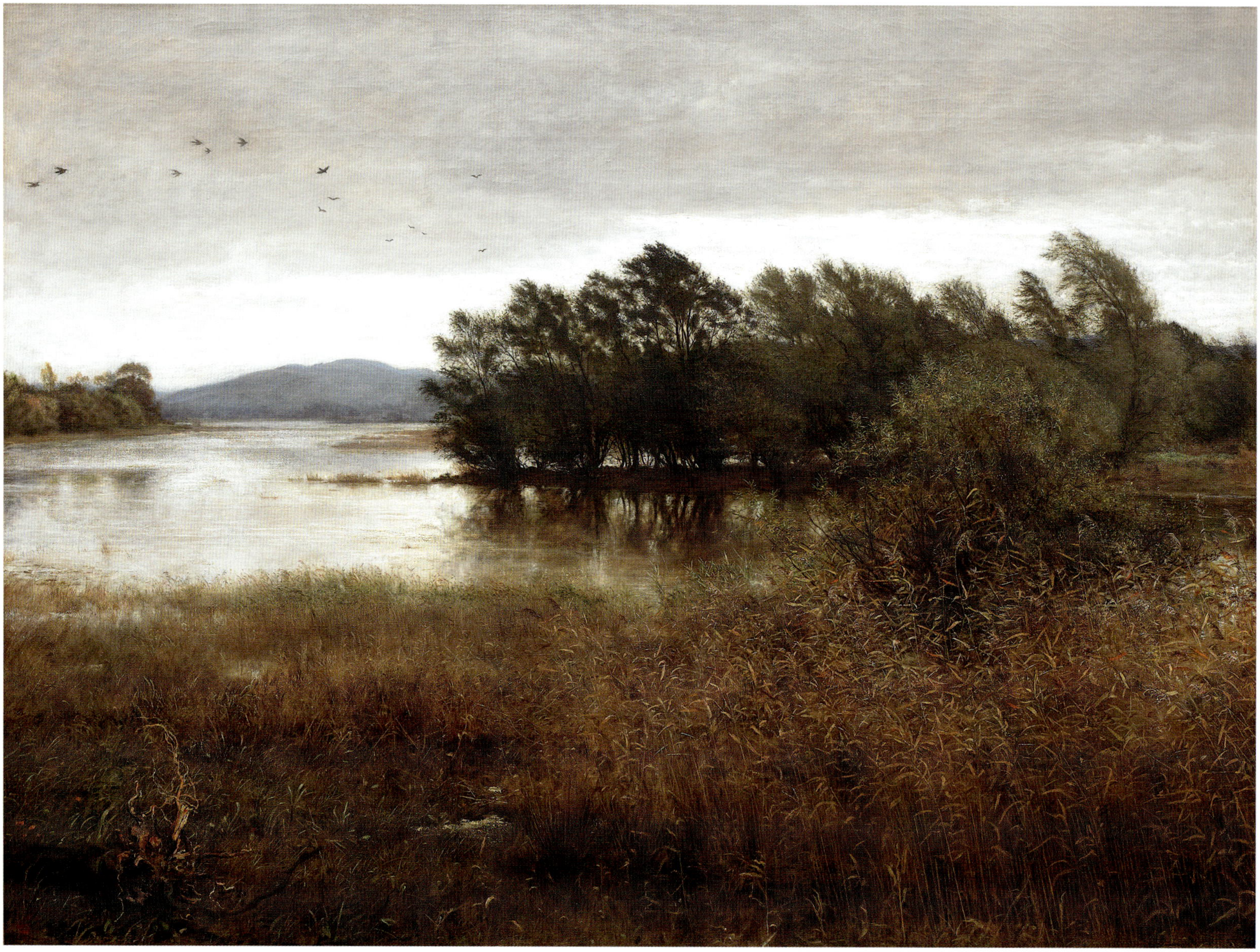

FIG. 30
John Everett Millais, *Chill October*, 1870. Oil on canvas, 186.7 × 141 cm. Private collection.

emotional state could shape or influence his visual perception. Carpenter thus concurred with Hamerton that "departure from *visual* truth, for the purpose of producing *ideal* truth," may be necessary for artists.[76] That is, fidelity to the truth of their feelings in response to nature might require that artists alter in their representations what they saw before them at any given moment. In short, the mind trumped the eye. Where did this leave the Pre-Raphaelite painter?

## Millais, 1870s–1890s: Transcript or Feeling?

In 1871, the critic and scholar Sidney Colvin, a fellow of Trinity College, Cambridge, wrote a profile of the artist John Everett Millais in Hamerton's journal the *Portfolio*. He opened the article by pronouncing the artist one of the greatest painters of modern art in Britain as well as one of the most popular. But, surprisingly, Colvin was very critical of Pre-Raphaelitism, on exactly the lines that Bain and Lewes had laid out. Colvin questioned whether it was feasible that artists should see and represent nature in universal agreement, observing that "nature in the mass—what any one of the sights of the world—at any moment is for any two or more of us is not a thing constant, identical, or upon which we can absolutely agree; it is a complicated presentation in the individual consciousness, arising from the contact of a given sum of outer impressions with a given sum of inner susceptibilities."[77]

For Colvin, the primary value of Pre-Raphaelitism lay in improving the artist's ability to represent particulars. He did admit his admiration for Millais's "gift in imitating in paint the thing before him." But Colvin reserved some of his highest praise for *Autumn Leaves*, which revealed Millais's ability (not always equally exercised) to capture "the poetical sentiment and sympathetic charm of a scene" and thus rise above mere imitation.[78] The painting rested upon highly individualized perception and knowledge of nature rather than on universally recognized truth to nature. Such opinions were clearly informed by Lewes's theories; Colvin knew the philosopher well. He later recalled that visiting the home of Lewes and George Eliot was one of the highlights of his Sundays from the late 1860s to the mid-1870s.[79]

At the conclusion of his profile of Millais, Colvin alerted readers that the artist would soon display "the new and eagerly-to-be-looked-for experiment of a large landscape"[80] This was a reference to Millais's *Chill October* (1870), a fully developed response to the question of what should constitute modern landscape painting and the first of the pure landscapes that Millais would produce over the next twenty years (fig. 30).[81]

*Chill October* was a heroic effort to paint directly from the motif, and in this respect Millais was faithful to his Pre-Raphaelite roots. To capture the scene of the backwater of the River Tay in Scotland, Millais built a special platform on which to place his easel. He recorded how bursts of wind from passing trains threatened the canvas and how the rising tide once carried away the platform. "I made no sketch for it," he claimed, "but painted every touch from Nature, on the canvas itself, under irritating trials of wind and rain."[82]

The canvas shows the River Tay stretching from the middle ground to the background, where gently rolling hills cast in bluish shadow occupy the distance. In the foreground, a cluster of dead brown grass echoes the movement of the shore-side grove of trees in the middle ground. A gray sky, relieved by a few wheeling birds at left, presses down on the scene. In its reference to the passing season, signaled by the fawn-colored plants in the foreground, the work is close in tone to *Autumn Leaves*. The presence of death was quite palpable in Millais's life at this time; his father had passed away earlier that year.

The *Times* critic, probably Tom Taylor, noting that this was the first time Millais had dispensed with figures, praised the artist for conveying feeling and narrative through pure landscape: "It is the sentiment of autumnal sadness, the story of recent storm and storm to come." For Taylor, this canvas epitomized excellence, and it was the first work he noticed in his review of that year's Royal Academy display. "It strikes home to the imagination with a power at once swift, straight, and strong," he wrote. "It does what all good landscape should do—embodies a sentiment and expresses a feeling."[83] In other words, Millais had achieved in *Chill October* what Brett had not in *Massa, Bay of Naples.* Noticeably absent from Taylor's review are the terms and phrases associated with Pre-Raphaelitism: transcript, detail, fact. The *Athenaeum* critic, probably F. G. Stephens, referred to the work as "noble and pathetic" and as "a poem in painting."[84]

Millais signaled his ambition as a landscape painter by submitting another landscape, *Flowing to the River* (1871), to the Royal Academy the following year. This canvas continues the somber mood established by *Chill October* and *Autumn Leaves* but also incorporates an explicit gesture to the art-historical past that positions Millais within the lineage of Britain's landscape painters. The painting is set in Scotland, near the Stormontfield salmon ponds north of Perth. Millais was an avid fisherman and hunter; his son records that while in Scotland his paintings "kept him fully occupied in the intervals of sport."[85] The foreground depicts the watery expanse of a bend in the river; a discarded cart occupies the far right corner of the canvas; and at left a fisherman, nearly all in white, stands on an outcropping produced by a man-made sluice. In the far distance we see the reason for the sluice, a mill, before which can be seen white figures loading a cart.[86] The ingredients of the scene—mill, river, fisherman, and cart—are reminiscent of Constable's early oeuvre, particularly such paintings as *Stratford Mill* (1819–20), *Dedham Lock and Mill* (1820), and *The Haywain* (1821). The association with Constable is not simply fortuitous; the Royal Academy's winter exhibition of 1871 (the second in a series of exhibitions devoted to the Old Masters) included six works by Constable, among them *The Haywain* and *Flatford Mill.*[87] Millais's handling of paint also evokes Constable. It is conspicuously thicker and more varied than in his earlier paintings and applied more broadly (as in the leaves of the tree in the left foreground or the impastoed touches registering the choppy water).

Millais inscribed his monogram and the date of the painting on the cart—drawing further attention to this object in the landscape—thus underscoring the connection between himself and Constable. Indeed, *Flowing to the River* seems an imagined projection forward in time of *The Haywain.* But now, rather than evoking the prosperity of the post-Napoleonic demand for grain and commenting on georgic labor, as Constable had done, Millais conveys a melancholy mood through the abeyance of the harvest season. The russet, yellow, brown, and green leaves that remain on the trees in Millais's landscape contrast vividly with the late summer greens in Constable's work. In *Flowing to the River*, low clouds create a gray expanse, a departure from the high-floating clouds in the blue sky of *The Haywain.* The autumnal sensibility adds a bittersweet tinge to the georgic activities—wheat being ground at the mill and the fisherman at the river—which will end shortly with the winter storms that the scudding clouds presage.

Reviewers were divided in their estimation of the painting. The *Art Journal* maintained that it was only a description of nature and lacked "any sentimental narrative," whereas the *Athenaeum* praised the subject. The *Spectator* considered the canvas merely a "clever sketch" for which Millais had not sufficiently studied nature or captured the requisite breadth, as Turner had done in his *Liber Studiorum*. The *Times*, by contrast, while referring to *Flowing to the River* as a "direct transcript" "of such bits of the outer world as a man might take in by looking through a window . . . recorded with the vision of an eye never surpassed for keenness," compared the canvas to idyllic poetry, such as that of Robert Burns or Tennyson.[88] This lack of consensus reveals that critics were unsure how to evaluate Millais's work: using models from the past, the Pre-Raphaelite paradigm, the artist's ability to create emotive responses to external nature, or analogies with poetry. What reviewers did consistently agree upon was that illusionistically mimicking nature should no longer be the only aim of art.

When Millais exhibited *Winter Fuel* at the Royal Academy in 1874, the *Times* reviewer, probably Tom Taylor again, took the opportunity to teach his audience how to read Millais's new work.[89] The scene depicts, in the foreground, a cart full of felled trees, some with leaves and branches still intact, set against a green field flanked by a craggy hillside in the background. The gray sky and brown leaves once again set an autumnal mood. A young girl wearing an eye-catching red hood is perched on the cart; her gaze and that of the dog at her feet are directed beyond the picture frame. In the immediate right foreground is an axe next to a dinner pail; presumably, the girl and dog await the return of the wielder of the axe. The *Times* critic's instructions to readers establish that viewers perceived a distinct shift in the painter's work that was of value to the art world at large:

> *The difference between all the marvelous detail in this picture and in the painter's earlier subjects, including similar fullness and intricacy of bough or leafage, is that here the parts are subordinated to the artistic effect and intentions of the whole, and that there they were not. In the earlier work the painter was painfully trying to paint the infinite details of nature as they were, that he might learn, in this the later work of the master, to paint the effect of the same details in the eye, to translate into what they look like by tones and touches in themselves utterly unlike the thing they are to help in representing. We can suggest no more useful lesson to the young painter than to look close into the execution of this carpet of heather, moss, and grasses in the fir wood, and then, retiring to the distance required to take in the picture as a whole, to mark the gradual change from what the close eye saw to what the eye, at successive distance, has told the brain it was seeing.*

This phrase "has told the brain it was seeing" is highly significant, indicating the effect of the new literature of psychology on popular criticism. The terms of evaluating art have shifted. Millais is now praised for achieving truth to mental impressions of the scene as opposed to truth to the scene itself; that is, he has created painted illusions of what the mind saw rather than what the eye saw. The painting, the critic concluded, "lifts the work of the realist into a new region altogether."[90]

In the same year, the psychologist James Sully published a new guide to aesthetics in his *Sensation and*

*Intuition: Studies in Psychology and Aesthetics.* This collection of revised essays concerns cognition and perception as well as evolutionary biology. In the early 1870s, Sully, who had studied philosophy and worked as a classics tutor, began to explore aesthetics through the developing field of psychology, with a particular interest in the associationist school. In 1871–72, he traveled to Berlin to study physiology with the physicians Emil du Bois-Reymond and Helmholtz. Sully's debts to Bain and Darwin, particularly his work on expression, which explored the links between behavior and expression, are clear in *Sensation and Intuition.*

In the volume's closing essay, "On the Possibility of a Science of Aesthetics," Sully argued that psychology and anthropology (the study of human development) should be applied to aesthetics. Through a series of logical steps, he derived the following criteria for an artwork of high quality: "its several features correspond to sentiments which are refined and complex, which involve a large number of distinct intellectual and emotional activities, and which are the product of numberless far-reaching, and oft-repeated emotional influences." To achieve these ends, artists could adopt a variety of approaches, which Sully ranked within an evolutionary framework. (Sully drew repeatedly from Darwin and Herbert Spencer, who argued that evolutionary models could be applied not only to biological species but also to human society.) The simplest, most childish or savage way would be to reproduce some absent effect that pleased—in short, to imitate or reconstruct nature with "scrupulous fidelity." To reach a higher form of pleasure, the artist must judiciously select scenes from nature and employ imagination and creativity. Sully denied that art was evolving from mimesis to creative expression; rather, different proportions of these two approaches were required, depending on the genre of art making. Once viewers determined the type of art being viewed, their appropriate expectations would fall in line. Therefore, "when the observer recognizes that the work before him aims pre-eminently at truth to nature, he is far less ready to indulge in the free flights of imagination."[91]

While Sully claimed that he was not ranking art hierarchically, his appellation of the terms *childish* and *savage* to acts of mimesis, particularly within the context of contemporary evolutionary theory, implied that such artworks failed to fulfill art's fullest potential. Art critics who subscribed to Sully's views and wished to laud landscape painting faced a difficult task. To what degree should they praise and thereby encourage mimesis, and to what degree should they advocate "free flights of imagination"?

Ruskin, in his *Academy Notes* of 1875, maintained a clear distinction between different types of landscape paintings. Those he classified as "Natural History" were based on the close study of nature, were significant for science, and were without "emotion, or definite purpose of expression." Those he classified as "Landscape" showed "the relation between Nature and man; and, in fine work, a particular tone of thought in the painter's mind respecting what he represents."[92] This dichotomy was in some ways a rehearsal of the distinction between Millais and Turner that Ruskin had presented earlier in "Pre-Raphaelitism," but here he assigned far less significance to works he reviewed under the heading of natural history. They offered little value: they did not teach the viewer to see nature in new ways, and they did not communicate knowledge about nature.

In this context, Ruskin was particularly critical of Millais's *The Fringe of the Moor* (1874), a depiction of rolling hills and cows grazing within walled and fenced fields. He described it as possessing "essential, though rude, and apparently motiveless, veracity." This mode of painting he judged "an insult to every master, principle, and feeling, reverenced, or experienced, in the schools of noble art, from its nativity to this hour."[93] This response also reveals that the capacity of Millais's landscapes to rouse emotional reactions was not consistent across all his canvases or all his critics.

While Ruskin's criticism was particularly harsh, especially in light of his position twenty-five years earlier, he was not alone. The *Times* likewise established a binary opposition between artists who painted whatever they saw, and those, more universally recognized as true artists, who could take what they observed and "build up . . . new scenes, unbeheld save by the inner eye of creative power." Millais fell into the former category. *The Fringe of the Moor* was a "literal and exact transcript of a scene which in some way appeals to him"; but, unlike *Chill October*, the newest painting had "nothing in it calculated to impress the imagination."[94] The critical tide was clearly turning, fed by new theories of psychology and perception, against the Pre-Raphaelite paradigm of universal truth to observed nature. And Millais was finding it difficult to repeat the earlier success of *Chill October*.

In this context, Millais's decision to select poetic subjects for his pure landscapes was a strategic response to his critics. Literary references would ensure that the painting carried associations beyond what was depicted. In 1890, Millais exhibited two important landscapes: *"The Moon is Up, and Yet it is not Night,"* quoting a line from Byron's *Childe Harold's Pilgrimage* (1818), and *Dew-Drenched Furze*.[95] Both paintings depict a landscape swathed in mist, transient effects that obscure the scene, as though Millais wanted to ensure that he could not resort to the trope of the scrupulously observed and easily identifiable detail. Both paintings also inspired a lengthy, powerfully descriptive review in the *Athenaeum*, probably by F. G. Stephens, who was enthused by the beauty he perceived in the scenes. At a time when the sheer quantity of artistic output on display in many exhibitions seemed to militate against exegetic narratives, Stephens poured out prose that almost seemed to compensate for the murky landscapes. Of *"The Moon is Up,"* he wrote, "The landscape proper comprises, in front, an open, level glade, covered with the burnt remains of the year's furze and fern, and bounded on either hand by clumps of oaks, graceful birches, and larches, while beyond these the darker pines rear themselves like tall phantoms, and loom in the vapour; in the mid-distance a wan reflection of the moon's disc betrays the lake which, so to say, lies *perdu* there; in the remoter space the shadow-like masses of uplands are dimly seen against the hardly less substantial sky."[96]

Stephens, of all the critics reviewing exhibitions that spring, seems to have grasped the metaphorical distance that Millais had traveled in his landscape practice, now seeking out beauty rather than the commonplace, now emphasizing the poetic rather than the literal, and now adopting a new technical means of representation distinct from his Pre-Raphaelite manner of handling paint. The use of paint is particularly discernible in *Dew-Drenched Furze*, its textured surface resulting from a relatively dry application of paint, particularly in the upper layers. Instead of a smooth, polished surface, as in Millais's canvases in the mid-1840s, one can see where the canvas weave caught his brush, when loaded with white paint for the highlights, as he dragged it across the surface. Such relatively broad handling foreclosed the possibility of the minute rendering of nature's details that characterized such earlier works as the portrait of Ruskin. Forms are now softened, suggesting atmospheric effect, their edges barely discernible rather than sharply demarcated. Stephens recognized this technique as new for the painter but also old in that it harked back to the Old Masters.[97]

These two paintings continue Millais's exploration of pathos in nature begun with *Autumn Leaves.* Both scenes are set in the fall and feature landscapes that have been burned to promote new growth the following spring. Both also depict transitional moments, the rising moon and sun, which, like the changing of the seasons, point to the transience of nature and the passing of time. This melancholic trope of the passing of time is the pathway, in Millais's landscape paintings, to the emotional or sentimental responses that some critics located in his work. This space of affective recognition and reaction described by the nascent field of psychology is also what many scientists had carved out for art. The physicist William Thomson, Lord Kelvin, in his toast to the Royal Academy in 1893, expressed fear that post-Darwinian science had rendered nature "God-forsaken" and "soulless . . . deprived of life." To address this problem, Kelvin proposed a partnership of sorts: while "Science brings us to the threshold of life . . . it is the beautiful function of art to give the soul of life to the representation of nature, to the threshold of which only can science lead us."[98]

Perhaps even more than *"The Moon is Up, and Yet it is not Night," Dew-Drenched Furze* invites contemplation of physical metamorphosis and larger conceptual changes in attitudes to nature. The misty veil connotes the artist's strained vision as he struggles to see nature clearly, very unlike the confident, all-seeing, innocent eye that Ruskin had championed forty years earlier. This reading is in keeping with the associations roused by the title *Dew-Drenched Furze*, taken from Tennyson's elegiac poem *In Memoriam A. H. H.*, written in tribute to his dead friend Arthur Hallam. Tennyson and Millais were good friends, and Millais's choice of a subject derived from Tennyson was also likely to curry public favor given the poet laureate's considerable reputation and cultural eminence at the time.

The narrator of *In Memoriam A. H. H.*, grieving his lost friend, describes going to nature to seek relief and comfort for his sorrow:

> *Calm is the morn without a sound,*
> *Calm as to suit a calmer grief,*
> *And only thro' the faded leaf*
> *The chestnut pattering to the ground:*
>
> *Calm and deep peace on this high wold,*
> *And on these dews that drench the furze,*
> *And all the silvery gossamers*
> *That twinkle into green and gold:*

But the narrator realizes that this surface beauty and dreamy calm is an illusion, because modern science has robbed nature of its divine presence:

> *Who trusted God was love indeed*
> *And love Creation's final law—*
> *Tho' Nature, red in tooth and claw*
> *With ravine, shriek'd against his creed—*
>
> . . . . . . . . . . .
>
> *O life as futile, then, as frail!*
> *O for thy voice to soothe and bless!*
> *What hope of answer, or redress?*
> *Behind the veil, behind the veil.*[99]

Tennyson mourns his friend and equally mourns the loss of nature as solace. While writing *In Memoriam A. H. H.*, Tennyson was reading Lyell's *Principles of*

*Geology* and Chambers's *Vestiges of the Natural History of Creation* (1844), which revealed that nature was governed not by God's divine and ordered providence but by the struggle for survival in which only the fittest prevail. And how could it be, Tennyson asks in anguish, that his friend, so full of promise, was not fit to survive? No answers are forthcoming—they are hidden behind a veil. Correspondingly, Millais has drawn a veil of moisture and refracted sunlight across his scene. The beguiling beauty of the obscured prospect framed by the curtains of fir trees is an invitation extended to us and then withdrawn. Upon closer examination of the scene, we can see that to enter the space (if we imaginatively desired), we would have to disturb the partridges hidden in the foreground grasses and then trudge through cobwebbed, damp, and thorny furze. It is a scene better contemplated than metaphorically traversed. As much as we may want to know nature and probe its mysteries, we are discouraged from pursuing answers. Nature is no longer the source of knowledge envisioned by the inductive philosophers.

### Brett, 1870s–1890s: Intellect?

Just as Millais negotiated with difficulty the changing dynamic between art and science and the rise of new theories of perception and aesthetic judgment, so too did John Brett. Millais, however, enjoyed a certain level of privilege because of his relatively rapid ascension into the ranks of the Royal Academy and the popularity of his figural subjects, whereas Brett's audience shrank over time and his fortunes declined. Publishing became an important strategy by which Brett attempted to shift both critical and public opinion, though arguably with little success, as his often polemical positions alienated some readers.

Brett evidently envisioned himself as equal to his fellow scientists in knowledge and visual acumen. In 1870, the astronomer Norman Lockyer invited him to join an expedition to Sicily to witness the total eclipse of the sun. Brett, who had been an avid observer of the skies from adolescence, kept a well-documented log of his findings.[100] In preparation for the trip, Lockyer asked Brett to supply the group with "leading questions . . . with reference to the aspect of the corona, such as might be useful to regulate the efforts of men who propose to make drawings."[101] Brett's instructions for those making drawings of the corona included paying attention to "particulars," such as "whether the outer border exhibits any coruscations or whether its definition is permanent and equally pronounced all around."[102] The artist also contributed to published reports of the expedition, describing the telescopes used and the solar phenomena he observed. He repeatedly noted that the event, as it unfolded, did not accord with his original expectations, and he speculated that previous accounts of eclipses were tainted by ignorance. His visual skills, honed through his experience as an artist, corrected such potential errors. "I believe that the wonderful character of the terrestrial appearances recorded during eclipses," he wrote, "is to a great extent the measure of the observer's ignorance of ordinary cloud phenomena. Such appearance (to judge from my own experience) would hardly strike a landscape-painter as very abnormal; but the truth is, that very few persons besides artists have ever paid enough attention to clouds and their effects upon landscape scenery to enable them to answer the most elementary questions respecting them."[103] In short, Brett was arguing for the methods of inductive science and the value of experience along the lines set forth by Herschel and Mill.

In 1871, Brett became a fellow of the Royal Astronomical Society (RAS). Following the eclipse expedition, he continued to submit essays to astronomy journals on such topics as the sun's limb, the lunar atmosphere, the Great Comet of 1882 (a sketch of which was published in the RAS's journal), Jupiter, Venus, and Mars.[104] He was credited with "keen sight" and skills of observation. But Brett's contributions were not always welcomed. His assertion that the dark spots observed at the poles of Mars were clouds "met with almost a storm of incredulity in some quarters when communicated to the Society."[105] Indeed, even earlier, Brett had lost Ruskin's support because of their conflicting views on a scientific matter. Ruskin explained the falling-out to Rossetti in a letter of 1865: "Brett told me, a year ago, that a statement of mine respecting a scientific matter (which I knew *à fond* before he was born) was 'bosh.' I told him in return he was a fool; he left the house, and I will not see him again 'until he is wiser.'"[106]

Brett's astronomical conclusions drew the particular ire of Richard Proctor, with whom Brett had repeatedly butted heads over scientific findings and the composition of the RAS's council, on which Brett served in 1876 and 1877.[107] Proctor went on in 1881 to found the journal *Knowledge*, which he hoped would displace *Nature* in the periodicals market and wrest authority away from the professional scientific elite.[108] Proctor's antagonism toward Brett was fueled by a desire to purge uninformed amateurism from the ranks of the RAS.[109]

Brett's connections within scientific circles extended beyond astronomy, although that discipline dominated. The guests he invited to exhibition openings, by the mid-1870s, included Lockyer and Huxley as well as Arthur Cowper Ranyard and William Henry Mahoney Christie, among other astronomers (whereas in 1864 the only scientist included was Francis Galton, Darwin's cousin, known for his contributions to statistics and eugenics). In 1876, Brett met Lewes at a dinner organized by the Physiological Society, which was founded that year to support experimental methods in physiology. In 1881, the philosopher and biologist Herbert Spencer sponsored Brett for membership in the Athenaeum Club, a gentlemen's club that had formed in the early nineteenth century and was associated with intellectuals such as Spencer, Michael Faraday, J. M. W. Turner, Charles Darwin, Charles Dickens, William Makepeace Thackeray, and Matthew Arnold.[110]

Brett clearly aimed to establish himself as a scientific authority on painting, as in his review of the Royal Academy exhibition of 1870 for the journal *Nature*. That *Nature*, which aimed at both scientists and the general public, devoted pages to reviewing the Royal Academy's annual exhibition indicates both the significance of this annual display and scientists' aspiration to exert cultural authority.[111] Lockyer founded *Nature* in 1869 as a means of placing "before the general public the grand results of Scientific Work and Scientific Discovery, and to urge the claims of Science to a more general recognition in Education and in Daily Life."[112] Lockyer believed passionately that science should play a crucial role in modern society, a conviction that undoubtedly appealed to the ambitious Brett, just as Lockyer's dogged "love of truth" did.[113] Yet, as David A. Roos has argued, the founding of the journal "coincided with the decisive recognition among scientists of the impossibility of maintaining a single norm of truth."[114] The aim of *Nature* was to transcend differences and create a forum, as Lockyer put it, in which "early information of all advances made in any branch of Natural Knowledge" could be shared with both the scientific community and the general public.[115]

Brett knew Lockyer through their shared interest in astronomy, and he clearly positioned his review to appeal to scientists.[116] He not only titled his essay "Natural Science at the Royal Academy" but also praised scientists as reasoned, orderly, and laborious—unlike art critics, whose discourse he compared to empty air, i.e., gas. Brett set himself the task of providing guidelines of aesthetic judgment for *Nature*'s audience of "clear-headed and simple-souled naturalists," and he offered recommendations for both artistic and scientific practice. He concluded by warning that "scientific men pay too little attention to the broader aspects of the visible world; while the artists on their part pass by the clear fountain of natural beauty, and content themselves with dreamily sipping lukewarm water from the corroded vessels of their forefathers"; in other words, they based their art too much on models from the past and not enough on direct observation of the external world.[117] To correct both kinds of folly, Brett, like many of the speakers at the Royal Academy banquets, recommended that artists and scientists come together to teach one another how to see and represent nature properly.

In a sketchbook of 1869, Brett had drafted several paragraphs on the nature of human vision. While unpublished and not particularly coherent, this passage nonetheless shows that Brett was concerned with the perception of color and the operations of the human eye—he speculated that vision was the result of the shock received by the eye from the emanation of light from an object.[118] These fragments suggest that Brett adhered to a mechanistic conception of vision, as opposed to the associationist model proposed by psychologists. But he did believe that visual perception was learned rather than innate, as demonstrated by his extensive sketchbook notations. In other words, Brett believed that the eye was trainable and that his task as a painter was to develop his eye into the most perfect instrument of reception. In this sense, Brett's practice fulfils what Lorraine Daston and Peter Galison term "noninterventionist," or mechanical, objectivity, which "requires painstaking care and exactitude, infinite patience, unflagging perseverance, preternatural sensory acuity, and an insatiable appetite for work."[119]

The concept of mechanical objectivity clearly informed the critical reception of Brett's submission to the Royal Academy of 1873, *A Morning Amongst the Granite Boulders*, which depicts a sandy beach dotted with boulders and seaweed-covered rocks revealed by low tide. In the distance, past the cresting waves, minute sailboats are highlighted against the gathering mist, while clouds hover over the horizon line and swathe an island at center right. The critic for the *Art Journal* described Brett's painting as "marvellous" and "powerful" for its veracity. But the reviewer also observed that "this view of sea and sand is wholly devoid of imaginative significance. There is no trace of human sensibility, no suggestion that the scene so clearly realised has influenced the mind of the painter. All that is shown is the scene itself, untouched and unchanged, and suggesting no human passion save a passion for mechanical fidelity."[120]

*Fraser's Magazine* characterized the work as possessing "unimaginative truth."[121] The art collector, photographer, and amateur scientist John Forbes White, writing for the *Contemporary Review*, praised the artist's skill and mastery in registering such details as the "crystals of felspar imbedded in the rocks," but complained that a "human impression" was missing, that the artist had failed to convey "the feeling of the scene having passed through the artist's mind and heart." White hoped that in the future Brett would combine his "stores of knowledge" with "the stamp of his own feeling for nature," in order to produce a great work of art.[122] According to the aesthetic framework proposed by Bain and Sully, Brett had failed to make a compelling selection from nature or to convey any emotive consequences, and his painting thus missed the mark.

The critic Harry Quilter, an increasingly important voice in art criticism in the last quarter of the century, made the same observations in an essay called "The Human Element in Landscape Painting," first published in 1879. He praised *A Morning Amongst the Granite Boulders* for its "literal reproduction of nature" but found the painting wanting in terms of lasting power and significance, especially when compared to Turner's *Temeraire* and *Polyphemus*.[123] In sum, critics repeatedly called explicitly for marks of Brett's subjective presence, cues that would trigger viewers' emotive associations and evidence the imaginative leaps taken. Mimesis alone was no longer sufficient rationale for a work of art, as Sully had insisted; Brett aimed at truth to nature, but his audience demanded "free flights of imagination."

Ruskin, when he reviewed the Royal Academy exhibition of 1875, denied Brett consideration as a landscape painter and reviewed his painting *Spires and Steeples of the Channel Islands* (1874) under the category of natural history, just as he had done with Millais's *Fringe of the Moor* that year. Like the reviewers of *A Morning Amongst the Granite Boulders*, however, he found merit in Brett's meticulous rendering and informed observation, pointing out that "the paint on the water surface in the bay is too loaded, but laid with extreme science in alterations of colour."[124]

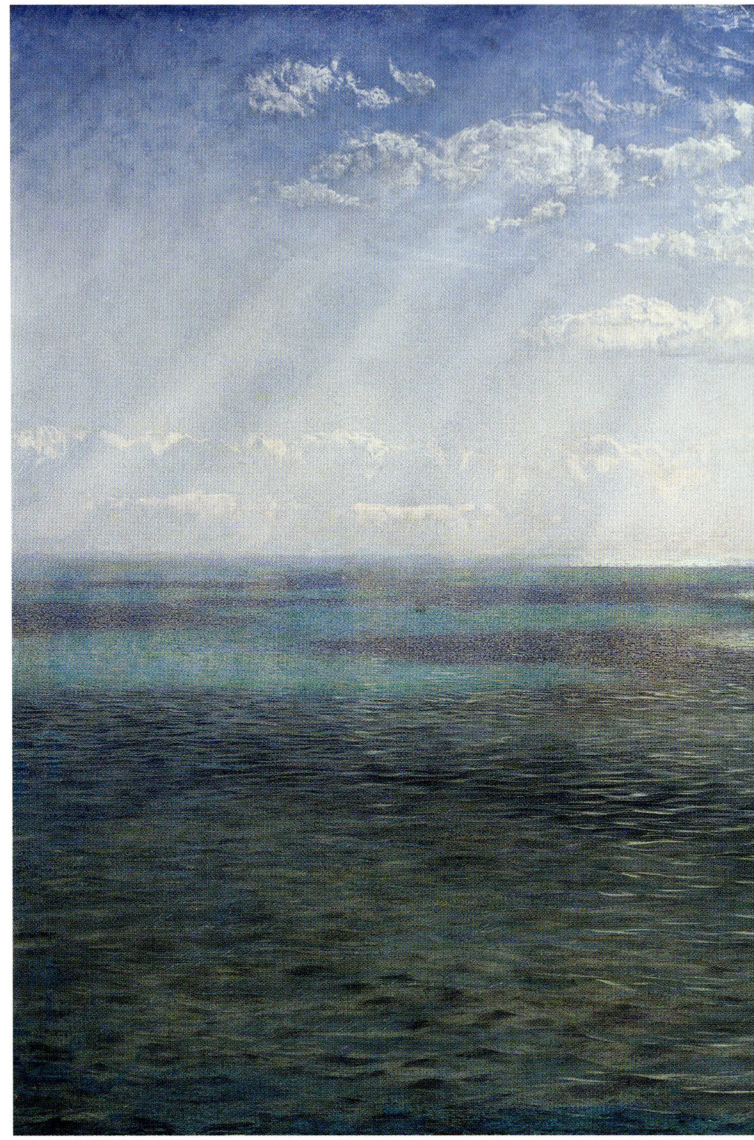

Ruskin's criticism of Brett's rendering of water draws attention to the painter's technique, which had, in fact, changed by this date from his earlier Pre-Raphaelite practice. Conservators at the Tate, when analyzing Brett's *British Channel Seen from the Dorsetshire Cliffs* (1871),

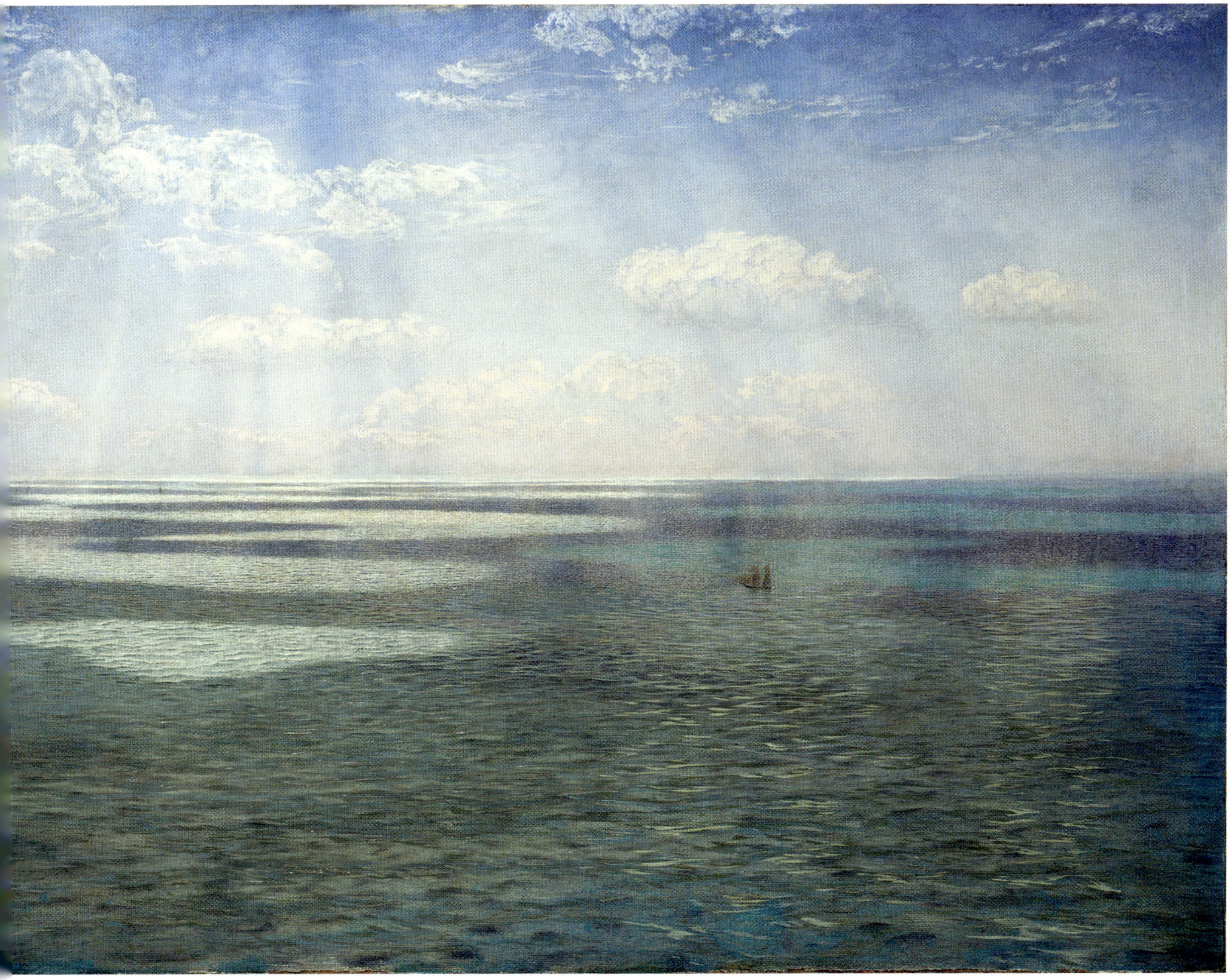

FIG. 31
John Brett, *The British Channel Seen from the Dorsetshire Cliffs*, 1871. Oil on canvas, support 1060 × 2127 mm. Presented by Mrs. Brett, 1902. 

made the invaluable discovery that the appearance of waves rippling the surface of the water, denoted so carefully in Brett's notebooks, was achieved not solely by individual brushstrokes or daubs of paint but also by what they characterize as "mechanical methods" (fig. 31).[125] This phrase describes how Brett used the preparation of the canvas to help achieve his final surface effects. He began his canvas, as he had in his earlier practice, by applying a coat of lead white over a pre-prepared white primed canvas. Now, however, this layer was then textured, perhaps by using a roller; Brett also used a knife to mark the surface of his canvases, as indicated by his notation in a sketchbook of 1863.[126] Brett appears to have deliberately allowed this layer to dry completely, so that when he applied the next layer of oil wash it would pull away slightly from the underlying layer (rather than adhering together as a wet-onto-wet application would allow). Brett's sketchbook notation also indicates that he allowed layers to dry before applying more paint. The reaction or resistance thus created would result in more surface texture, which could be uniformly achieved across the entire canvas and thereby save considerable time in producing such effects as rippling waves. For the sky, he then laid in a layer of lead white and ultramarine, which he could wipe away judiciously to create clouds. To color the sea, he used washes and more opaque touches of cobalt blue, viridian, and white paint, dabbing more (rather than stroking on the paint) as he reached the horizon line to suggest the dissolving effect of light on the water.[127]

What this means is that the faithfulness to nature that Brett achieved in his canvases, a feature of his work on which critics commented repeatedly, was not the result of purely mimetic pictorial means—the delineation, for example, of each wave. He also adopted illusionistic strategies dependent on scientific knowledge of the physical behavior of light. By creating a fine texture on the surface of his canvas and using more opaque paints, Brett scattered the light, enhancing the illusion that we are gazing on the choppy surface of the sunlit water.[128]

Lockyer endorsed this attention to the interaction of light and atmosphere with color. In 1878, Lockyer gave Brett's canvas *The Cornish Lions* a positive review in his essay on the Royal Academy's annual exhibition. According to Lockyer, the work was a "remarkable picture: the colours and the atmospheric absorption, and therefore transformation of the colour, perfect." Throughout the article, titled "Physical Science for Artists," Lockyer framed his analysis of specific paintings within the context of a larger exegesis on the state of contemporary art in relation to physical science—in particular, optics. He argued that most artists were woefully ignorant of the laws governing the behavior of natural phenomena, and he judged the success of paintings on the basis of their adherence to known scientific principles. For example, he divided the exhibited landscape paintings into two groups: those that accurately represented the sky and those that demonstrated cognizance that light is composed of particles that are dispersed or absorbed (to which class Brett's painting belonged).[129] What is particularly noteworthy is Lockyer's attempt to ground art criticism in scientific practice, which was entirely in keeping with his ambitions for science as a voice of cultural authority. But Lockyer was also out of step with prominent strands of art criticism that valorized individual subjectivity over inductive accuracy.

I should make clear, however, that the division between critics who advocated an art rooted in inductive science and the observation of nature and those who conceived art's role as the expression of individual consciousness and poetic sentiment does not map cleanly onto the dichotomy of art critics versus scientists. For example, the art critic and writer William Wilthew Fenn, in his 1881 essay "Fact in Landscape," welcomed what he described as a fact-based approach to landscape painting as Britain's contribution to modern civilization, and he heralded the contributions of Brett and Hunt to this development.[130] The painter J. E. Hodgson praised Brett's painting *Britannia's Realm* (fig. 32), acquired by the Chantrey Bequest in 1880, for eschewing emotion in favor of focusing on external phenomena: "There is no other sentiment, and the artist intended there should be none, but a feeling of vastness and light—a suggestion of the infinite complexity of Nature."[131] What is most significant here is the absence of consensus and the implicit tensions concerning what defined excellence in art.

Like *The British Channel Seen from the Dorsetshire Cliffs*, *Britannia's Realm* depicts a broad panoramic expanse of water. The scene of watery surface and cloud-filled sky spreads out, uninterrupted by any evidence of shoreline. Patches of bright light, indicating breaks in the clouds, play over the surface of the water, which is further interpolated by a variety of hues of blue, green, and purple rendered more vivid by the delicate insertion of contrasting touches of yellow and orange. The waves are registered by intricate patterns of parallel and intersecting lines of paint that stand out against the smoother texture of the underbelly of the clouds overhead. Just as the color of the water is carefully represented to indicate the passages of light and shadow caused by the interaction of sunlight and the moving waves, the clouds are modulated to indicate meteorological diversity, with purple-tinged tufts closest to the watery surface and white wisps whisking through the upper atmosphere.

The title conveys imperialist ambitions, and the work defines the British Empire as a watery expanse with no boundaries. Edward T. Cook, in his handbook to the Tate Gallery, explained that "through Brett's vast windows . . . we may look over the lovely expanse of shimmering placid blue—Britannia's realm and all the glory of it."[132] The language of possession and control is echoed by Brett's painting technique, in which every element is rendered with equal precision and control from foreground to background. In short, Brett's paintings evoked mastery, both political and personal, over nature. But Hodgson insisted that Brett intended no sentiment, only a feeling of vastness and light. The pursuit of truth to external nature—mimetic faithfulness to the chosen scene—has foreclosed any other narrative possibility for the work. Thus what was also at stake in this debate about truth to nature was how meaning for art was produced. Could and should the painting convey further associations? Hodgson answered no, whereas Colvin, writing about Millais's landscapes, answered with an emphatic yes.

In the same month that Brett's canvas made its public debut, the landscape painter Alfred William Hunt published "Modern English Landscape Painting," an essay marked by declarations, retractions, an odd doubling back, and fissures that revealed the author's struggle to reach decisive conclusions about the present state and future of landscape painting. Hunt was part of Ruskin's circle at midcentury, and his detailed studies of closely observed natural phenomena led to the nickname "Bird's Nest" Hunt. What is particularly remarkable about Hunt's essay is that he recommends an artistic practice that he himself never achieved and that was at odds with his public reputation.

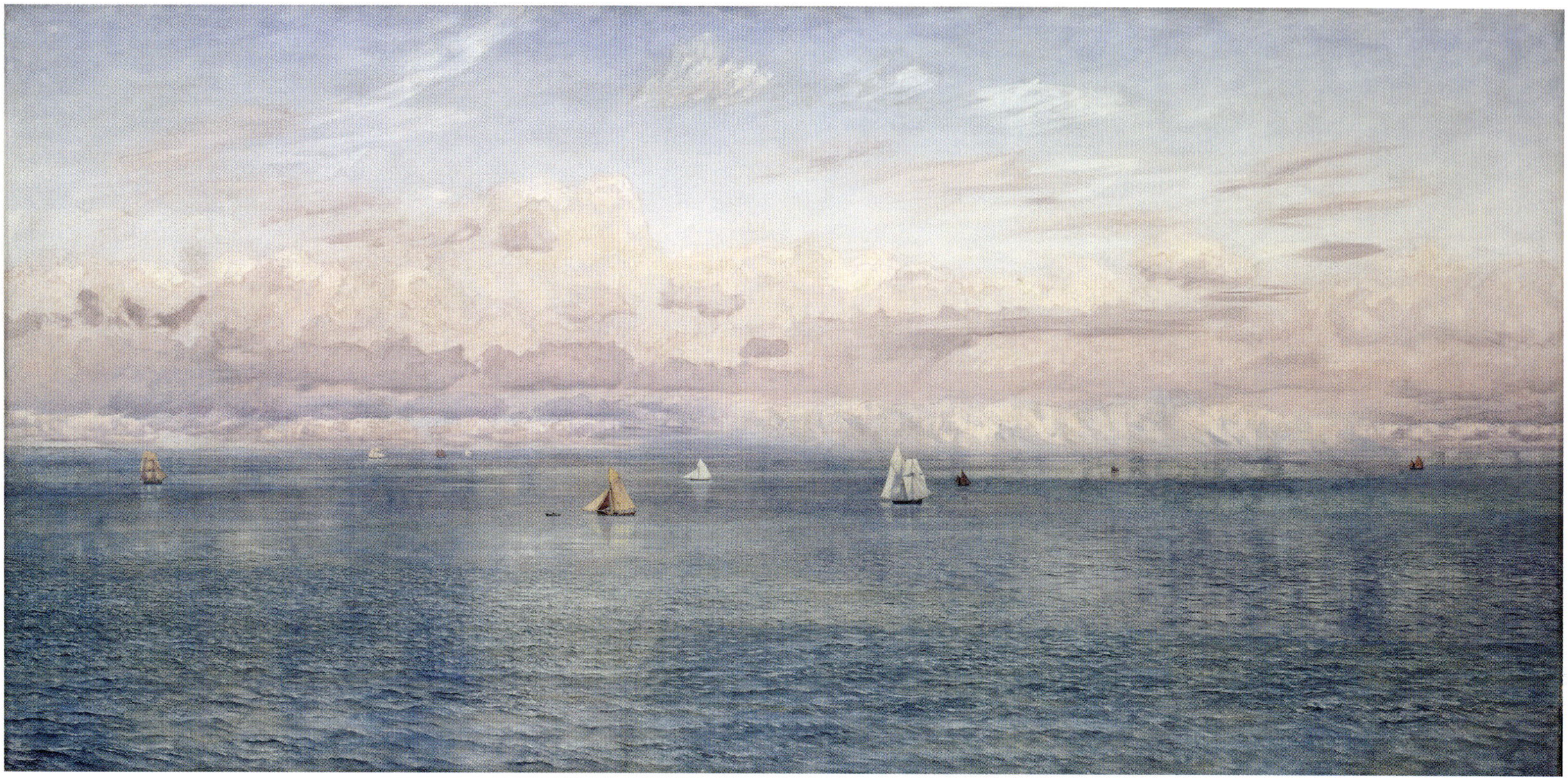

FIG. 32
John Brett, *Britannia's Realm*, 1880. Oil on canvas, support 1054 × 2121 mm. Presented by the Trustees of the Chantrey Bequest, 1880. © Tate, London 2015.

Like Fenn before him, Hunt recognized close observation of nature, fact, detail, truth, and realism as hallmarks of a new, post-Turner direction in British landscape painting, and in this context he praised the Pre-Raphaelites. Yet despite his forceful celebration of the modern age of landscape to which he clearly belonged, Hunt expressed wistful regret for the demise of the tropes of the pastoral and the sublime, and of the "old style," characterized by "clearness, simplicity, swiftness of execution, poetical conception, and unity of design." With a tinge of melancholy, he described how the past had been swept away by "a new zeal in the pursuit of truth" driven by "the inroads of the scientific spirit, the invention of photography, the writings of John Ruskin, [and] sympathy with the Pre-Raphaelite movement." "The most elaborate transcript of the forms and colours of nature, without the expression by design of the sympathy or poetical feeling excited by them," Hunt concluded, "is rightly to be called a study, not a picture of a landscape." Whereas F. G. Stephens had used the word *transcript* as a term of praise for Pre-Raphaelite canvases, as well as a mark of modernity, Hunt's use of the term connotes limits and even reproach, if not a dead end. The task of the landscape painter, he explained, was to "make us feel the hold which this deep mysterious beauty of nature has taken of his heart and soul"—in short, the emotive power of the scene.[133]

Two years later, an anonymous writer in *Blackwood's Magazine* repeated many of the points made in Hunt's essay but attributed the perceived decline in the quality of landscape painting more definitively to science: "science has been positively mischief making" because it lacked imagination. The author warned that science was "cold to sentiment," and that its "scrutinizing vision 'would peep and botanise upon a mother's grave.'" Because of science, "intellect has usurped the place of emotions."[134] This was a reiteration of the dichotomy of knowledge found in the later Royal Academy banquet speeches.

Such pronouncements led some in the materialist faction to dig their heels in even deeper. Lockyer, in an essay in *Nature* called "Science and Art" that appeared one year after the *Blackwood's* essay, reasserted his belief that landscape paintings should be judged purely on their accuracy in depicting the color of the sky, clouds, distance and atmosphere, sunsets, moons, water, reflection, snowstorms, and other natural phenomena. In the fall of 1886, Brett wrote a lengthy vituperative essay to accompany his exhibition of views of the Scottish coast at the Fine Art Society, in which he reconfirmed the value of accurate observation.[135] He indirectly refuted the formula that equated art with feeling or emotion and science with thought or intellect. Indeed, he argued that scientists, when compared to experienced landscape painters (such as himself), were not sufficiently knowledgeable about external nature or sufficiently rigorous in their methodologies.

The Fine Art Society, a commercial art firm that cultivated an audience of wealthy elites and projected an image of erudition, typically commissioned outside authors to write essays for their exhibition catalogues (Ruskin wrote on the work of Samuel Prout and William Hunt in 1880). But in this instance Brett insisted upon providing his own exegesis. Indeed, he noted in the catalogue his strenuous objections to most art criticism because most critics "have no technical education, and but little acquaintance with nature." The aim of his exhibition, he explained, was to introduce his sketching practice to the public and to refute critics' frequent denigration of his work as overly laborious and detailed. Above all, Brett wished to be praised for the optical correctness of his work.[136]

Brett used his essay as a platform from which to preach how his practice might be adopted by others. The artist's memory should be so rigorously developed that a painting executed "without any direct reference to nature, should be more like the real scene than even your studies done on the spot." He polemicized against art based on imagination or limited to "some hint . . . just sufficient to revive the pleasant recollection of an early experience which was already registered in his brain." Landscape painters, he advised, should begin by learning how to draw and paint the human figure in order to gain skills in recognizing errors in anatomical accuracy. They should then acquire skills in geometry, perspective, and optics before venturing out into nature and learning how to translate what they see into observations. To complement this knowledge of geometry, optics, and perspective, the artist should learn the scientific laws governing rocks and the sea.[137]

But, paradoxically, Brett also dissuaded his readers from seeking knowledge in science. For example, he insisted that he learned more about clouds from his fellow landscape painter Peter Graham than "from all the transactions of the Royal Society put together." He disparaged the work of geologists, which was filled with jargon. He claimed that astronomy texts were riddled with unsupported observations.[138] The only exceptions were the writings of the Herschels (father and son) and Huxley, as well as Brett's own scientific contributions. He reemphasized the artist's need for rigorous observation.

Brett intended his exhibition as a retort not only to critics who pushed for art informed by emotion, feeling, and poetry, and to philosophers who argued for associationist models of perception and cognition, but also to artists such as James McNeill Whistler, who was pursuing new directions in modernist painting, particularly practices associated with France. Whistler, an American-born artist, studied in Paris at the École impériale and the studio of Marc-Charles-Gabriel Gleyre before moving to London in the 1860s. He earned a place among the generation of painters interested in reinventing painting by depicting modern life and realist subjects. His first major canvas exhibited in London, *At the Piano* (1858–59), which depicts his half sister seated at the keyboard while her daughter looks on, was a modern-life subject and also registered his early interest in anchoring his paintings in a strong compositional order. Mother, in black, is balanced against daughter, in white, the two almost forming the opposite sides of an isosceles triangle. Their heads bisect painting frames that repeat the paired pattern, while the wainscoting and the piano establish a horizontal line that extends the composition beyond the pyramid formed by the figures. Over the course of the 1860s, Whistler became greatly interested in the notion of decoration and the concept of beauty for beauty's sake, given form in such paintings as *Symphony in White, No. 2: The Little White Girl* (1864) and in his nocturnes. These oils captured the transforming effects of twilight, fog, and other atmospheric effects on the Thames in broad strokes of highly liquid paint that mitigated against capturing detail and instead emphasized flattened pictorial patterns. It was one of these nocturnes, *Nocturne in Black and Gold: The Falling Rocket* (ca. 1875), that Ruskin famously objected to when it was exhibited at the Grosvenor Gallery in 1877. Whistler sued Ruskin for libel, and while he eventually won the case in court (which he strategically used to advance his concept of painting), he was bankrupted by the costs of the trial. He rebuilt his finances and his career initially through a commission from the Fine Art Society to produce a series of etchings of Venice. Whistler quickly grasped the benefits of controlling how his work would be seen, and he deliberately staged his one-person exhibitions, as well as group shows in which he participated, in the 1880s and 1890s. For example, for his exhibition *Notes—Harmonies—Nocturnes*, held at Dowdeswell's Gallery in 1884, he orchestrated a wall scheme of flesh color and gray that was carried over to the uniform of the gallery's guard.

In this exhibition, Whistler included small-scale landscape oil sketches with which he achieved, as Anna Gruetzner Robins has described it, "a new kind of painting" and "a new way of looking at the motif" (figs. 33, 34). These innovative directions hinged upon many of the scientific findings discussed thus far, including Helmholtz's theories of vision, which led Whistler to emphasize the flatness of pictorial space, masses of form rendered without detail, and "low-key tonalities combined with an accent of brighter colour." In his earlier nocturnes, Whistler claimed that he sought an innocent

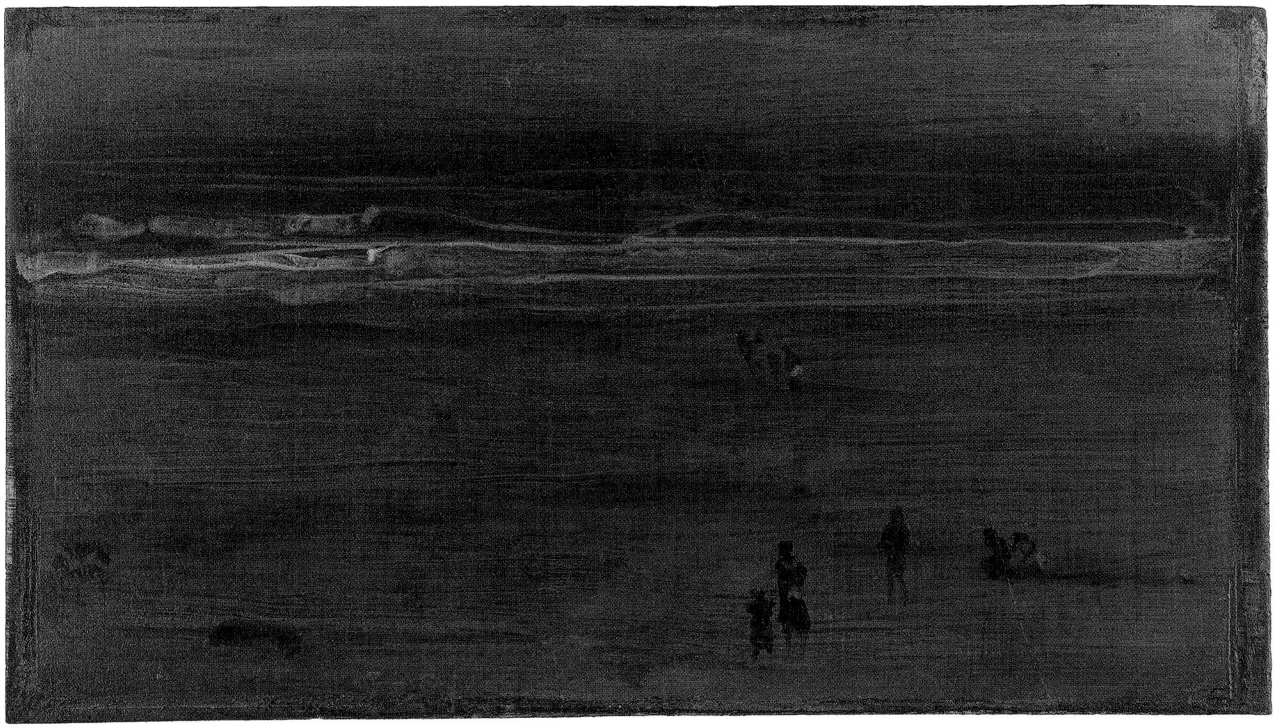

FIG. 33
James McNeill Whistler (1834–1903), *The Sea and the Sand*, 1884. Oil on wood panel, 13.4 × 23.4 cm. Freer Gallery of Art, Smithsonian Institution, Washington, D.C. Gift of Charles Lang Freer, F1902.151a–b.

eye—but his process was not to focus exclusively on the act of seeing, as Ruskin suggested, but instead was achieved by "emptying his mind," making it a blank before he painted. For the later oil sketches he painted directly from the scene, but he developed various exercises through which to rethink how to represent nature, arriving at a means of creating "a distilled image of successive impressions acquired from direct observation."[139]

Judicious and even aggressive editing of the scene as found in nature was Whistler's platform statement by the 1880s. In both paint and print, Whistler wholeheartedly pursued the distillation of the motif. His follower Mortimer Menpes records that Whistler "maintained that to see nature as nature absolute, one must, figuratively speaking, go into the coal-cellar and look at her through a chink, it would then, he declared, be proved incontrovertibly that nature was composed of five tones, and no more—that distance, detail, gradations of tint, even of form, were degrading and false alike either in nature and [*sic*] in art."[140]

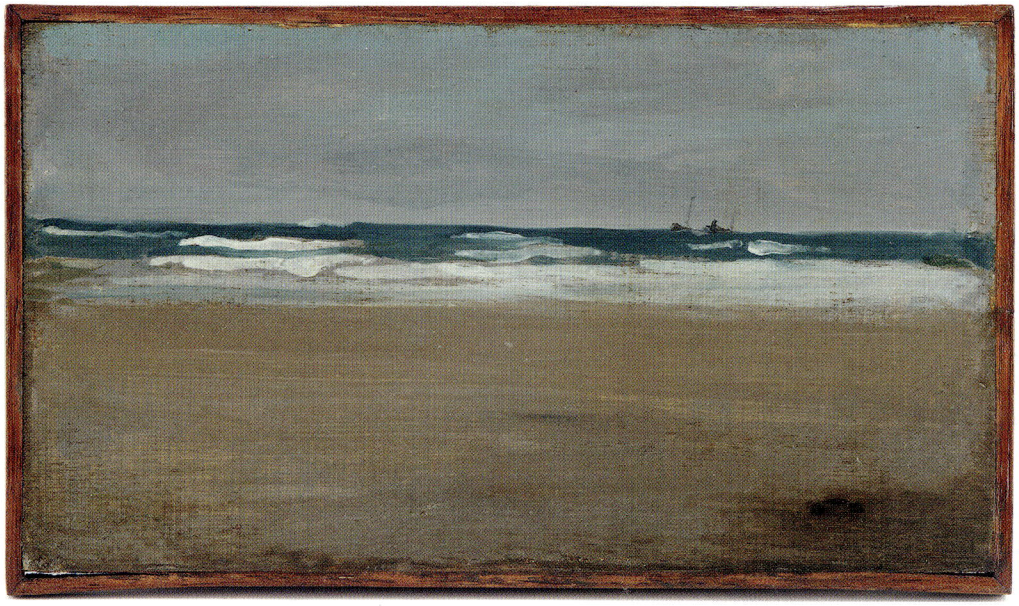

FIG. 34
James McNeill Whistler, *The Angry Sea*, 1883 or 1884. Oil on wood panel, 12.4 × 21.7 cm. Freer Gallery of Art, Smithsonian Institution, Washington, D.C. Gift of Charles Lang Freer, F1904.76a–c.

Whistler was willing to forgo what he had learned in order to achieve truth to nature. The artist, according to the critic Frederick Wedmore, "aims generally to record what the eye actually sees, and not what the mind knows the eye to see," a point perhaps fed to Wedmore by Whistler himself.[141] This act of seeing anew required a degree of detachment. Indeed, writing from Cornwall in 1884, Whistler proclaimed disdain for nature: "The Country wearies me. . . . Well—for dulness [*sic*], this place is simply amazing!—nothing but Nature about—and Nature is but a poor creature after all—as I have often told you—poor country certainly—and, artistically, often offensive."[142]

In 1885, in his "Ten O'Clock" lecture, Whistler explicitly challenged Ruskin's advice that the artist reject nothing, select nothing, and scorn nothing. "Nature contains the elements of color and form of all pictures," he wrote, "as the keyboard contains the notes of all music—but the artist is born to pick, and choose, and group with science, these elements, that the result may be beautiful."[143] P. G. Hamerton adopted this analogy in a series of essays on imagination in landscape painting published in the months preceding Brett's exhibition.[144] Hamerton distinguished between many forms and uses of imagination, but it is possible to identify a general pattern whereby he acknowledged nature as the source of all ideas for the landscape painter but then emphasized the artist's acts of perception. In his opening essay, "Is the Landscape Painter's Imagination of a Special Kind?," Hamerton described how the artist should search nature for objects and effects that provide analogies to the "moods of man," a strategy highly reminiscent of Constable. But the artist should not merely copy the scene; imaginative power required that the artist selectively choose and combine from nature what was desired for the final composition.

Hamerton acknowledged that artists were not unique in working from imagination; scientists also drew upon it to develop new insights. But what distinguished the artist's use of imagination, according to Hamerton, was how it was coupled with feeling, and how spectators in turn responded most strongly to works that communicated the artist's emotions, offered an "appeal to imaginative sympathy" (as in scenes of rural labor), or conjured reverie. The ultimate merit of a painting, he claimed, lay not in how well it captured nature but in its capacity to trigger a response in the viewer. "Effect is the supreme power in landscape-painting," he maintained, "that it arouses or soothes the feelings like music," adding that the "scientific conception of nature is always without effect."[145] Here we have the associationist theories of Sully and other psychologists, now fully realized in an artistic treatise.

In keeping with his earlier interest in memory, Hamerton argued that using imagination also entitled the artist to recall absent scenes in the mind, and he referred to the research of Francis Galton to verify his claims about the clarity and persistence of memory.[146] Galton, perhaps better known today for his work in eugenics, was also a pioneer in the field of psychology, particularly in developing experiments to assess human abilities. His *Inquiries into Human Faculty and Its Development* was published in 1883, and this text appears to have informed Hamerton's thinking—in particular, the section devoted to "mental imagery," in which Galton described his inquiries into individual perception, that is, the human ability to imagine a scene. Using questionnaires and statistical computation, Galton concluded that the ability to recall or conjure images varied extensively among individuals. While his investigations led him to question whether artists could create mental imagery sufficiently "clear" to draw from, he recognized that the "visualising faculty" could be "developed by education," citing the successes of the French teacher Horace Lecoq de Boisbaudran, director of the École impériale in Paris from 1866 to 1869.[147] This was probably welcome news to Hamerton, who continued to advocate for training memory and devoted a chapter to the subject in *Imagination in Landscape Painting*, in which he also cited the work of Lecoq de Boisbaudran.

The final chapter of Hamerton's series, "Substance and Effect," carried the logic of his arguments to their ultimate conclusion—a form of art without reference to nature. If the artist, Hamerton explained, dispensed with a scientific conception of nature, and if, in addition, the "artist mind loses the sense of substance," a painting would become "eye music," "no longer a study of tangible things at all," and concerned "only with sequences and relations." While this point is reminiscent of Whistler's keyboard analogy, in Whistler's schema the painter still retained elements of nature, whereas Hamerton proposed "colour-music without meaning," that is, severing the relationship between nature and painting, or nearly so: "When we come to this kind of imagination, in which substance is either banished altogether or reduced to a minimum, whilst the delicacies of colour are retained, the only intelligent way of considering it is to think of it as an art existing on its own basis, which is almost, though not quite, independent of Nature."[148] Arguably, Hamerton's vision for art was not realized for another thirty years, with the rise of abstract painting in the early twentieth century, but what is noteworthy here is that Hamerton clearly asserted a new means by which art should be judged—no longer in comparison to nature but instead by how well it evoked the personal feelings of the artist and the viewer. The universalist paradigm of inductive science has been replaced by a highly individualistic approach informed by new research in psychology.

Whistler's and Hamerton's theories of art were an affront to Brett (as well as to Ruskin). Brett could never envision an art independent of nature. He maintained that the value of a painting lay in its optical accuracy. Furthermore, Brett's admiration of finish was diametrically opposed to Whistler's painting practice, with its thin, fluid washes of paint that nearly dissolved forms. In the libel trial of 1877, the labor Whistler invested in his canvases became a subject of inquiry. He was asked whether he charged two hundred guineas for a mere two days of work, to which he famously replied that he charged that amount for the knowledge of a lifetime, which informed his work.[149] Brett, too, wanted each of his canvases to embody a lifetime of knowledge, but knowledge rooted in strenuously observing and recording landscapes. Brett's Fine Art Society exhibition can therefore be understood as his attempt to reclaim the authentic, scientific response to nature from Whistler, who was described in the 1880s as "the most serious scientist in his art."[150]

One might have anticipated that the journal *Nature* would greet Brett's exhibition, with its rhetorical roots in inductive science, enthusiastically. But Brett's charges against scientists put off the journal's critic. Scientists, the journal responded, "are, no doubt, often wrong. But at least they take the trouble to try to be right. Their greatest aim is to get at the truth, and they welcome whatever will lead them nearer to that goal." They might even be willing to learn from Brett, but only "if he has anything to teach them." But this, the journal argued, was not the case. Brett was capable only of "feeble little scientific bantling," and, perhaps worse, his paintings were not consistently accurate.[151] In an age in which scientists increasingly prized professionalism, Brett's claims of scientific expertise had exceeded *Nature*'s boundaries of acceptance, just as they had done for notable members of the Astronomical Society.

As Christiana Payne explains, Brett's exhibition was a failure financially, and it began a period of precipitous decline in popular and critical reception of his work and a corresponding drop in sales. Nonetheless, the artist continued to exhibit and to publish, although he suffered a breakdown in 1894.[152] Close reading of his publications reveals that he sometimes attempted to accommodate new directions in landscape art. Brett's lecture and subsequent article "The Relation of the Pictorial to the Decorative Arts" (1888), for example, conjured up Hamerton and Bain in an account of "the power" of the pictorial arts in "kindling our imagination."[153] Traces of competing philosophies were even more apparent in Brett's lecture, delivered to the Camera Club in 1889, "The Relation of Photography to the Pictorial Art." At pains to distinguish painting from photography, yet cognizant that the concept of truth to nature seemingly bound them together, Brett underscored the artist's ability to idealize—to remove from nature her "flaws and blemishes" in order to realize the highest level of beauty. Moreover, he acknowledged that artists often worked from memory and that the resulting work was "liable to be biased by the personal taste of the performer."[154]

While this rhetoric suggests that Brett had absorbed some of the findings of leading scientists of perception, his painting practice remained largely the same, with perhaps slightly more individuated brushwork. He continued to work from both pencil and oil sketches and to assemble the final work from selected motifs while attempting to stay true to the original scene, as in *Echoes of a Far-Off Storm* (1890). This painting recalls his seascapes of a decade earlier in the broad panoramic sweep of clouds, and continues his attention to geological formations and reflecting surfaces. By this time, Brett's fortunes had turned decisively; this large-scale painting failed to find a buyer, as did many others toward the end of his career.[155]

Brett's writing culminated in a lengthy essay titled "Realism in Painting" that appeared in the *Contemporary Review* in 1899, more than fifty years after the birth of Pre-Raphaelitism. Here, Brett circled around the question of optical accuracy. Like others before him, he explained that the origins of modern realism in art, which he associated with the faithful representation of the facts of nature, lay in modern science and had first been manifested by the Pre-Raphaelites. Brett opened the door to associationist models of perception by acknowledging that facts can result in sentiment: "However prosaic the facts may be in themselves, they may be so disposed or arranged as to develop a fascination over the mind, which is called Poetic."[156] Brett seems to have moved beyond considering the reception of a painting as a purely optical exercise by encouraging the artist to exert influence on the mental interpretation of a scene.

But when Brett turned to actual examples of realist painting, he praised Turner's *Frosty Morning* (exhibited in 1813) as "unmitigated realism," with "no attempt made to excite our imagination," whereas he criticized Turner's *Ulysses Deriding Polyphemus* (1829) as fantastical. He pushed aside associationism by claiming that the true value of realist painting lay in its ability to stand up to a "direct comparison with nature." He returned to his model of optical accuracy in his condemnation of impressionism, which he could only explain as the result of the poor eyesight of both artists and audiences. And he advised young painters, those whom Ruskin had commanded in *Modern Painters*, "to represent the real in its naked reality."[157]

This seemingly uncomplicated assertion, in its resolute belief in an identifiable external reality, swept away all the debates about the role of imagination and cognition in both perception and representation. But Brett's written record, in its anxieties and assertions, leaves a profound trace of those debates. That he chose to wage battle with photography and impressionism is a reflection of the ways in which these fields were critical arenas for debates about the relationship between art and science in the 1880s and 1890s, the subject of the next chapter.

What is particularly intriguing about Brett's career is that when he came of age as a painter, his practice was firmly in step with the advice of the leading art critic, John Ruskin, and closely aligned with scientific naturalism, which was heavily invested with cultural authority. By the close of his career, he was out of step with such critics as Quilter and Hamerton and had antagonized his scientific colleagues. Millais, who shared similar origins, had fared better with his landscapes, redolent of poetic effects, but had not escaped occasional criticism for failing to provide anything more significant than a transcript of nature. But this did not mean that the close relationship between art and science so repeatedly predicted by the speakers at the Royal Academy had been severed. Rather than build on a shared engagement with inductive methodology and scientific naturalism, the relationship was reconfigured around questions of perception and cognition that were of increasing concern to the nascent field of psychology.

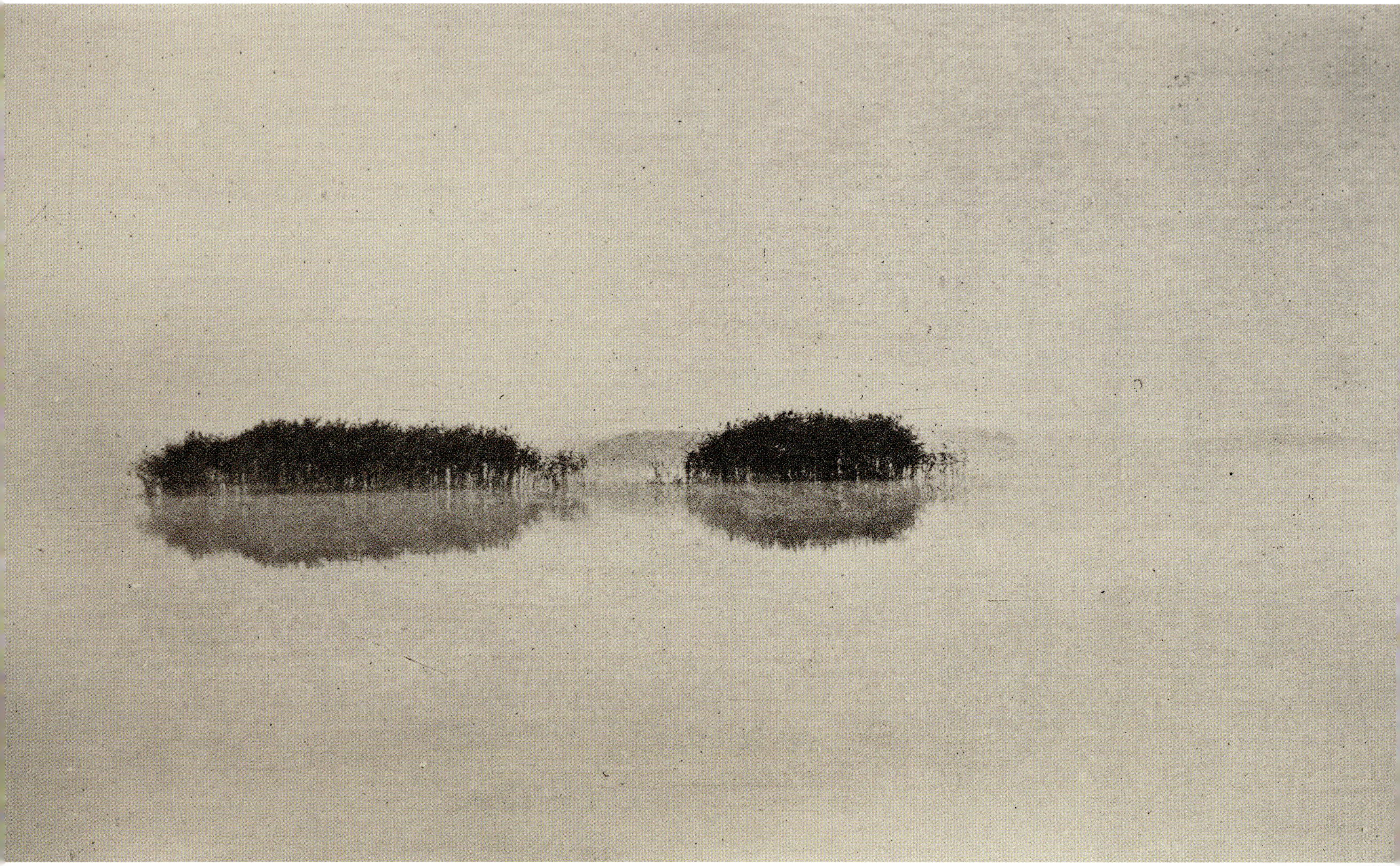

FIG. 35
Peter Henry Emerson (1856–1936), *The Lone Lagoon*, 1895. Photogravure, 7.5 × 12.5 cm. The J. Paul Getty Museum, Los Angeles.

{ CHAPTER THREE }

# P. H. Emerson and George Clausen

## *Renouncing the Quest*

The image, impressed on paper, floats on a field of white (fig. 35). Plate marks help to distinguish the photogravure print from its support. More difficult to distinguish is the horizon line of the image; where does water meet sky? Clusters of trees, suggesting islands floating in the water, provide the only clue; their dark clumps, silhouetted against the sky, are mirrored on the hazy watery surface. Reading left and right from these twinned masses, the rest of the image dissolves into an undistinguished whitish-gray tone. Markers of scale or other clues to the greater environment are absent. Didactic information withheld, the image appears as an aesthetic exercise, distinct in function from Talbot's photographs. The history of the photograph as a model application of inductive methodology or a purveyor of the picturesque is suppressed. Instead, the image engages in dialogue with its painterly counterparts, such as Millais's *Chill October*, with its watery expanse and melancholy mood, or Whistler's nocturnes, with their rendering of the transformative effects of obscuring atmosphere.

How and why did photography, as practiced by Peter Henry Emerson, stray so far from its roots? How does its evolution shed light on the increasingly fraught relationship between art and science and the shared project of seeking truth to nature?

A medical doctor by training, Emerson began taking photographs as a sort of ethnographic pursuit, documenting the peasants of the Norfolk Broads in east England. This background put Emerson in an ideal position to offer a bridge between art and science; yet, although he repeatedly weighed in on the issue, his position was inconsistent. As Ellen Handy has observed, "successive attempts to oppose or to synthesize the two resulted in an unevenly dialectical process which favoured sometimes contradiction and sometimes a transcendental fusion."[1] Mary Marien has described Emerson as "an intellectual *flâneur*, strolling the avenues for what he felt modern science had to say about photography."[2] Early in his practice, Emerson confidently declared photography an unmediated medium that could closely approximate the operations of the human eye; he advocated naturalism as both subject and pictorial mode. Then, seemingly abruptly, he upended his own carefully constructed position, unable to reconcile the truth claims of the increasingly divergent practices of art and science. Rather than fault Emerson, a temptation given his pugnacious rhetoric and tone of moral certitude, we should see these dialectical maneuvers and "bricolage" as indicative of a rapidly changing intellectual landscape. His shift from closely observed representations of natural scenes to evocative studies was not a paradoxical gesture but of a piece with larger scientific and aesthetic discourses.

Emerson often derived lessons from contemporary British art; he clearly favored the New English Art Club, an exhibiting society founded in 1886 by young artists committed to open-air painting and eager to champion techniques, such as a highly visible facture, that many of them had learned through training in France. The increasing critical attention garnered by this loosely affiliated group of painters, as well as by Whistler and his circle, provoked a sense of crisis within British painting in the closing decades of the nineteenth century, as artists who took their cue from progressive French painting attempted to put in place new models of practice under the labels of naturalism, realism, and impressionism. Strong voices within the British art world, such as those of the painters William Powell Frith and Edward Burne-Jones, fought strenuously against what they perceived as a foreign incursion. Moreover, painters who were initially important advocates of naturalism, most particularly George Clausen, eventually renounced it in the late 1880s and early 1890s. Clausen was corresponding with Emerson at the time, and their letters reveal their shared struggles with issues of visual perception and pictorial representation.

Focusing on Emerson and Clausen, this chapter explores how the concept of truth to nature in pictorial representation changed in the last decades of the nineteenth century, as new criteria for artistic quality and changing scientific paradigms, spurred on by psychology, emerged. I have chosen Emerson and Clausen not only for their epistolary exchanges but also because they were highly regarded by their peers as exemplars of a new direction in photography and painting: what was most frequently called naturalism. Their eventual questioning of this artistic practice thus had a substantial impact on their respective communities. Moreover, in addition to significant bodies of work in photography and painting, respectively, Emerson and Clausen left to us important published texts that indicate their desire to exert broader influence and reveal the intellectual problems shared by painting and photography.

## Emerson and Truth to Nature

Emerson's publications, in which he typically combined image and text, reveal his studious attempts to advance his knowledge of art and to turn photography away from its dependence on the tourist picturesque and reinvest it with scientific authority. In so doing, he thrust photography into debates about what constituted truth to nature. Concomitantly, he entered into arguments concerning the relationship between the operations of the eye and mind and the production of art.

Emerson, unlike the gentlemen-scientists of an earlier generation, initially pursued a profession. Born in Cuba, he passed much of his childhood there and then in Wilmington, Delaware. Upon his father's death, when Emerson was in his early teens, he moved to England with his English-born mother. His studies at King's College, London, in preparation for joining the Royal College of Surgeons, never captured his interest fully, and he poured his energies into photography, having learned it from his physics tutor Ernest Griffiths. Wealth inherited from his father enabled him to set aside his professional training to pursue his polymathic interests, which eventually included writing, boating, and billiards in addition to photography.

Yet his medical interests persisted in his concern with vision and perception. Moreover, this early training introduced Emerson to ongoing debates about the value of a scientific practice rooted in observation and dedicated to the pursuit of facts and laws. This theme emerges in his first book, a fictionalized treatment of his early years in medical training, which features Paul Ray, a Protestant and "a staunch believer in everything," who is contrasted with his fellow medical student Reve, a young man who doubts everything, subjecting it to rigorous inquiry.[3] Darwinism is a strong subtext in this narrative, but Emerson suggests that even the clergy are interested in scientific inquiry, as he describes a visiting missionary who responds with delight to Ray's invitation to peer through his microscope.[4]

Here, in anecdotal fashion, Emerson lays out the tensions of the age: Did knowledge derive from visual observation facilitated by instruments and established scientific and philosophical principles, or from some intangible presence requiring belief and faith? Moreover, Emerson models multiple ways of gazing upon and visually comprehending and learning from nature: through the microscope or modes of artistic representation, i.e., the picturesque. While in the Lake District, Reve and Ray admire the view as a "lovely natural picture gallery."[5]

Within a few years of publishing *Paul Ray at the Hospital*, Emerson took up photography wholeheartedly. Yet he did so from a distinctly amateur position, that is, as a sort of gentlemanly pursuit, as opposed to a strictly commercial venture. By the time Emerson entered the field of photography, it had developed a number of important structures by which it could sustain itself and by which knowledge could be developed and shared. Amateur clubs and societies had formed, among them the Exchange Club, the Amateur Photographic Association, and the Photographic Society of London (renamed the Photographic Society of Great Britain in 1874); the latter two groups organized exhibitions that helped to secure recognition of the artistic significance of the medium.[6] Emerson joined the Photographic Society, helped to found the Camera Club in 1885, and submitted works to photographic competitions sponsored by corporations eager to raise the visibility of photography.

Publications as well as exhibitions proved key means by which practitioners could disseminate ideas and advice about photography, ranging from the construction of cameras and lenses and the chemistry of developing to the place of the medium in the panoply of the arts. Photography, in short, was of a piece with the explosion of serial literature that took place over the course of the nineteenth century. The medium possessed specialized journals, such as the *Amateur Photographer* (founded in 1884), *British Journal of Photography* (founded in 1860), *Journal of the Photographic Society* (founded in 1853), and *Photographic Quarterly* (founded in 1889), and was the subject of many articles in general interest periodicals such as the *Spectator* and the *Saturday Review*.

Within a few years of focusing his attention on photography, Emerson upped the ante on his practice by giving lectures, publishing essays and articles, and authoring photographically illustrated books. His early publications served as platforms for his advocacy of the Ruskinian practice "to go direct to Nature, to live with her, and study her," as Emerson phrased it.[7] In these publications, Emerson was also preoccupied with establishing his presence in a field already dominated by well-recognized photographer-writers such as Henry Peach Robinson, who argued strenuously for art photography and adopted combination printing in order to produce his figural narrative subjects.

Emerson's fine art texts indicate that he sought an audience beyond that of the photographic world. Reviews of his books in such periodicals as the *Spectator* and the *Photographic News* reflect the success of this plan. Yet the small editions of his books and their relatively high cost meant that a limited body of readers could patronize his work. Only about 750 copies of *Pictures of East Anglian Life* (1888) were issued, a stark contrast to the vast quantities of images produced by the major commercial firms that served the tourist industry with topographic views, such as those by Francis Frith, James Valentine, and George Washington Wilson—Wilson was producing more than half a million prints a year by the mid-1860s.[8] Nonetheless, Emerson's books gave the photographer control over his program; in the periodical press, his essay might appear cheek by jowl with an opposing opinion. Moreover, Emerson's books doubled as treatises intended to foster a community of practice, as well as providing portfolio collections of his photographs.

By the time Emerson took up the medium, photography had changed considerably from the days of Talbot, facilitating the mass production of prints. In 1854, Talbot lost a decisive lawsuit against the operator of a London photography studio, Martin Laroche, for infringement of his patent, which Talbot argued extended to the new process of wet plate collodion. Frederick Archer, who had published his findings on this process in 1851, recognized that keeping a solution of iodized collodion wet on the glass plate during the exposure process produced a highly detailed negative that required relatively less exposure time than other processes. Paper prints produced from the glass-plate negative (typically made with albumen paper, which has a smooth, glossy surface because the egg white used in the formula masks the texture of the paper) offered sharply defined images, in contrast to the more softly rendered forms associated with the paper negative. Because the collodion process tended to be more reliable and consistent than paper, it quickly took hold in the commercial sphere, particularly once Talbot's patent was lifted. By the early 1880s, technological innovations allowed photographers to use dry plates coated with light-sensitive silver salts suspended in gelatin that permitted even shorter exposure times.

Yet other innovations returned photography to the realm of handcraftmanship, as in the case of the combination print, in which photographers combined multiple negatives, most often one for the sky, which could easily be overexposed, and another for the landscape. The photographer Henry Peach Robinson realized that this technique could be adopted for figural subjects and began to produce complex photos akin to genre paintings, for example, *Fading Away* (1858), which depicts a young girl, presumably on her deathbed, surrounded by members of her family. Robinson promoted his vision for a photographic practice based on the principles of painting in his popular book *Pictorial Effect in Photography* (1869). Active in the Photographic Society and a successful commercial photographer, Robinson was an elder statesman when Emerson entered the field.

The book that earned Emerson his reputation, *Life and Landscape on the Norfolk Broads* (1886), was, in the words of the art historian John Taylor, "the last great book of the 19$^{th}$ century to be illustrated with original photographs."[9] Emerson used the platinum printing process to produce the forty platinotypes contained in the volume. The platinum process had been developed in the 1870s by William Willis and required contact printing, meaning that the print was based directly on the negative and the photographer could not manipulate the print during the developing process. The metals in the final print produced a distinctive aesthetic; these photographs usually have a silvery-gray tonality, with subtle gradations well suited to capturing atmospheric effects. Since albumen is not involved, the texture of the paper is also visible in the print, which can lessen the sharp distinction of forms.

Emerson also made use of the photogravure process, which had been developed in the late 1870s. Like the platinum print, a photogravure is contact printed. But otherwise the processes are significantly different; most important, the photogravure entails transferring the photograph to a copper printing plate onto which the image is etched. The final print is executed from the inked plate, and this method is conducive to producing large editions on a printing press. The paper print of a photogravure, like an etching, is impressed with the plate mark, thus enhancing the impression that the photograph is akin to other reproductive print media.

The portfolios that Emerson produced around this time, which include *Pictures from Life in Field and Fen* (1887) (twenty photogravures), *Idyls of the Norfolk Broads* (1887) (twelve platinotypes), co-authored with Thomas Frederick Goodall, and *Pictures of East Anglian Life* (1888) (thirty-two photogravures), are exceptional not only for the quality of the photographs but also for the textual rhetoric employed. The books derive largely from Emerson's time on the Norfolk Broads in East Anglia—he had first gone yachting there in the summer of 1885.

These books are part of a lineage of natural history and topographical narratives that can be traced back to at least the eighteenth century and the work of the Reverend Gilbert White, whose *Natural History and Antiquities of Selborne* (1789) offered a detailed account of White's local environment based on his close observations and remained popular well into the nineteenth century.[10] Even more than in the eighteenth century, accounts of the natural history of Britain published in the second half of the nineteenth century were attuned to literate urban audiences curious about the work and social life of the rural population and the landscape, which seemed to be disappearing in the face of the rapid growth of cities and industry. Emerson's early portfolios are also important indices of the issues and concerns regarding the nature of truth in photography, which Emerson attempted to resolve later in his *Naturalistic Photography* (1890), a highly significant statement on naturalism.

To produce several of these portfolios from 1886 to 1888, Emerson worked closely with the painter T. F. Goodall, whom he had met during his first trip to the Norfolk Broads in 1885 and with whom he returned shortly thereafter to explore the watery landscape. Goodall was affiliated with the New English Art Club (NEAC), which included George Clausen, Stanhope Forbes, John Lavery, John Singer Sargent, and Philip Wilson Steer. The group eventually developed a formal protocol of membership by election and a committee structure for administering the organization, selecting paintings for the organization's biannual exhibitions, and organizing the displays. The goal was to create a mutually supportive community of painters, but the group split into factions over time with the infusion of new members, such as Walter Sickert, who joined in 1888, and Spencer Gore, Harold Gilman, and Augustus John, who joined in the 1890s.

In the early days, many members of the group were drawn together in admiration of the plein-air painting of Jules Bastien-Lepage.[11] Bastien-Lepage was a French painter of rustic genre scenes whose work was popularly received in France and Great Britain, which he visited frequently, particularly in the 1880s. His work drew considerable attention in 1880, when the Grosvenor Gallery devoted a small section of its summer exhibition to the painter. The display included *Les Foins (Hay Making)* (1878), first shown at the Salon of 1878, which depicts two rural laborers, a male and female, dressed in rough clothes, clearly exhausted after a stint of hard work. The male figure lies in the grass, his hat covering his face as he presumably naps. The female figure, situated in the immediate foreground, caused the most difficulty for critics, who were disturbed by her awkward pose, her legs stretched out before her and her arms hanging slack in her lap, as well as her blank visage, staring gaze, and slightly open mouth. Labor has reduced her to this numbed state, which the critic for the *Magazine of Art* described as the "passive dreaming of an animal."[12] This critical reaction was, in fact, an indication of the growing interest in Bastien-Lepage's rustic naturalism, which found a ready market that encouraged British painters to follow in his path in both peasant subjects and attention to the broader rural landscape and its seasonal changes.

The NEAC members, like the Pre-Raphaelite Brotherhood before them, were perceived as reformers intent on escaping from conventional art making, now, ironically, associated with artists like Brett. In the eyes of critics, the NEAC was strongly linked with representing nature in a fresh manner; the *Spectator* cheered that "the members of the 'New English Art Club' are dealing the deathblow to many antiquated and trivial ways of regarding Nature, and reproducing it in pictures." The painters, by and large, successfully arrived at a means of conveying an artist's personal perspective on nature while still adhering to the highly respected practice of sustained visual observation, a model that several critics had hoped for when reviewing the work of Brett and Millais.[13] Indeed, it is worth remembering that the NEAC was formed the year that Brett exhibited his sketches at the Fine Art Society, which the older painter offered as an authoritative guide to scientifically based painting. The NEAC, according to the *Glasgow Herald*, was also committed to truth to nature, but it eschewed the representation of carefully rendered details in favor of pursuing forces not readily visible to the untrained or insensitive eye and requiring highly individualized and attenuated perception to discern:

> *It differs from every one of its rivals, from the ponderous Royal Academy to the pretentious and futile Nineteenth Century Art Society, in the important respect that the exhibitors have one aim in common—each attempts to express himself absolutely, individually, to record, in his own degree and according to his measure of insight, the secrets of Nature which have become revealed to him, and to do this without heed to convention, indifferent, where needs be, of the most sacred tradition. Each is an experimenter, constantly curious. . . . The artist should have but one idea—truth to nature; but one passion—life. Of course, much clap-trap is talked about truth to nature; for every convention-bound dauber asserts, and perhaps believes, that nature's lessons can be taught professionally, and that mere pictorial reproduction of details is equivalent to obtaining the spirit of one of nature's not numerous secrets. There are many who truly love nature; there are some who perceive*

*her haunting, yet evanescent, features; but there are very few who can perceive beyond her outer aspect, her veil, as it were, and depict what they see there.*[14]

Given this ambitious agenda and emphasis upon individuality, it is not too surprising that it is difficult to locate consensus on the NEAC aesthetic when it was put into practice. It is far easier to identify particular paintings championed by critics in their exhibition reviews, as in the case of the *Bohemian*'s praise for Goodall's painting *The Bow Net* (1886), exhibited in 1887 (fig. 36). The journal referred to the scene of fishing in Norfolk as the "best landscape and figure painting in the exhibition. River scene painted early in the year with a very bright sunny interpretation of nature. No portion of the work is scamped or partakes of the nature of an impression"—a dig at the rising tide of Whistlerian or French impressionist–influenced paintings that would soon overtake the NEAC.[15]

Goodall's canvas suggests that the painter worked it up while observing the scene: the shadow cast over the face of the young woman at the oars registers the passage of light, as does the watery reflection of the two figures seated in the boat. Moreover, the painter's decision to focus on indigenous fishing techniques conveys a sense of authenticity for his urban audience. Unlike the Pre-Raphaelite clarity of form and brilliance of complementary color, the scene is suffused with a slightly hazy atmosphere, caused by the sun's beating on the water, that softens the edges of forms and cools the harmonizing palette. The critical acclaim that greeted the painting, along with Goodall's interest in working out of doors, explains why Goodall's paintings would have appealed to Emerson and his project to document the Norfolk Broads.[16]

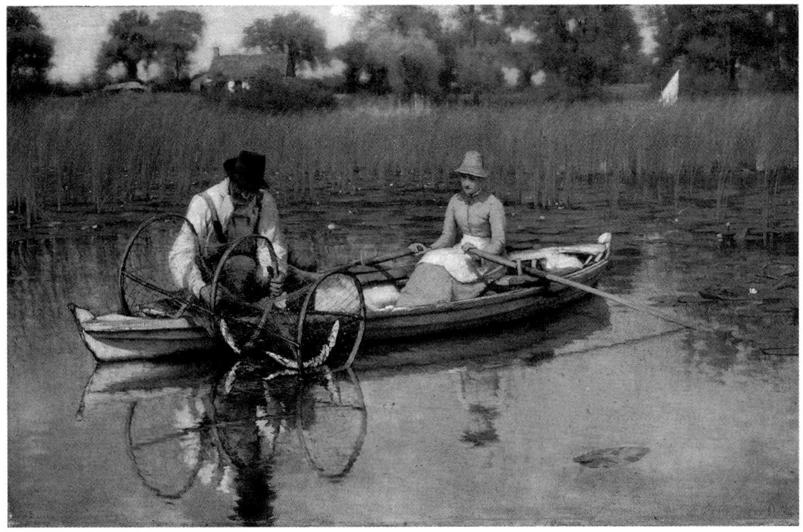

FIG. 36
Thomas Frederick Goodall, *The Bow Net*, 1886. Oil on canvas, 83.8 × 127 cm. Courtesy of National Museums Liverpool (Walker Art Gallery).

In *Life and Landscape on the Norfolk Broads*, Emerson took pains to ensure that the photographs (reproduced as platinotypes) were read as truthful documents, making the case in the text that accompanied the images. The plates *An Autumn Morning* (plate 38) and *The Fringe of the Marsh* (plate 39) (fig. 37), for example, were described by Goodall as "literal transcripts" of scenes inherently artistic or paintable: "designs presented by nature, ready made." Goodall also insisted that their work was stripped of conventionality and forswore both the mistakes of the past and "traditional artificial rules."[17]

But despite these ostensible rhetorical links to Pre-Raphaelitism, rather than establish an affiliation with the older painters, Emerson and Goodall indicted them for their attempts at optical accuracy. By focusing on details, Goodall argued, they had missed "entirely

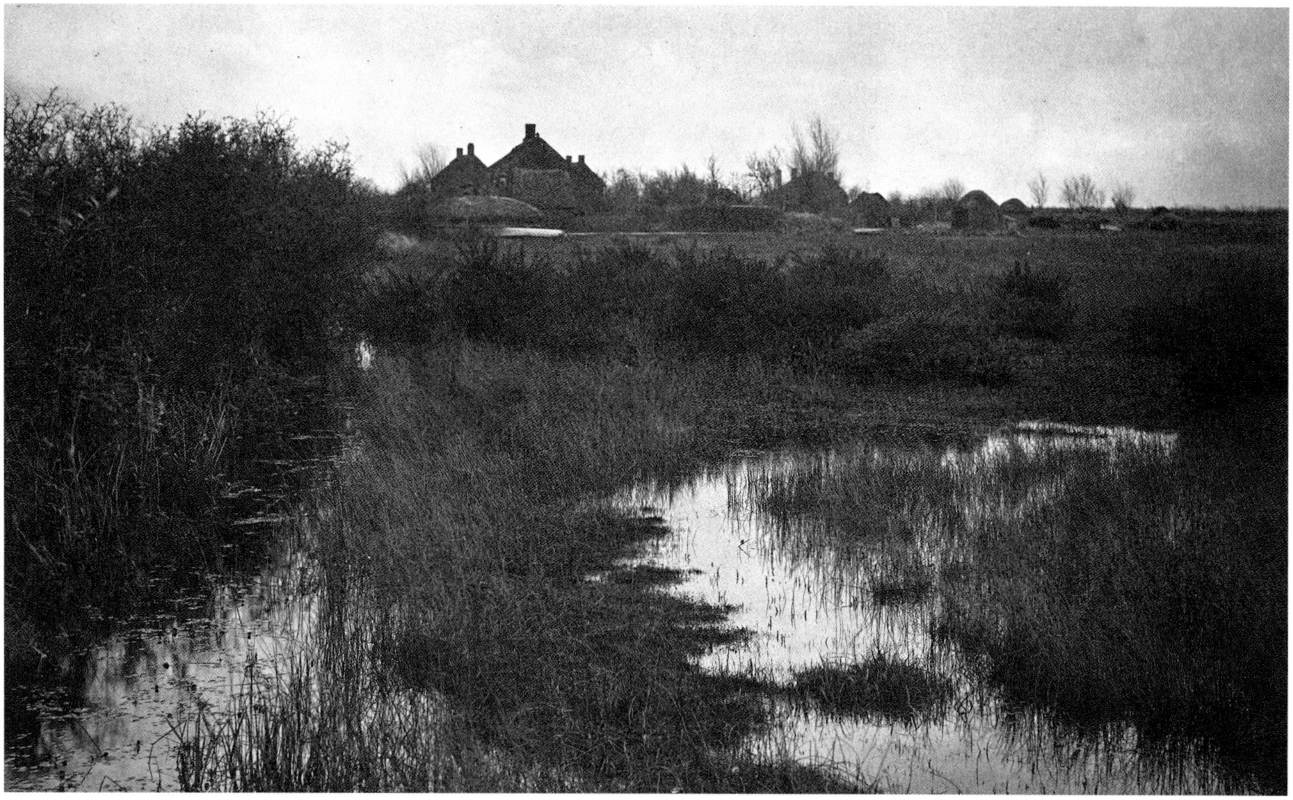

FIG. 37
Peter Henry Emerson, *The Fringe of the Marsh*, 1885–86. Platinum print, 17.6 × 28.7 cm. In P. H. Emerson and T. F. Goodall, *Life and Landscape on the Norfolk Broads* (London: S. Low, Marston, Searle & Rivington, 1886). Gernsheim Collection, Harry Ransom Center, The University of Texas at Austin.

the big truths of nature; the paintings of this school being perfectly flat and unreal in consequence."[18] Clearly, Goodall had in mind a different measure of truth than Ruskin had established in *Modern Painters* and *The Elements of Drawing*.

Emerson expanded on this critique of the Pre-Raphaelites in his essay "Photography: A Pictorial Art" (1886), which also made clear his desire to challenge Henry Peach Robinson's authority in the realm of fine art photography. Emerson charged the Pre-Raphaelites with fallacy and falsehood because they "imitated the pigments, not the light," and failed to register space, atmosphere, and color precisely—the very qualities for which Norman Lockyer had recommended Brett's paintings in 1878.[19]

Emerson's critique was supported by scientific analysis derived from Hermann von Helmholtz, in particular his essay "On the Relation of Optics to Painting"

(1871), which was part of his ongoing investigation of the physiology of perception.[20] Helmholtz was concerned with the question of whether artists could precisely imitate the physical behavior of light with respect to color. What pictorial strategies could artists adopt to translate the laws of physics and physiology into pictorial representation? Emerson quoted Helmholtz directly: "If the artist is to imitate exactly the impression which the object produces on our eye, he ought to be able to dispose of lightness and darkness equal to that which nature offers"; that is, the painting would achieve a one-to-one correspondence with nature. But Helmholtz acknowledged that painters are unable to achieve the same levels of brightness that exist in nature. This led him to inquire, in Emerson's paraphrase, "How, then, is it that there is any similarity between the picture and the reality?" The answer lay in the "physiological process of fatigue; to wit, the effect of loud noises and bright lights on hearing and seeing." That is, the artist does not need to duplicate the scene itself, with its "actual luminous phenomena," but only the effect or impression such phenomena have upon vision. The painter should thus avoid transcribing in favor of translating, rendering "his impression into another scale," as Helmholtz put it.[21] This still qualified as truth because—Emerson again quoted Helmholtz—"perfect artistic painting is only reached when we have succeeded in imitating the action of light upon the eye," an achievement that Emerson claimed had eluded the Pre-Raphaelites.[22]

Emerson recognized that photographers could not enter this debate about color, but he insisted that they were comparable to painters in that both were attempting "to force a scene on our perceptions as a plane surface."[23] Emerson's thinking was in keeping with a developing discourse, promoted heavily by Henry Peach Robinson and also articulated by members of the Photographic Society, which held that "photographers can think with their cameras as well as painters can with their brushes or sculptors with their chisels."[24] Emerson even argued that the camera was more adept than the tools of painting at rendering three dimensions in two. He insisted, therefore, that photographers should draw lessons from Helmholtz and, in the absence of effective means of rendering color, pay particular attention to atmospheric turbidity—what Helmholtz, building on Tyndall's work, described as "the scattering of light by atmospheric turbidity, more generally moisture."[25] These points were significant in an age when photographers sometimes resorted to using multiple negatives to overcome the failure of the plate to register differentiations in the sky.

To this argument about why a truly scientific and truthful image of nature had not yet been achieved in art, Emerson added the crucial caveat that the photographer had "ample room for selection, judgment, and posing." Here, Emerson offered photographers advice similar to what Alexander Bain, James Sully, and Philip Gilbert Hamerton had laid out for fine artists. "Photography: A Pictorial Art" concluded triumphantly that "a finished photograph is a work of Art" and that "Art has at last found a scientific basis."[26]

Emerson's claim is striking, and I want to draw particular attention to his pronouncement that he had finally found a solid basis for art in science. His claim was all the more conspicuous given how often this assertion had been made in the preceding fifty years. The repeated attempt to situate art within a scientific framework reflects the cultural authority that science, particularly scientific naturalism, continued to hold in the last decades of the nineteenth century, despite challenges from such diverse

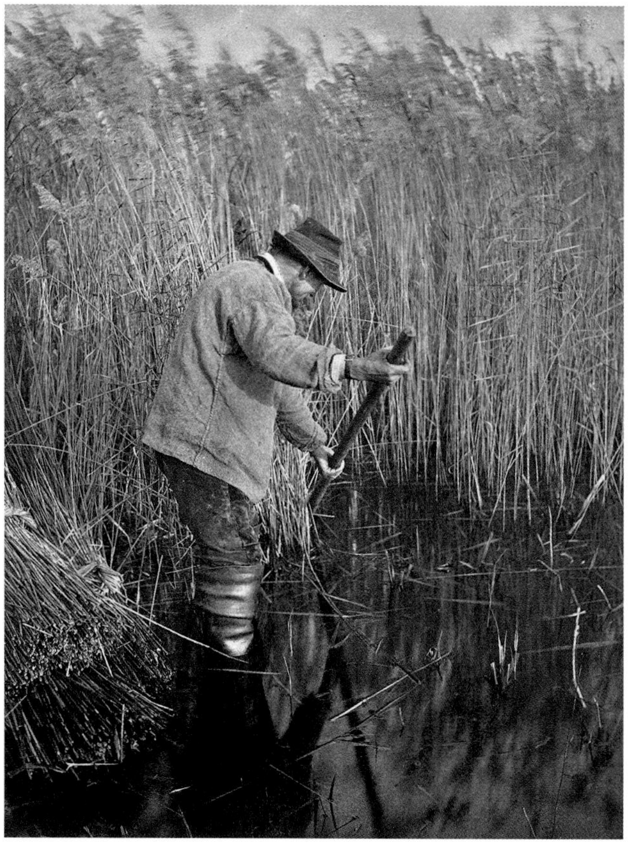

FIG. 38
Peter Henry Emerson, *A Reed-Cutter at Work*, 1886. Platinum print, 27.6 × 20.3 cm. In P. H. Emerson and T. F. Goodall, *Life and Landscape on the Norfolk Broads* (London: S. Low, Marston, Searle & Rivington, 1886). The J. Paul Getty Museum, Los Angeles.

quarters as the church, spiritualists, psychologists, and writers of popular scientific literature. Entering the debate about the relationship between art and science also revealed Emerson's intellectual ambitions. It allowed him, for example, to carry on a conversation in writing with such figures as Helmholtz and Tyndall.

If the Pre-Raphaelites did not constitute the modern school that should function as a guide for photography, then who did? It was British plein-air painters, such as Stanhope Forbes and George Clausen and their genealogical forebears, Théodore Rousseau and Jean-Baptiste-Camille Corot (French painters associated with the Barbizon school and the forest of Fontainebleau), who brought a new level of naturalism to landscape painting—and, most critically, John Constable, to whom Emerson, like many art critics of the day, attributed the innovations of the Barbizon school.[27] In a published lecture of 1886, Emerson reached back further in time and claimed that the notion of going directly to nature had begun with the seventeenth-century Dutch landscape painter Jacob van Ruisdael and his student Meindert Hobbema, even if their work was "trammeled by conventionality."[28]

Despite Emerson and Goodall's critique of Pre-Raphaelitism, they were driven by a Ruskinian devotion to nature, and despite their obvious debt to Helmholtz, they perpetuated an implicit belief in the superiority of bodily or optical vision. Complicating this agenda was the fact that the camera was a mechanical instrument. Rather than rely on this feature to make their arguments regarding truth to nature, they sought to suppress it. They believed that they could transform the camera into the equivalent of the human eye. To prevent the fatal flaw of Pre-Raphaelitism—overly detailed scenes of nature—Goodall instructed the photographer to deliberately "select his lens and adjust his focus so as to get an impression as nearly identical with the visual one as possible; and he must print in such tone as will give the closest approximation to the values in nature."[29] Whereas Talbot delighted in the possibility that the camera could exceed the capacities of human vision, Goodall and Emerson wanted to foreclose upon this possibility. Emerson's photographs from the 1880s demonstrate that he had begun to employ differential focus, by which a small area of the image—typically the chief subject—was sharply focused, while the rest was allowed to fall out of focus.[30]

But, according to Goodall and Emerson, focus and depth of field were only part of the process of making the artistic photograph; the other crucial ingredient was selecting the proper subject. The choice involved shifting attention from the depicted scene to its maker. Goodall asserted that a successful artist should possess an "inner consciousness" attuned to nature. The artist was a highly evolved individual who had passed beyond the "childish" stage of admiring "gaudy and unnatural production" and had developed the "power to observe" in order to recognize the beauty of nature. This act of recognition required imagination, according to Goodall, but should not be tainted by emotion or associationism: "it is much more difficult to realize the beauty of actual life and nature, than to attract by a subject which appeals to the emotions by the associations it suggests."[31] Discipline was required.

Goodall's discussion of artistic practice in *Life and Landscape on the Norfolk Broads* was interspersed with Emerson's photographs *A Rushy Shore*, *'Twixt Land and Water*, *Evening*, *An Autumn Morning*, and *The Fringe of the Marsh*, images that are much more challenging to weave into the rhetoric of truth to nature than, for example, *A Reed-Cutter at Work* (fig. 38) or *Setting up the Bow-Net*. The titles of the last two photographs describe activities that occupy the full scope of the compositions, and the images contain far more descriptive information than is found in the accompanying texts. By contrast, *A Rushy Shore*, *'Twixt Land and Water*, *Evening*, *An Autumn Morning*, and *The Fringe of the Marsh* (fig. 37) do not elucidate any particular aspect of the habitus of those making their living from the land and water of the Norfolk Broads.

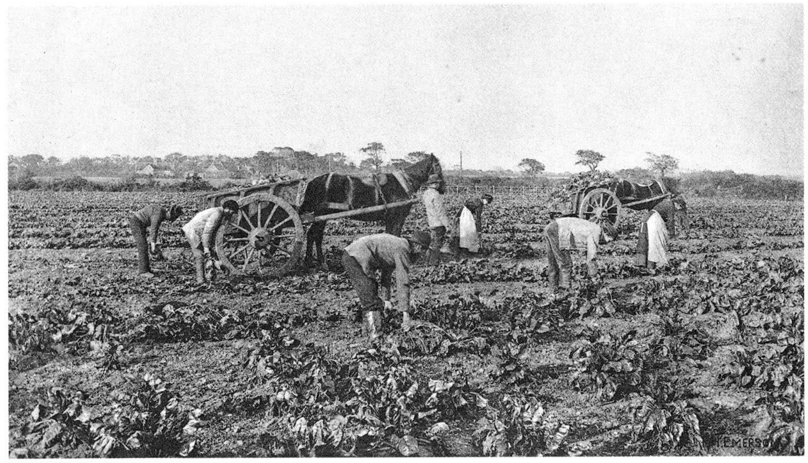

FIG. 39
Peter Henry Emerson, *The Mangold Harvest*, 1887. Photogravure, 16.5 × 29.1 cm. In Emerson, *Pictures from Life in Field and Fen* (London: George Bell and Sons, 1887). The J. Paul Getty Museum, Los Angeles.

Instead, these photographs convey atmospheric scenes evocative of preexisting artistic tropes. *'Twixt Land and Water* and *The Fringe of the Marsh* are reminiscent of B. W. Leader's *February Fill Dyke* (ca. 1880–81), a popular Royal Academy submission that was widely disseminated in print form. Perhaps more striking are analogies with Millais's *Chill October*, in the shared somber mood, heightened horizon line, and composition divided between watery expanse, trees, and marsh. Goodall invited comparison between *Evening* and the work of Corot: "some slender alders spring gracefully from the marshy ground; the mystery of this dreamy corner reminds one of Corot."[32] Emerson's photograph *The Mangold Harvest* (fig. 39) from *Pictures from Life in Field and Fen* (1887) easily connotes the work of Bastien-Lepage, whom Emerson refers to in this text as "the greatest painter that had ever lived, because the greatest naturalist."[33]

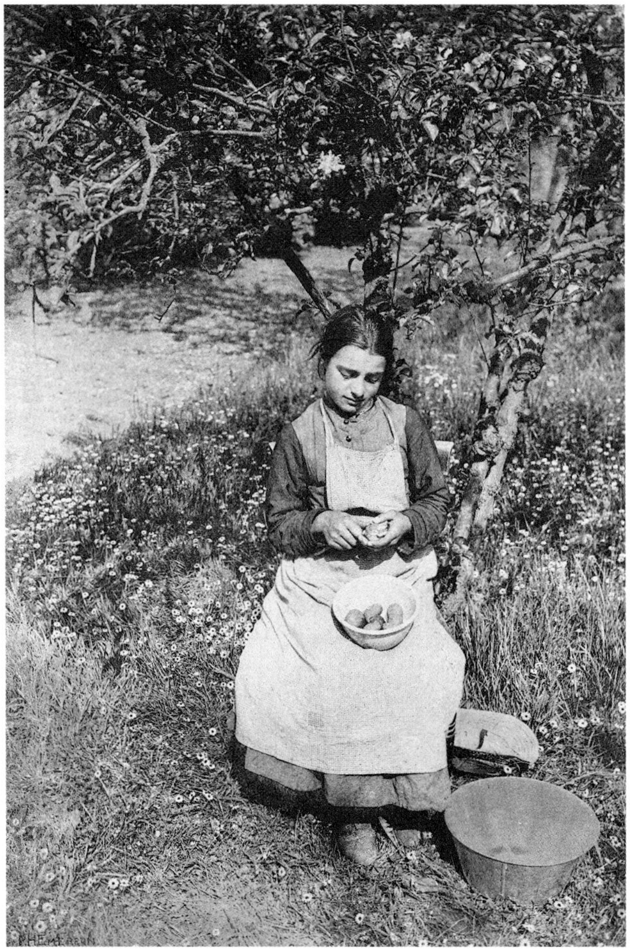

FIG. 40
Peter Henry Emerson, *A Spring Idyl*, 1887. Photogravure, 26.8 × 17.9 cm. In Emerson, *Pictures from Life in Field and Fen* (London: George Bell and Sons, 1887). The J. Paul Getty Museum, Los Angeles.

As the photo-historian Jennifer Green notes, "the countryside to Emerson is already 'landscape,' pictorial, two-dimensional."[34] Robinson had also promoted this way of seeing the external world in *Pictorial Effect in Photography*, and in many ways this mode of viewing the landscape as picture was an extension of the earlier picturesque.[35] As in the picturesque, Emerson and Goodall aimed to naturalize their processes. They argued that their work expressed the "natural language" of art.[36]

Yet this artistic borrowing produces a tension in Emerson's body of Norfolk images. While Emerson claimed that his intent was to provide his readers with an insider's insights about life on the Norfolk Broads, the images often pull the viewer away from contemplating the visual information conveyed and toward the artistic precedents being quoted. Moreover, as Jennifer Green has noted with respect to *The Mangold Harvest* and other images in *Pictures from Life in Field and Fen*, they "aestheticize labor to such a degree that not only the identity of the workers is obscured but also the precise nature of their work."[37] In addition, in many cases, the carefully arranged poses of the rural laborers, as in *A Spring Idyl* (fig. 40), register the presence of the photographer and undermine the author's claims of naturalism.[38]

Yet Emerson's failure, initially, to perceive how his photographic representations contradicted his own truth-to-nature rhetoric reveals the power of Goodall's logic that art was a natural language. This argument, as already suggested, required that Goodall and Emerson construct a particular genealogy of "true" art—eschewing the Pre-Raphaelites in favor of seventeenth-century Dutch painters, Constable, and the Barbizon school. This artistic lineage provided Emerson and Goodall with both legitimacy and aesthetic knowledge. If art was a "natural language," as Goodall claimed, then the contradiction of basing supposedly unmediated images on artistic precedents disappeared, because true art was, and always had been, synonymous with nature.

Borrowing a page from Millais, Emerson further enhanced the artistic legitimacy of his photographs by

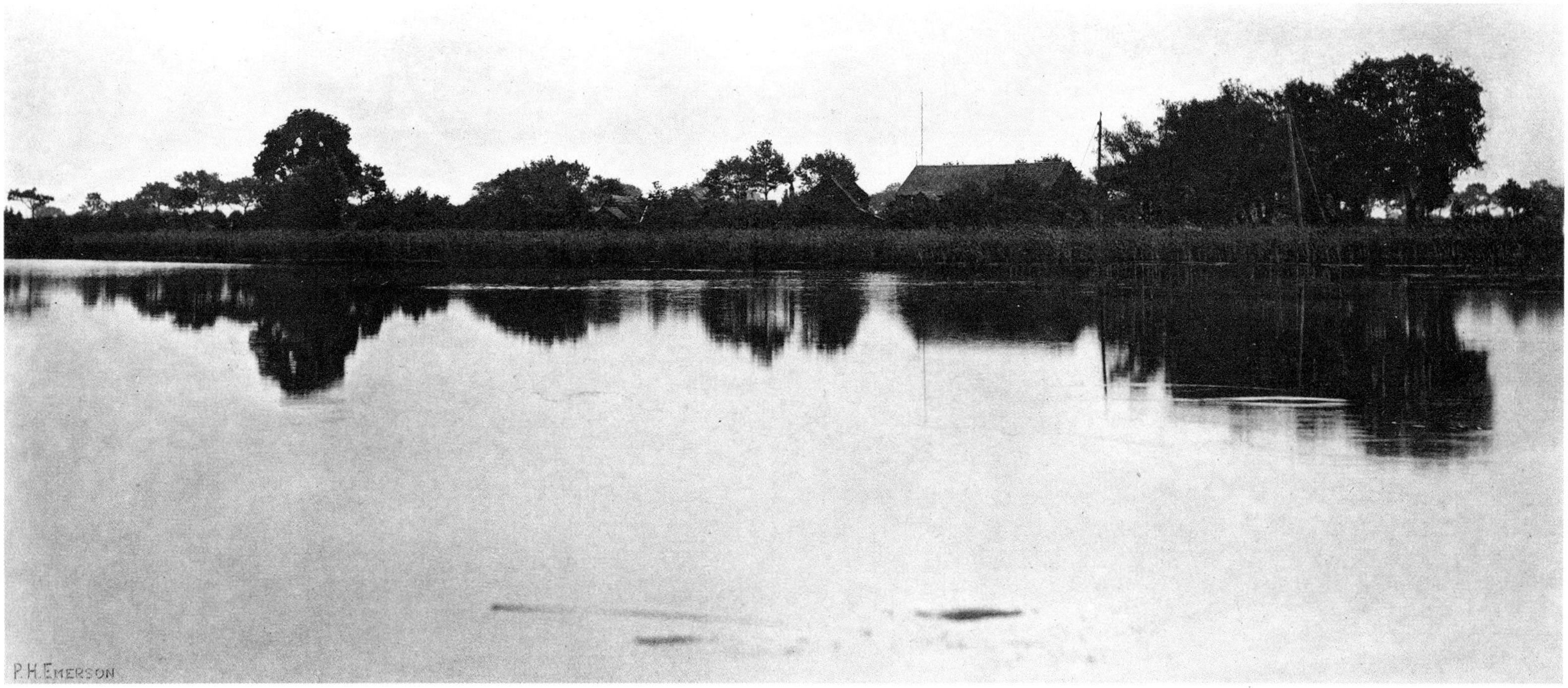


FIG. 41
Peter Henry Emerson, *The Reedy Marge*, 1887. Photogravure, 11.9 × 27.5 cm. The J. Paul Getty Museum, Los Angeles.

linking them with poetry, which created yet another layer of connotation and mediation despite Goodall's disdain for associationism. Emerson prefaced *Pictures from Life in Field and Fen* with a Robert Browning poem, the closing lines of which reinforce his precept that art was the source of truth: "That Art remains the one way possible / Of speaking truth, to mouths like mine, at least." In *Idyls of the Norfolk Broads* (1887), Emerson evoked the pastoral—"High up in the velvety brown branches of the poplars and elms the gentle zephyr soughs and sighs its Aeolian cadence, and passing on ripples the stilly surface of the mere with silvery streaks." Poetic references continue; with respect to the photograph *The Reedy Marge* (fig. 41), Emerson states, "We name this plate as nature-loving Wordsworth named the border of one of his beloved lakes."[39] Emerson also included a lengthy excerpt from aesthete poet Algernon Swinburne's "By the North Sea" (1880); here, the poem functions more explicitly as a framing element, so that the reception of Emerson's photographs is doubly shaped by the photographer's text and Swinburne's poem.

The passage that Emerson selected from "By the North Sea" was based on the landscape around Dunwich, Suffolk. Swinburne describes the marshy scenery through a series of sensuous phrases: "Tall the plumage of the rush-flower tosses, / Sharp and soft in many a

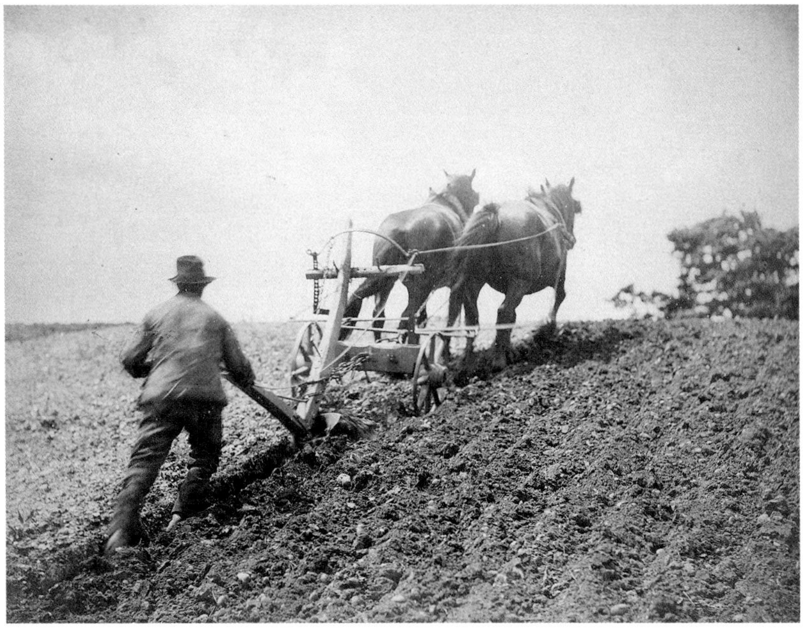

FIG. 42
Peter Henry Emerson, *A Stiff Pull*, 1885. Albumen silver print, 22.9 × 29.1 cm. The J. Paul Getty Museum, Los Angeles.

curve and line / . . . Streak on streak of glimmering sea-shine crosses / All the land sea-saturate as with wine."[40] The potential appeal of this poem for Emerson is readily apparent; the watery location is akin to what he experienced in Norfolk, and, perhaps more significantly, the poem describes a journey of personal transformation triggered by the speaker's changing perception of sea and coastline. The opening stanza describes a gloomy locale, and references to death abound: "A land that is lonelier than ruin; / A sea that is stranger than death." But gradually the narrator begins to envision the landscape differently—assigning new associations and values—and the first hint of this alteration occurs just before the passage that Emerson excerpted: "Slowly, gladly, full of peace and wonder / Grows his heart who journeys here alone." This metamorphosis requires that the speaker rethink the savagery of the sea; in the words of scholar Kerry McSweeney, "he has simply come to accept the scene he confronts—the way things are for him—as the place where he has to be, and as the place which can satisfy and solace him." Thus the poem engages major themes in Swinburne's corpus that could also describe Emerson's project: "the necessity of rejecting all supernatural beliefs and hopes in favour of a purely naturalistic vision of life."[41]

Emerson's textual referents in *Idyls of the Norfolk Broads* reveal another side of the photographer's corpus, distinguishable from his dispassionate ethnography—an engagement with beauty, sentiment, and sensuality that infuses a number of images. *Flowers of the Mere*, for example, depicts a young girl examining the petals of a water lily; she extracts pleasure, rather than moral or fiscal benefit, from nature. *Water Babies* shows young nude boys walking out of the water toward the marshy bank. It echoes Goodall's frontispiece as well as the work of Henry Scott Tuke, whose sea-bathing boys delight in the warmth of the sun and conjure an adolescent existence spent close to nature and separate from the adult world of commerce and materialism.

When *Idyls on the Norfolk Broads* is compared with Emerson's next published collection, *Pictures of East Anglian Life* (1888), it becomes clear that Emerson was struggling with his aim to achieve truth to nature as expressed in a "natural language." In this portfolio, even more than in his earlier collections, different modes of representing nature can be distinguished. Overall, *Pictures of East Anglian Life* is strongly embedded in the discourses of natural history. Emerson wanted to immerse his readers in a sense of place—the ways of life and habits of those who live there—and to teach them to see the potential beauty in the quotidian, such as a horse-drawn ferry. Indeed, Emerson declared in the preface, "I have endeavoured in the plates to express sympathetically various phases of peasant and fisherfolk life and landscape which have appealed to me in Nature by their sentiment or poetry."[42] Here, Emerson pointed to the decisive act of selection that he and Goodall described earlier.

At the same time, Emerson's text participated in the emergent discourses of ethnography and anthropology by which the ordering gaze was turned increasingly to the human inhabitants of the landscape and the organization of society. Emerson set himself the task of categorizing peasants and fisherfolk around such topics as character and intelligence, religion, superstitions, witchcraft, ghost stories, medicine, politics, habits and customs, sports and pastimes, and the relations between farmer and landlord. Evolutionary science had fueled interest in studies of peoples and races; in 1844, the year that Chambers published *Vestiges of the Natural History of Creation*, Dr. Thomas Hodgkin and others founded the Ethnological Society, and the group eventually supported Darwin's theory of evolution by natural selection. In 1863, an anti-Darwin faction of the group set up the rival Anthropological Society. In 1871, largely through Thomas Henry Huxley's efforts, the groups were reunited as the Anthropological Institute of Great Britain and Ireland. Popular writers also took up the topic of anthropology, as in John G. Wood's *Natural History of Man*, first published between 1868 and 1870.

Emerson's aim in *Pictures of East Anglian Life* was "to produce truthful pictures of East Anglian Peasant and Fisherfolk Life, and of the landscape in which such life is lived."[43] The art historian Charles Palermo argues that Emerson "first believed that correspondence with the

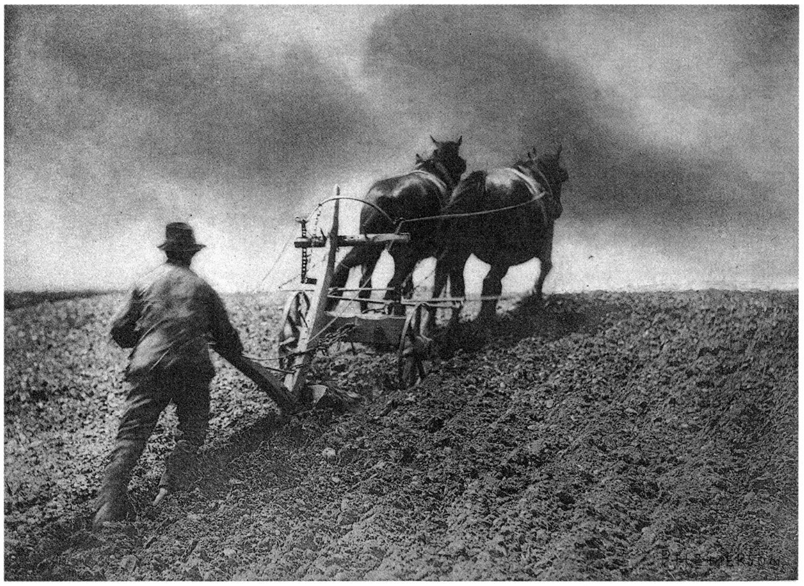

FIG. 43
Peter Henry Emerson, *A Stiff Pull [Suffolk]*, 1888. Photogravure, 20.8 × 28.6 cm. The J. Paul Getty Museum, Los Angeles.

workmanlike ways of his subjects was the key condition of truth in art."[44] But Emerson's truth claims were undermined by his admission that he had manipulated several of his plates. He clearly altered *A Stiff Pull [Suffolk]*, for example, as can be seen by comparing the photogravure that appeared in *Pictures of East Anglian Life* (plate IV) (fig. 43) with an earlier, albumen version published in *A Selection of Prize Pictures from the London Stereoscopic Photographic Company's Amateur Photographic Exhibition* (1886) (fig. 42): the later print has been trimmed; the trees along the horizon line at right have been eradicated; and two negatives were used—one for the toned clouds and another for the horses, plow, and plowman—registered by the slight halation along the horizon line and around the figures set against the sky. These changes isolate the human figure and horses; they no longer compete with the trees for visual attention and are highlighted against the sky. The open-ended horizon generates the sense that the furrows and, by implication, the attendant labor, are unending. The clouds, borrowing a concept from Constable, add an emotional valence. In sum, Emerson's changes enhance the narrative capacity of the image.

Further contradictions can be found in the collection. For example, the frontispiece, *In the Haysel*, is firmly in the georgic tradition. A single man leads a horse-drawn cart along the edge of a narrow canal, the orthogonals of which direct attention from the open foreground to the far distance, suggesting that the fecundity of the scene extends beyond the horizon. Yet plate II, *The Poacher*, reverses this trajectory. Poaching is counter to the georgic as it is immoral rather than moral labor. Furthermore, the landscape is emptied out, and it is not clear to what the man and dog in the foreground direct their attention. Plate III, *Leafless March*, shifts again, no longer focusing on the productivity of the land but instead conveying a dreary mood.

The text likewise embodies different voices, from the detached observer who catalogues the landscape and its society to the sensuous interpreter of the scene who evokes the wonders of the changing seasons. At certain junctures, Emerson is aware that he is employing distinct, sometimes divergent, discourses. After describing how in deepest winter "man alone is full of hope" and can envision "life-giving spring . . . with the warm sunshine, balmy showers, and gentle zephyrs," he commands the reader, "now let us look at things from a utilitarian point of view," and he quantifies the assets of the countryside as both grazing and hunting territory. In other passages, he appears unaware of the disjuncture he produces, as when he first describes the painter's exultation at the "exquisite scenes" produced by mist hanging over the marsh and then quickly shifts gears, criticizing the decoration of an inn he visited—a critique redolent of class antagonism.[45] Emerson's language cultivates intimacy and claims sympathy, but it also produces detachment and even antagonism. This seemingly unresolved dialectic provokes larger questions concerning the project of truth to nature.

## George Clausen, Naturalism, Realism, and Impressionism: Forming a New Language of Truth to Nature

Questions of how to "produce truthful pictures" also concerned George Clausen. As a young boy, Clausen had been apprenticed to an interior-decorating firm where his father worked. In the evenings, he attended classes at the Government Art Training School in South Kensington. On the advice of the painter Edwin Long, for whom he had undertaken decorative work, Clausen traveled to the Low Countries in the mid-1870s and began to paint genre scenes set in recognizably Dutch landscapes and populated by figures in local costume. These paintings were generally well received, but around 1880 Clausen began to favor studies of fashionably dressed women in interiors, paintings that demonstrate his knowledge of the artistic vocabulary of aestheticism, and modern urban subjects, as in his paintings *Schoolgirls* (1880) and *Spring Morning: Haverstock Hill* (1881).[46]

But Clausen's decision to move to Childwick Green, Hertfordshire, in 1881 stimulated another dramatic shift in his work. The rural landscape—the tenant farmers and rural laborers around the estate of Childwickbury, where Clausen and his family took up residence—became the focus of his work, a maneuver akin to Emerson's time on the Norfolk Broads and in keeping with the move away from urban subjects associated with the rise of artists' colonies, such as Pont-Aven in France and Newlyn in Britain, in the late nineteenth century.[47] Clausen's new rural work fell largely into two categories, as Anna Gruetzner Robins explains. One consisted of oil paintings depicting pretty young rustic women interrupting their work for moments of reverie, as in *Haying* (1882). These subjects helped to invent the myth of the rural idyll at a time of momentous social and political change in the rural English countryside, the result of shifting landholding patterns that undercut tenant farmers and threatened to displace rural laborers. The second featured scenes of rural labor that depicted the harshness of working the land, as in *Winter Work* (1883–84); these were ambitious canvases intended to compete with the major works of French naturalist painters like Bastien-Lepage. Indeed, to produce these paintings, Clausen looked to Bastien-Lepage and Jean-François Millet, whose works could be seen in London at this time and were clearly an inspiration for a number of Clausen's paintings.[48] Clausen sketched from life, using laborers from the local estate, and also used photographs that he had taken using a Royal Academy camera, designed to allow relatively short exposure times and therefore well suited to working quickly. He occasionally worked up a painting largely out of doors (as opposed to working from sketches and photographs in the studio), as in the case of *Day Dreams* (1883), which took a year to paint and was profoundly difficult because of the changing patterns of sunlight and shade.[49] In the fall of 1883, Clausen spent several months at the Académie Julian, a private art school founded by Rodolphe Julian in Paris in 1868, an experience that confirmed his new direction and strengthened his commitment to rustic naturalism.

Clausen's mode of painting was by now markedly distinct from that of the Pre-Raphaelites. Rather than apply glazes, Clausen used thick, opaque paints that he sometimes applied with a brush, leaving distinct passages of dragged paint, and sometimes with a palette knife, registered by passages of smoother but dense paint. Clausen's choice of relatively viscous paint meant that the edges of forms could not be crisply delineated but were registered by touches of built-up paint that often helped to fuse the figure with the ground and, to use the vocabulary of Tyndall and Emerson, suggested the turbidity of the atmosphere.

In 1883, Clausen left Childwick Green for St. Albans, Hertfordshire; two years later, he and his family moved to Cookham Dean on the Thames. Around this time he joined the NEAC, becoming one of its early members. For the second NEAC exhibition, in 1887, he submitted what the art historian Kenneth McConkey describes as his "most significant new work," *The Stone Pickers* (1887) (fig. 44), a representation of two women removing stones from a field in preparation for plowing.[50] The *Pall Mall Gazette* described it as a "study of peasantry as seen through . . . Anglo-French spectacles."[51] The painting conflates the two categories of Clausen's paintings identified by Robins, as the young woman standing in the foreground gazes down and to her right, her head slightly downturned, with a pensive or dreamy visage as she empties the stones from her burlap apron to the ground. Wildflowers are scattered in the field, and a food basket lies on the ground near the young woman's feet, but neither flowers nor basket seem to attract the woman's gaze. Behind her stoops another woman, her head bowed nearly to the ground,

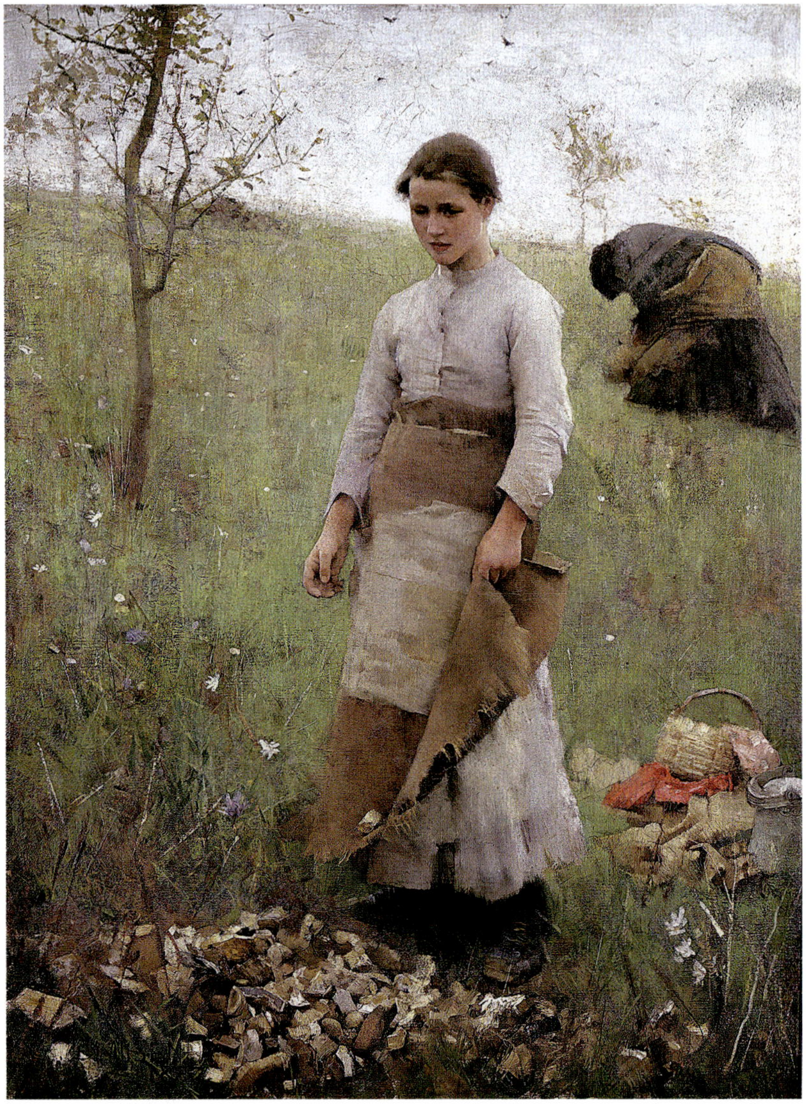

FIG. 44
George Clausen (1852–1944), *The Stone Pickers*, 1887. Oil on canvas, 107.6 × 79.2 cm. Laing Art Gallery, Newcastle upon Tyne, U.K.

and we assume that the young woman will shortly turn to join her. The gray sky against which the young trees stand silhouetted casts a bleak mood over the scene.

The year after this painting was exhibited at the NEAC, Clausen published a tribute to Bastien-Lepage in the *Scottish Art Review*, which had recently been founded by the critic Robert Alan Mowbray Stevenson, one of a new school of art critics in Britain that included George Moore, Elizabeth Pennell, Walter Sickert, and Frederick Wedmore.[52] These critics looked favorably upon the NEAC and those painters affiliated with French impressionism, and often criticized the Royal Academy for failing to keep pace with changes in the art world.

In his essay, Clausen sought to distinguish Bastien-Lepage's contribution to the concept of realism. He contrasted Bastien-Lepage's work with that of Millet, who, he explained, subordinated "the facts of nature to the expression of sentiment," whereas Bastien-Lepage emphasized facts over sentiment (fig. 45). Bastien-Lepage's achievement was to create "a new and living interest" in his canvases by insisting "on the claims of smaller things—commonly slurred over and suppressed—to a full and complete realisation." This produced an "exquisite and subtle truth" that also reflected the painter's dedication to nature, direct study from the motif, and sympathy for rural subjects.[53] Clausen hit upon an absolutely crucial point here; at the time, the value of the fact or detail was very much up for debate in British artistic circles. Emerson, for example, questioned it, and Whistler clearly rejected it.

Yet, within a year, Clausen qualified his praise of his mentor in a letter to his dealer, David Croal Thomson of the Goupil Gallery. Clausen had recently visited the Exposition Universelle in Paris and had been stunned by

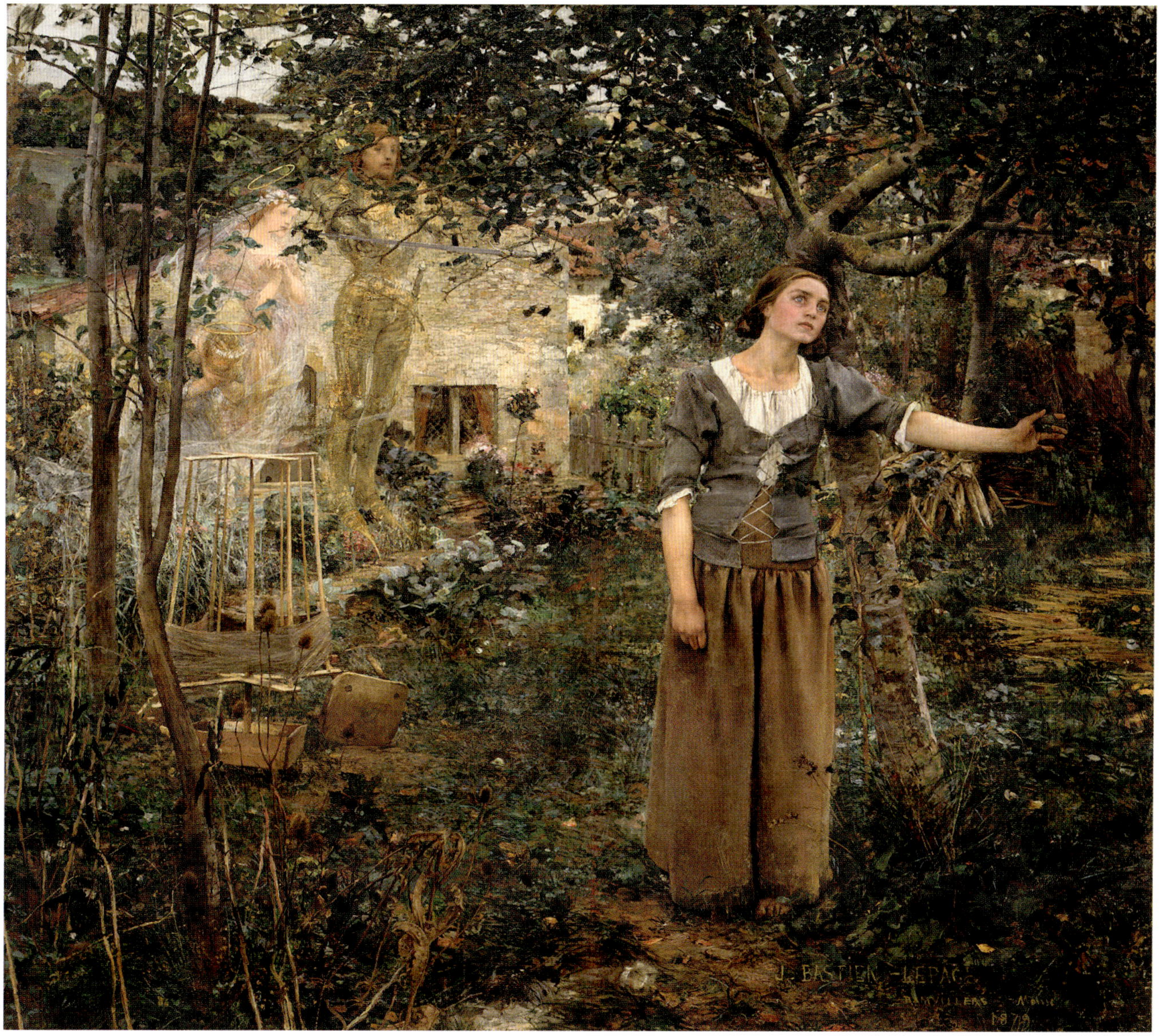

FIG. 45
Jules Bastien-Lepage (1848–1884), *Joan of Arc*, 1879. Oil on canvas, 254 × 279.4 cm. The Metropolitan Museum of Art. Gift of Erwin Davis, 1889 (89.21.1).

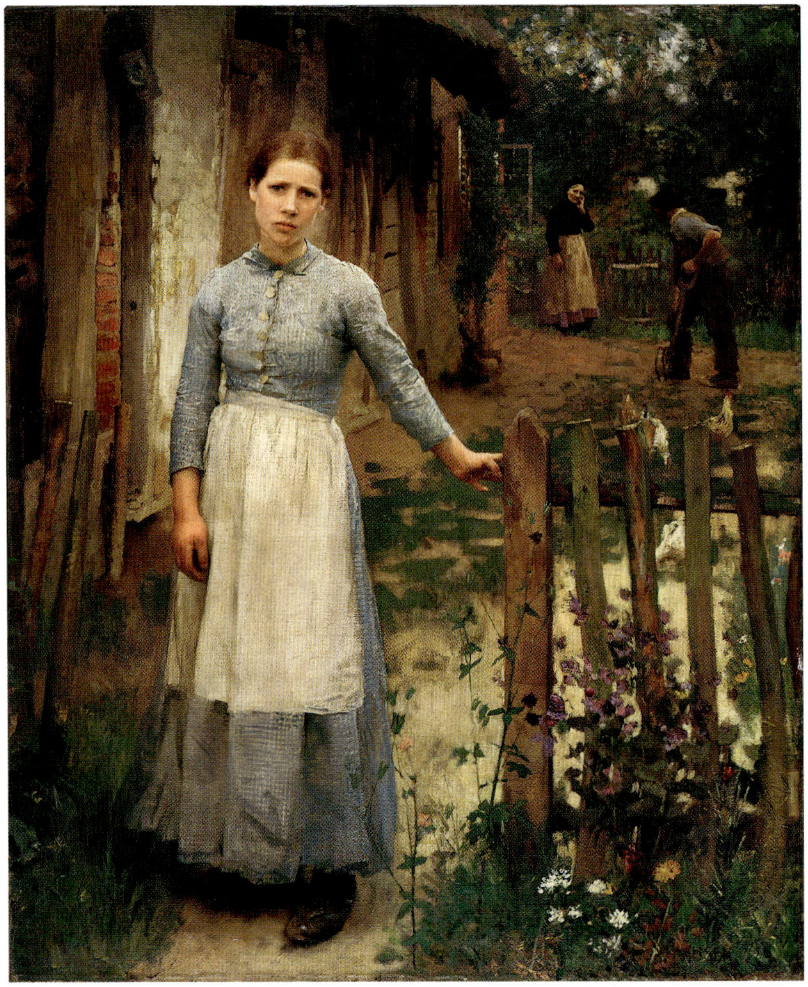

FIG. 46
George Clausen, *The Girl at the Gate*, 1889. Oil on canvas, support 1714 × 1384 mm. Presented by the Trustees of the Chantrey Bequest, 1890. © Tate, London 2015.

the work of Gustave Courbet. He described Courbet's painting as "to my mind perhaps the greatest thing in the show." He continued, detecting the source of Courbet's success, "Lepage who puts in everything . . . as much as he can & Courbet who leaves out as much as he can—thereby gaining, as only the essentials remain."[54] Suddenly, well-rendered details had lost their appeal. So what, for Clausen, was realism to be?

The difficulty of this question was exacerbated by the antagonism of several British artists and critics toward progressive French painting. In 1887, for example, Harry Quilter wrote a long essay called "The Tendencies of French Art" for the *Contemporary Review* in which he drew attention to aspects of French painting that he believed British artists should not emulate. He claimed that the Barbizon school of painters demanded that the audience "disregard all the ideas of a subject, all desire for beautiful arrangement, all notions of composition, and simply accept as the one sufficient subject of a picture, a piece gouged out of Nature, as it were with a cheese-scoop." The impressionists, and the "English modification of [their work, of] which Mr. Whistler's painting is the most notable example," believed solely in "truthfulness to the impression of any given scene. . . . To paint that which is impressed upon the retina within the space during which an eye can be rapidly opened and shut—this is the end to which the artist's effort should be directed." Quilter despaired because this formula did not allow the "mind . . . to exert its influence" and appeared to dispense with "beautiful form . . . glow and depth of colour, . . . dignity of aim . . . tenderness and poetry of meaning . . . patient, industrious endeavour to depict every portion of its subject with completion and loveliness."[55]

The genre and history painter William Powell Frith echoed this view of impressionism, which he dismissed as a craze in an article published in the *Magazine of Art* in April 1888. Whereas Brett had accused the impressionists of faulty eyesight, Frith claimed that they must have malformed minds if, indeed, their paintings reproduced the impressions that nature had made upon them. In a subsequent article, Frith warned of the dangers of realism, of imitating details beyond an acceptable level and detracting from the narrative at hand, as the Pre-Raphaelites did. But even more pernicious, according to Frith, were certain kinds of foreign realist painting "where the accidents of nature—such as dirt, distortion, and exceptionable ugliness of type—are insisted upon." This was a "revolting practice" that Frith cautioned young English art students to avoid.[56]

Yet the reception of Clausen's next major artistic statement, the painting *The Girl at the Gate* (1889), exhibited at the Grosvenor Gallery in 1890, reveals that not all critics concurred with Quilter and Frith (fig. 46). The painting depicts a girl standing at a wooden gate adjacent to a ramshackle cottage cobbled together from brick, wood planks, timbers, and stucco. Behind her, at the opposite end of the cottage, stands an older woman, her hand at her cheek, presumably in conversation with a man, his back to us, who leans on a pitchfork. Chickens scurry in the yard. The young woman, wearing a blue-checked dress and a white apron, meets our gaze, but the critic Claude Phillips, writing for the *Art Journal*, described her expression as "an abstracted and somewhat joyless look into vacancy." While Phillips noted the ways in which Clausen had achieved decorative harmony through the balance of the blue dress against the girl's pale skin and light brown hair, he described the canvas overall as "a pathetic and unexaggerated study of every-day country life." This reading of realism was complemented by links the critic established between Clausen and his mentors Bastien-Lepage and Millet.[57]

Clausen's painting received further critical recognition when it was purchased for the nation through the auspices of the Chantrey Bequest, then under the management of the Royal Academy. John Singer Sargent's painting *Carnation, Lily, Lily, Rose* (1885–86), a painting strongly associated with French impressionism by British critics, had similarly been purchased through the Chantrey Bequest in 1887, and these purchases were probably the result of an effort by the Royal Academy's president, Frederic Leighton, to demonstrate openness and amenability to new directions in art.[58] But despite Clausen's major achievement, *The Girl at the Gate* marked another turning point in the painter's career, what McConkey describes as "a moment when consciousness of the need for change rose up like a tidal wave."[59] That need for change, I will argue, was one that he shared with P. H. Emerson and was arguably triggered, in part, by the publication of Emerson's *Naturalistic Photography*.

## Emerson: Naturalistic Photography

In 1889, Emerson revamped and consolidated his agenda in his treatise *Naturalistic Photography*. Ellen Handy describes the text as a combination of "a rambling history of art from a 'naturalistic point of view,' minutely detailed technical instruction in photographic practice, and assorted harangues on nature, and on art and its scientific basis."[60] The last point, most significantly, forced Emerson to reconsider the question of truth to nature. Signs of his changing attitude can be found in the essay "Science and Art," which he published in the *Photographic News* around the time that *Naturalistic Photography* appeared. Like many of the scientists at the Royal Academy banquets, Emerson sharply distinguished between art and science in this essay, associating art with "the emotional side of man's nature" and science with "his intellectual side." While he perceived a connection between the two fields—they both deployed facts—they did so in such different ways that he described them as "poles asunder." Therefore, once an art student had acquired sufficient scientific knowledge to "express what he wishes," Emerson advised the student to keep science at bay. In fact, he argued strenuously that science could have ill effects on art students, leading them, in the pursuit of truth, to be overly attuned to the details of nature to the exclusion of all else and making them indifferent to the means of expression. Achieving aesthetic pleasure, the true aim of art, would be denied them.[61] This essay indicates clearly that Emerson was no longer interested in placing art on a scientific foundation, as he had done in his earlier essay "Photography: A Pictorial Art."

In *Naturalistic Photography*, Emerson again demonstrated an interest in patrolling boundaries by dividing photography into three categories: art, science, and industrial.[62] His text focused largely on art photography, in which "the aim of the work is to give aesthetic pleasure *alone*," in contrast to science photography, which aimed "to investigate the phenomena of nature," and utilitarian industrial photography. To shore up his category of art photography, Emerson once again constructed a genealogy, offering the lengthiest treatment in his corpus thus far. He drew extensively from a history of art by the German art historians Alfred Woltmann, professor at the Imperial University of Strasbourg, and Karl Woermann, director of the Old Masters Picture Gallery in Dresden. Their history of painting had been translated into English by Sidney Colvin (vol. 1) and Clara Bell (vol. 2), sister of the painter Edward Poynter. Woltmann's history of art was informed by a cyclical or evolutionary model of progress in art, with naturalism at the apex, which Emerson likewise adopted for his text. From this history Emerson derived his conclusion that the best art expressed "an impression of nature as nearly as possible similar to that made on the retina of the human eye." In other words, "a picture should be a translation of a scene as seen by the normal human eye."[63]

Elsewhere in his treatise, Emerson explained that the artist's task was to offer "not a mere transcript of the object, but a translation of his impression into another scale of sensitiveness." Again, Emerson was rephrasing Helmholtz, who stated, "The artist cannot transcribe Nature; he must translate her."[64] The choice of the word *translation* is important here, for it distinguishes Emerson's project from Talbot's. In his *Pencil of Nature*, Talbot had described the camera and the photograph in this way: "The object glass is the *eye* of the instrument—the sensitive paper may be compared to the *retina*"—a view in keeping with scientific theories of contemporary scholars like David Brewster, who argued that images were "painted on the retina," suggesting a direct correspondence between external nature and human perception. But Helmholtz had revised this model of vision by arguing that "visual perception is not produced directly in each retina, but only in the brain itself by the means of impressions transmitted to it from both eyes."

Emerson was clearly aware of Helmholtz's research into perception; his thinking was also informed by recent investigations concerning the behavior of the human eye by Helmholtz and Tyndall. In *Naturalistic Photography* he explained that the eye was essentially a flawed instrument because of such phenomena as dispersion, spherical aberration, astigmatism, and blind spots.[65] To this list of phenomena that produced imperfect vision, Emerson added his previous analysis, drawn from Helmholtz, of the fatiguing effect of light on the eye. He concluded, in a section titled "Art Principles Deducted from the Data Cited," that "the human eye does not see nature exactly as she is, but sees instead a number of signs which represent nature, signs which the eye grows accustomed to, and which from habit we call nature herself."[66] This statement echoes Whewell's observation regarding the role of mental interpretation in visual observation—"most men are unconscious of this perpetual habit of reading the language of the external world, and translating as they read."[67] Helmholtz, through his physiological studies, had updated Whewell's cautionary tale; as he explained in his essay on sight, "none of our sensations give us anything more than 'signs' for external objects and movements, and . . . we can only learn how to interpret these signs by means of experience and practice."[68]

Emerson marshaled this evidence, drawn from art history and science, to argue that an artistic photograph should not strive for optical accuracy or evenly distributed sharpness and attention to detail. He derived a new set of visual markers to delineate truth to nature. Photographers, he advised, should pitch their dark tones to approximate what they saw in nature but then calibrate their bright tones in relation to the dark, rather than judge them against nature. Avoiding strong contrast was also an excellent means by which to suggest the atmosphere (or turbidity) that modulated the perception of objects. Atmosphere, like aerial distance, could also be indicated by softening the outlines of images, suppressing details, and deploying differential focus.[69]

Emerson dubbed his project "naturalism," which he described as "the true and natural expression of an impression of nature by an art."[70] Naturalism, he insisted, should not be confused with mimesis, in that the artist should "interpret," not copy, what he saw in nature. To illustrate his point, Emerson provided an anecdote of a realist and a naturalist rendering a tree. "The realist makes no analysis. . . . He will, in so far as he is able, paint all the veins of the leaves as they really are, not as *they look* as a whole." The naturalist, by contrast, "would endeavour to express the impression of the tree as it *appeared* to him when standing a hundred yards off, the tree taken as a whole, and as it looked, modified as it would be by various phenomena and accidental circumstances."[71]

Emerson thus developed a dramatically new way to articulate truth to nature that was nearly diametrically opposed to that offered by Talbot, Ruskin, and the early Pre-Raphaelites. The journey from the origins of the truth-to-nature project in inductive methodology to Emerson's naturalism was charted by changes in both science and aesthetics that contributed to the ways in which Brett and Millais rethought their art production and that informed the mixed reception of their work, and that spurred Emerson to consolidate his theories of photography.

It should be emphasized, however, that none of the artists discussed thus far followed the full logic of Hamerton's vision of an art independent of nature. Although Emerson acknowledged the role of individual subjectivity in perception, he still believed that nature offered an ultimate standard of truth. Indeed, he criticized impressionism for abandoning such a standard and allowing each artist to become "a standard for himself." "The subject in nature," Emerson explained, "must be the thing which strikes the man [the photographer] and moves him to render it, not the plate he has to fill."[72] Furthermore, Emerson clearly believed that once photographers possessed the requisite scientific knowledge, no further danger would be posed by individual subjectivity, for they would know how to adjust their processes accordingly. Like an inductive scientist explaining how to conduct an experiment in order to arrive at verifiable and repeatable experiments, Emerson sought to train photographers in how to render their impressions of nature in ways that would be recognized as truthful according to new understandings of how light, air, and even the human eye operated. But the very sciences that he relied on to arrive at these instructions also told him that perception was contingent. Therefore, we can understand Emerson's lengthy, and perhaps anxious, treatise as an attempt to get around that contingency.

Emerson's text sparked a wave of critical reviews in both leading photography journals and general interest periodicals. Perhaps the most significant and perceptive was by Henry Peach Robinson, published anonymously in the *Amateur Photographer* shortly after the book's release. Robinson attended to the fine points of argument perhaps even more closely than Emerson did—for example, in discussing Emerson's definition of naturalism and the ways in which Emerson sought to bolster his case by citing painters and paintings. "There is the fallacy here," Robinson observed, "of supposing that nature is the standard to the naturalistic school [of painting]; it is their method of *seeing* nature they always refer to." With respect to the technique of differential focus, Robinson argued persuasively that Emerson, like naturalistic painters, ignored the fact that the human eye shifted focus "so automatically and instantaneously . . . that we are not conscious of it." The review's great significance, however, lay in its resounding conclusion. Robinson referred to Emerson's system as a "pernicious" "false doctrine" and a "disease" requiring a "disinfectant" to prevent its spread.[73] The threat posed by Emerson's text was such that Robinson felt compelled to ensure its dismissal by whatever rhetorical means possible.

Emerson pugnaciously defended the book, largely by attacking colleagues and editors in the photographic press.[74] His other major tactic was to cultivate artists he admired by sending them his book, a practice he followed throughout his career. While the artists' written responses might never appear in print, they did sometimes shore up Emerson's confidence, as in the case of the artist James Charles. Charles had trained at the Royal Academy, studying with the realist painter Luke Fildes, and then at the Académie Julian in Paris, where he studied plein-air painting. Landscape subjects emerged as the focus of his corpus. Emerson must have been gratified that Charles replied to the gift of *Naturalistic Photography*, "your views on artistic matters I thoroughly agree with." More significantly, Charles affirmed the terms Emerson used to describe his project (a point of dispute in the photographic press) as well as his new methodology: "I am convinced we are on the right road, our aim is not as you say realism, that is detestable, but is an attempt to convey to other minds the impression and delight we receive on contemplating nature. There is no such thing as realism if we were to study a thousand years it would be impossible to put down everything there is in one scene in nature, but is possible to give the sense of it."[75] Charles recognized the same struggle that A. W. Hunt faced when attempting to convey the complex fullness of nature, and he was convinced by Emerson's solution of representing the sensations or impressions of nature received by the artist's mind rather than scrupulously mimicking the numerous details of the scene. (This solution, as we have seen, was endorsed by psychology.)

Perhaps it was with this confirmation of his beliefs still fresh in his mind that Emerson agreed to participate in an April 1889 public debate with the painter

John Brett at the Camera Club. In the end, Emerson did not appear, but his paper was read following that of Brett, who had some knowledge of photography as an amateur photographer himself. In his paper, Brett took up the larger question that had preoccupied Emerson—was photography a fine art equivalent to painting?—and answered decisively in the negative. Photography, he insisted, was best suited as an artist's tool, for it could not idealize or improve upon nature.[76] Emerson's paper sidestepped Brett's argument about photography's failure to achieve the status of fine art and instead took up the question of the relationship between art and science. Quoting Huxley, Emerson defined science as "all accurate knowledge" derived from "*observation* and *experiment*." Art, by contrast, required the judicious selection of facts in order to achieve the desired expression and aesthetic pleasure—points that echoed Bain and Sully.[77]

Thus, surprisingly, Brett and Emerson reached similar conclusions regarding truth to nature in art, agreeing that the characteristic or the ideal should be pursued, not the individual, in all its detail and accidental variety. The discussion of the two papers focused largely on Brett's paper, and the objections that listeners raised make clear that exactitude, fidelity, and detail still defined the truth-to-nature project for this group of photographers, despite Emerson's attempts to reconceptualize the field.[78]

It was perhaps with relief, then, that Emerson greeted Clausen's engaged response to his work; Clausen wrote lengthy epistles to the artist for several years. Clausen responded very positively to *Naturalistic Photography* overall, although, like Brett, he was reluctant to recognize photography as a fine art because of its inherently mechanical nature, which he believed interrupted the circuit between the artist's mind and the resulting artwork, whereas painters and sculptors, through the physicality of their craft, forged a direct correspondence between eye, mind, hand, and object. Yet Clausen recognized the value in many of Emerson's points. He too condemned excessive detail in photography and argued that "the province of art was not the representation of facts, but the appearance of things," a distinction that resonates with James Charles's logic.[79] For Clausen and Charles, truth was measured not by adherence to minute details but by effectiveness in capturing a general impression. The whole outweighed the part; the overall effect was more critical than the detail.

The epistolary exchange between Emerson and Clausen mirrored the debates in contemporary photography journals. To demonstrate this point before turning more fully to Clausen's own reconceptualization of realism, I want to examine briefly a selection of articles published in the *Photographic Quarterly* between 1889 and 1892, the high point of Emerson's impact on the field. The educator Graham Balfour, for example, cautioned against excessive detail in photography and the mechanical transcription of nature and argued that the artist's task was to evoke "his own personality." By contrast, the art critic James Stanley Little claimed that photography could never be an art form because it was the camera, not the artist's soul or ego, that perceived the scene. Moreover, he concluded, "art is born of the marriage of fact with idealism, and idealism owes as much to the vagaries of memory as it does to the conceits of anticipation."[80] That is, art was dependent upon human emotion and mental processes, which could not be rendered by the camera.

Henry Peach Robinson seized upon this definition of art and reclaimed it for his brand of photography. The photographer, he advised, should be well steeped in the traditions and principles of art making so that he could select from nature appropriately artistic scenes, an argument that echoes Gilpin's view on the picturesque.[81] The optician and instrument designer H. Dennis Taylor ingeniously sidestepped the whole debate about what the photograph should include or exclude, pointing out that if human vision was flawed, it would be so when looking at either nature or the photograph of it—so no need to be concerned about detail or focus, since the "eye . . . interpreter to the mind . . . reports to the brain in its own *fragmentary fashion*." H. Ernest Murchison pointed out that the photographer could manipulate the plate or print after taking the image in order to align it with his remembered or imagined experience.[82] In short, despite Emerson's efforts to establish a definitive platform for the field, the goals and methods of photography were still very much up for debate.

Emerson's treatise, then, failed to build consensus in the field. Indeed, by pointing out the impossibility of seeing and thus representing "nature exactly as she is," Emerson complicated the questions not only of how one should photograph nature and what constituted truth to nature, but also whether photography was a fine art.

## Emerson's Crisis

Emerson attempted to put an end to these debates in 1891 with his infamous black-bordered pamphlet *The Death of Naturalistic Photography*, a refutation of naturalism. Before issuing his pamphlet, in January 1891, he circulated his essay "A Renunciation" to editors of photographic journals. Emerson gave little premonition of this about-face in his book of the year before, *Wild Life on a Tidal Water*, which chronicles several months spent with Goodall on a houseboat on Breydon Water, Norfolk, in 1887.

In justifying his new renunciation of naturalism, Emerson explained that he had recently gained knowledge that forced him to rethink the question: "Misgivings seized me after conversations with a great artist after the Paris Exhibition; these were strengthened by the appearance of certain researches in psychology, and Hurter and Driffield's papers; and finally the exhibition of Katsushika Hokusai's work and a study of the National Gallery Pictures."[83]

Taking the points in the order given, it is quite likely that the exhibition to which Emerson refers is the Exposition Universelle, and the great artist, George Clausen.[84] This was the moment at which Clausen had declared his admiration for Courbet and, in particular, his ability to represent only what was essential for a scene.

Given the plethora of new publications in psychology at this time, it is difficult to pinpoint with certainty which texts Emerson might have read. In his lecture to the Camera Club, Emerson referred to Herbert Spencer, an important contributor to psychology, but Spencer's *Principles of Psychology* had first been published in 1855. Emerson's correspondence with the sculptor J. H. Thomas in 1890 refers to French experiments in hypnotism, perhaps those of Jean-Martin Charcot, which Emerson speculated would open "up a vast field to the psychologist."[85] Charcot was investigating hysteria at the time, using hypnosis as a tool to demonstrate the links between the psychological and the physiological, which perhaps suggested to Emerson the potential depths of unconscious mental processing.

In 1890, Ferdinand Hurter and Vero C. Driffield, a chemist and an engineer, respectively, at Gaskell Deacon

& Company in England, published "Photochemical Investigations and a New Method of Determination at the Sensitiveness of Photographic Plates." Emerson subsequently corresponded with the pair on their experiments and conclusions. He summarized the impact of these discussions on his thinking in a letter of 13 January 1891, in which he explained to Driffield, "your investigations shew, I think—how little control the photographer really has—they of course shew 'how much' that control is but that 'how much' seems to me to be a very little much. For example, I cannot modify, emphasise, or suppress as I wish, a fatal limitation."[86]

The Japanese Edo period printmaker Hokusai, known for his ukiyo-e subjects, presumably revealed to Emerson new ways of representing nature pictorially. Hokusai had made a decisive impression on Whistler, who was attracted to the Japanese artist's stylized representation of landscape: simple bold forms reinforced by broad planes of bright colors and relatively shallow pictorial space. Indeed, elsewhere in his essay "A Renunciation," Emerson cited Whistler as another influence on his rethinking: "Art—as Whistler said—*is not* nature, is not necessarily the reproduction or translation; much, so very much, that is good art, some of the very best, is not nature at all, not even based upon it."[87]

With respect to the National Gallery of Art in London, no evidence has yet surfaced that allows us to identify which works contributed to Emerson's thinking at this time, but his earlier reference to seventeenth-century Dutch landscape painting and nineteenth-century French Barbizon art is suggestive of his interests.

Via this litany of diverse influences, Emerson found his former certainties upended. He came to recognize that his recommendation that photographers select from nature would encourage them to manipulate either the negative or the print, as Murchison had suggested, which he rejected as inauthentic. Photography, Emerson concluded, was only capable of "a true and literal scientific transcript of nature." He threw his previous theories "upon the dust heap" and disclaimed any label—rejecting "idealist, realist, naturalist" and "*photographic impressionist.*"[88] He had reached the end of the road.

Emerson had been trained to recognize truth and science as leitmotifs of the modern age. For art to be modern, he believed that it had to be based upon truth and science. He attempted to achieve this for photography but discovered that photography then no longer remained art—he could not reconcile truth to nature, science, or art.[89] He wrote to the sculptor J. H. Thomas, "I have withdrawn Nat. Photog.—I am no longer a follower of any school. . . . I don't even think I must be perfectly true to Nature."[90] This was a radical about-face from Emerson's beginnings as a photographer.

Despite Emerson's claim that he would withdraw from the field after publishing his renunciation, he entered the fray again in June, this time with a treatise, co-written with Goodall, that focused on perspective drawing and vision and aimed at psychologists and physiologists as well as artists. Emerson and Goodall set out to demonstrate that "a mathematical drawing of various subjects in the field of vision (*i.e.* angular measurements), as received on a glass plate (ordinary perspective drawing) or upon the screen of a camera," was not truthful via "a series of provisional propositions, experiments, proofs and deductions."[91] Like Ruskin before them, they proposed a series of exercises by which the observer could see nature freshly, untainted by conventional rules of art making.

Influenced by Darwinism, Emerson and Goodall argued that human vision was not physically accurate.

For example, to gain an advantage in the struggle for survival, human sight would "try and draw distant objects nearer and to enlarge them." What we actually "see," they explained, is an optical illusion.[92] Therefore, perspective drawings, to be "true," should be based on human vision rather than on mathematical principles. Truth should be measured against perception, not physical nature. Here, finally, Emerson had crystalized his argument, at just the moment in which he renounced photography.

In the relatively short period of six years, Emerson had gone from asserting that photography was a fine art (in his 1886 essay "Photography: A Pictorial Art"), to denying that photography was art (in "Photography Not Art," which appeared in January 1892). The latter essay echoes Clausen's chief complaint—that photography was the product of a machine that intervened between the artist's mind and the resulting image.[93] Emerson modified this extreme denial of his former position somewhat in an essay of 1893, "Naturalistic Photography."[94]

Crucially, this 1893 essay offered Emerson an opportunity to clarify his evolving notion of truth to nature. It was originally delivered as a paper, read by T. R. Dallmeyer, to the Photographic Society of Great Britain. It also functioned as Emerson's explanation for withdrawing his book *Naturalistic Photography*; he told the gathering that "my researches into the subject of vision and perspective drawing were mainly responsible for that act." He then took his audience through his findings. He began by explaining that rather than a singular truth to nature, as he had always implied before, "there were two truths to nature—the perspective or mathematical truth and the psychological or visual truth."[95] The latter was Emerson's main concern, and he believed he had developed the best technical means to achieve it in photography through differential focusing, which he distinguished from Julie Margaret Cameron's method of deliberately throwing the photograph out of focus.

He also made clear to his listeners that he had relinquished his earlier position that nature was the ultimate standard or authority against which art should be judged. Instead, he explained, "there is no *absolute* truth to Nature from the visual standpoint, for as each man's sight is different, the only absolute truth to Nature for each man is his own view of her." In recognition of this fact, which Emerson claimed he had acquired from reading a recent textbook on psychology, he asserted that what photographers should strive for is the "*illusion*" of naturalism.[96] Emerson was now committing to print ideas he shared with Charles and Clausen.

Emerson had evidently been influenced by William James's *Psychology* (1892), an abridged version of his *Principles of Psychology*; he quoted a passage (without attribution) from James's discussion of the perception of space to shore up his point regarding the lack of absolute truth to nature.[97] James, after considering a career as a painter, studied medicine at Harvard. While he never practiced as a medical doctor, his interest in physiology eventually led him to psychology, and he was deeply familiar with British scholars on the subject, such as Herbert Spencer, Alexander Bain, and George Henry Lewes. James, in his *Principles of Psychology,* established his interest in probing the materialist nature of consciousness and perception (he described psychology as a "natural science," as opposed to a philosophy). He rooted his analysis in the study of the nervous system: "1) The fibres which carry currents in; 2) The organs of central redirection of them; and 3) The fibres which carry them out." Sensation was correlated with incoming currents, reflection with redirections, and motion with the outgoing message. Sensation, combined with "*reproductive brain-processes*," produced the content of perception (what formerly was understood as "association radiating from the sense-impression").[101] Like many of his predecessors, James was also engaged by questions of automatic or unconscious mental processing and the autonomy of the human will. He argued that the human mind can acquire habits of perception that explain why the viewer can perceive the two flat pictures shown in the stereoscope as presenting a solid ("there is a whole batch of illusions which come from optical sensations interpreted by us in accordance with our usual rule, although they are now produced by an unusual object"). "*Whilst part of what we perceive comes through our senses from the object before us,*" James concluded, "*another part* (and it may be the larger part) *always comes out of our own mind.*"[98]

James's text had a profound effect on Emerson, and in his paper before the Photographic Society he offered a series of seventeen propositions, derived largely from psychology (namely, from James), and from Charles Darwin's theory that perception evolved in animal species over time. Emerson's logic is not entirely clear, but he slowly built a case for regarding art as highly personal, an index to the creator's mind rather than external nature. The photographer, according to Emerson, had only a limited repertoire of means to exert individual expression; for example, a photographer could take fifty views of the same subject and produce the same results, whereas the painter, according to Emerson, would always produce something different, "for the *mind* would work differently on each so-called 'replica' of the original." Therefore, Emerson recommended that photographers aim to produce either decorative works, in which the chief aim was beauty, or scientifically accurate images. But neither would be true art.[99] So, while denying that art should be influenced by science, Emerson used science, i.e., psychology, to negate photography as art. He refuted the truth claims of the medium and instead opened a new path for photography as a decorative assemblage that implied a highly personal perception of and response to the natural scene. It was a path that the next generation of photographers associated with pictorialism would follow, despite Emerson's dismissal of the artistic possibilities of the medium.

Emerson's rhetoric was preceded by his practice, a body of work published in 1893 under the title *English Lagoons* but begun in September 1890, when he took up residence on his houseboat, *Maid of the Mist*. The accompanying images take elements of the quotidian landscape—houses, cows, fields, inns, and the like—as subjects but render them in transformative conditions, such as veils of mist, transitory times of day, or changeful weather phenomena. *Marsh Leaves* (1895), illustrated with photo etchings, confirms the trajectory of *English Lagoons*. The cover, with its embossed stylized rendering of marsh plants, sets the tone for the images that follow. The scenes, often photographed at ephemeral moments, are imbued with a hazy atmosphere that obscures the figural elements, underscoring both Emerson's differential focus and his deference to turbidity. In the first plate, *A Winter's Sunrise* (fig. 47), the camera points directly into the haloed sun, partially obscured by rising mist and narrow bands of clouds. The middle ground of the image, dominated by the wooden stile, dark against the frost-laden grasses and misty horizon, dominates the image so that the foreground balances harmoniously with the whitish-gray distance beyond. This grounding of the image in the middle, with expanses of grayish-white

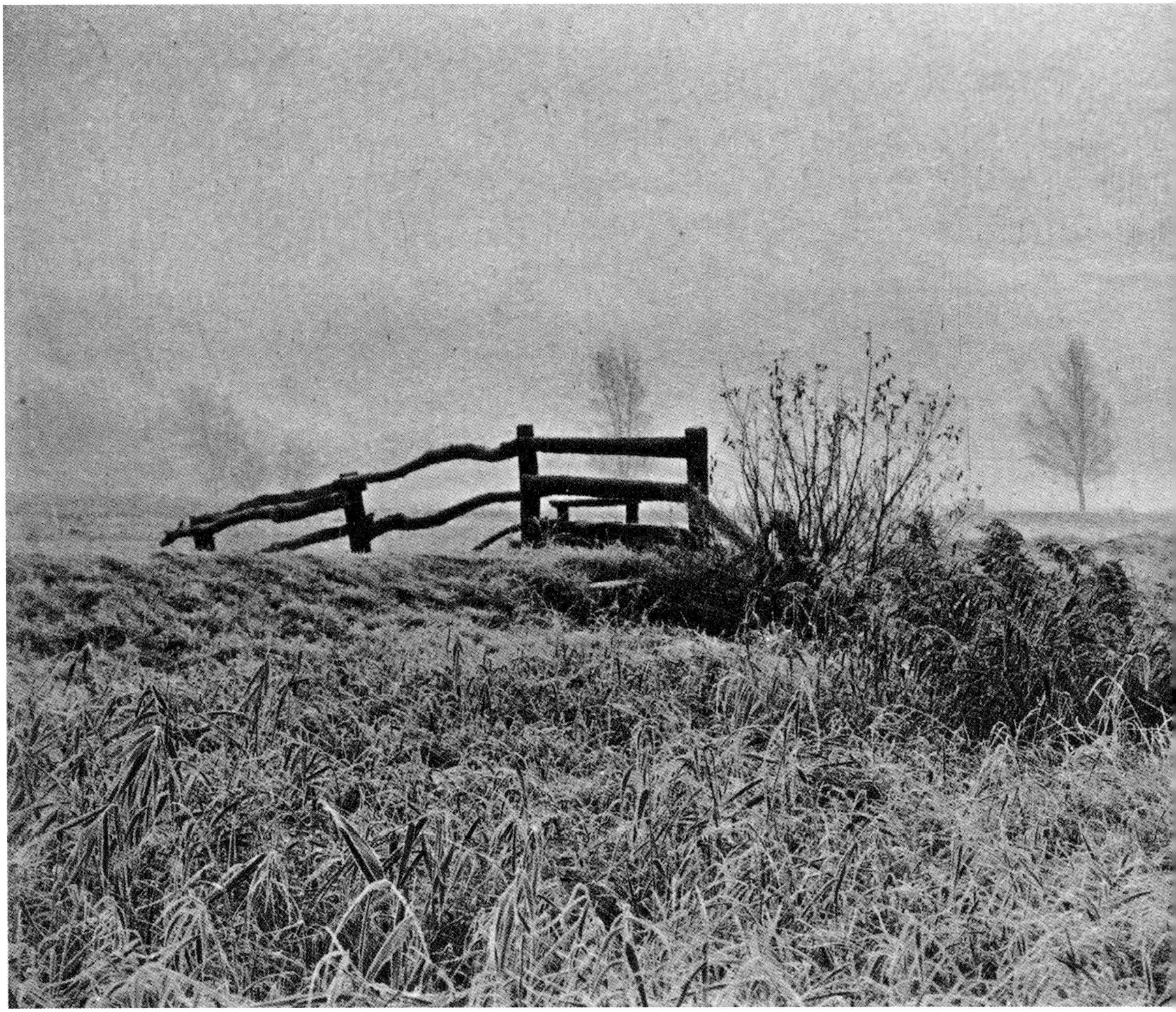

FIG. 47
Peter Henry Emerson, *A Winter's Sunrise*, 1895. Photogravure, 6.8 × 12.9 cm. The J. Paul Getty Museum, Los Angeles.

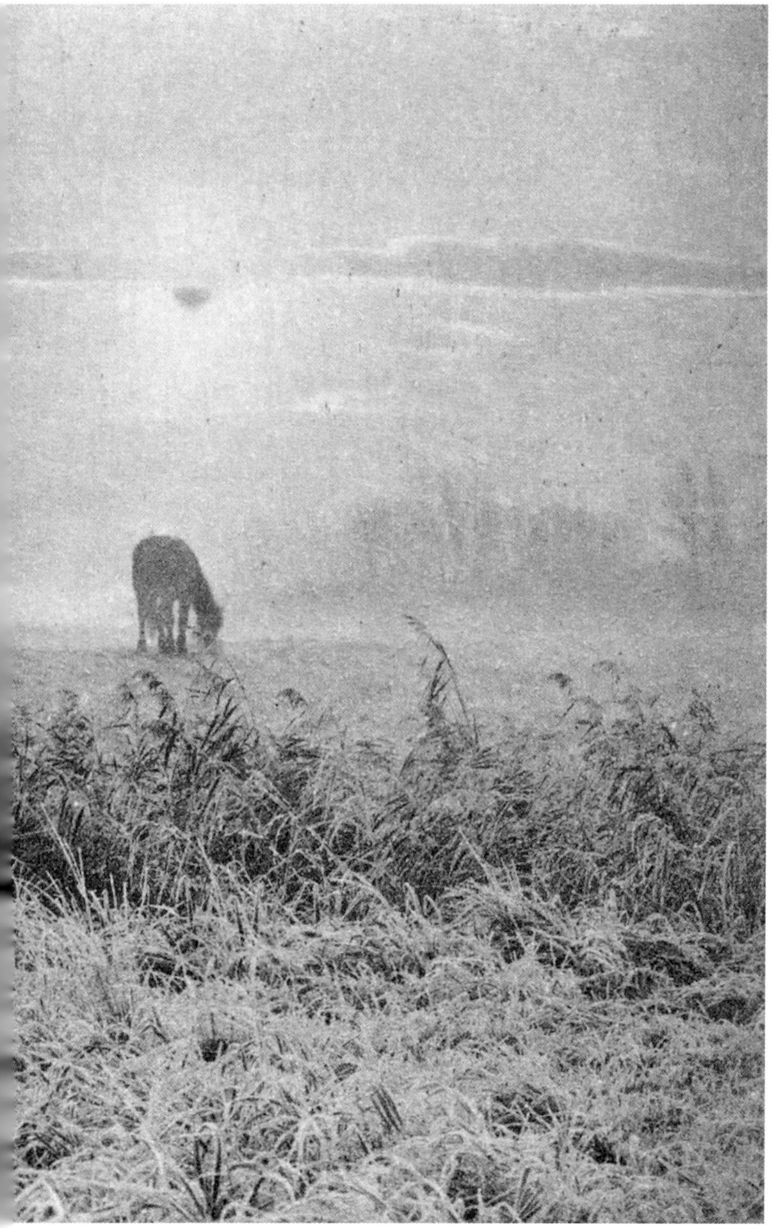

above and below, flattens the picture space, rendering the scene in decorative patterns, an effect emphasized by the trees and animals silhouetted by the rising sun. The accompanying text combines anecdotes of Emerson's encounters with "natives," poetic descriptions of landscape, and the recounting of a dream. The text thus reinforces the implication of the photographs—that knowledge of the landscape is unstable, ever shifting—and also underscores that the photographs were the products of highly personalized encounters with the scene, putting the perceiving subject at the heart of the project.[100]

## Clausen's Crisis

Just as Emerson rethought his practice in the late 1880s and early 1890s, so did Clausen, often in dialogue with Emerson, whom he had probably first met in 1886 when he sold a small portrait to the photographer.[101] Emerson evidently sent the painter many of his writings; in 1889, Clausen wrote to thank Emerson for his essay "Science and Art." Emerson also sent him *Naturalistic Photography*, and Clausen responded that he concurred with Emerson's "condemnation of the false (full of details) photographs. It needed an artist—one who saw that the province of art was not the representation of facts, but of the appearance of things—to point this out." Clausen also shared with Emerson a series of diagrams and exercises showing how to approximate human vision in pictorial representation. "My inclination now," he wrote, "is . . . to trust more and more to pure impression—the first unconscious impression." He acknowledged that it was "very difficult to consciously do this and to be able to get behind yourself so to speak and to find out why it is that things impress us in a certain way. I think one may reach solid ground that way."[102] Truthfulness, for Clausen, was increasingly becoming a measure of what was left out, rather than what was left in.

Clausen continued his inquiry into perception in another letter to Emerson, in which he acknowledged Emerson's congratulations that his painting *The Girl at the Gate* had been purchased for the nation through the Chantrey Bequest. Despite this major accomplishment, Clausen mused, "I don't know if it is that I am advancing (I hope it is) or going back: but I find I am beginning to see many things differently to what I used to." He continued, "I want to get at a broad view of the question: to find out what is essential & vital & to hold on to that and & let the rest go."[103]

The question of what constitutes the essential continued to plague Clausen, as evidenced in another letter to Emerson, written in response to Emerson's treatise on perspective drawing. The letter indicates that the two men had also been engaged in a debate about whether photography or painting could produce truthful images of nature. Clausen, perhaps not surprisingly, came down in favor of painting because the task of working up a canvas forced the artist to engage with nature over an extended period of time. Through this long period of observation and deep familiarity with nature, the artist could identify what was most significant in the scene. He declared to Emerson, "You must take what the photo gives, you have only an instant's touch with nature—in the act of exposing and in that its a hundred to one against hitting the exact moment." He added, "But I am in touch all the time. I can go again and again—and get a firm grip of the essential things—so that I can in a sense make nature mine—at least its only my fault that I can't." Clausen then asked Emerson, "What is to be relied on? After all nothing but the human eye: and with that one must take all the other variations of individuals—so that one scheme of colour is right to one man, another to another—it expresses for him nature—I don't mean to say that anything is anyhow and that a man may draw and colour as he pleases and achieve a worthy result in doing so—but only that within certain limits. There do exist many different interpretations of nature—each having a hold of some truth. Don't try to make art 'scientific.'"[104]

While Clausen seems to rebuke Emerson for attempting to use science to establish firm categories of what could constitute art and truth, he also propelled his colleague toward the conclusion that Emerson would find verified in psychology, namely, the individuality of perception. In his description of his painting practice, Clausen held on to the notion that it is the artist's duty to go to nature. But he did not suggest that the artist should studiously observe in order to gain further knowledge and then craft carefully detailed depictions. Rather, the task is to determine what is most fundamental, to extract from the scene that which initially captured the artist's attention.

These questions about the artist's role occurred alongside Clausen's rethinking of his artistic affiliations. In June 1890, he resigned from the NEAC, complaining to Thomson that the artists in the group associated with impressionism were too concerned with advertising themselves.[105] Presumably, he was referring to Philip Wilson Steer and Walter Sickert, who had taken over the Selection and Hanging Committee of the organization in 1888.

Steer had trained both in Britain, at the school of art affiliated with the Science and Art Department at South Kensington, and in Paris, at the Académie Julian and then at the École des beaux arts. When he returned to England, his paintings demonstrated his knowledge of rural naturalism, most particularly the work of Bastien-Lepage, as well as the decorative aestheticism associated with Whistler. By the late 1880s, Steer's work revealed a close affiliation with the work of the French impressionists in the use of broken brushwork and bright, complementary colors. He was particularly drawn to Claude Monet's paintings, which appeared in the 1877 winter exhibition of the Society of British Artists. Steer's paintings received a mixed critical reception at the early exhibitions of the NEAC. The *Pall Mall Gazette*, for example, described his contributions to the second NEAC exhibition as "interesting and challenging"; Steer's study of a French garden exemplified "the eccentric manner just now affected by some of the 'Independent Artists' [i.e., the French impressionists] in the Champs Elysees."[106]

Sickert had initially aspired to be an actor, but in 1881 he relinquished the stage in favor of training at the Slade School of Art. In March 1882, he departed to study with Whistler and joined the "school" of painters and printmakers that the older artist gathered around him.[107] Sickert, in addition to working as an artist, was also a critic, understanding like Whistler before him the power of public discourse. He used his reviews to advance the cause of his faction of painters within the NEAC.[108] Like Steer, Sickert admired the work of Degas (whom he met through an introduction from Whistler) and publicly acknowledged the French printmaker, painter, and sculptor as "one of the greatest artists the world has ever seen."[109] Degas's impact can be seen in the complicated compositions of Sickert's music hall pictures of the late 1880s and early 1890s, which drew on Whistler's atmospheric effects while adopting unsettling points of view married with social observations.[110] These subjects drew the condemnation of some critics; the *Pall Mall Gazette* referred to Sickert's music hall scene at the 1889 NEAC exhibition as an "outrage on art" because of its supposedly vulgar subject and "contempt for drawing."[111]

As Anne Gruetzner Robins explains, for the 1888 NEAC exhibition, Steer and Sickert chose their most important works and discouraged colleagues from doing the same, thus garnering the best places on the crowded walls.[112] This tactic paid off; the *Belfast News Letter* noticed that "the place of honor at the foot of the room is awarded to a *tour de force*—a large canvas, by P. Wilson Steer, entitled 'A Summer's Evening.'"[113] It eventually became apparent that the NEAC had largely split into two factions: the rural naturalists and the London impressionists, a name derived from an exhibition that Sickert organized at the Goupil Gallery in December 1889.[114]

Although the Goupil Gallery represented Clausen, he was not among the London impressionists, which included Steer, Sickert, Sidney Starr, Frederick Brown, Paul Maitland, and Theodore Roussel.[115] While Clausen was not included in this important show, Sickert's essay for the exhibition catalogue provides crucial insight into the crisis of naturalism in which Clausen was embroiled. In his essay, Sickert first tackled the question of how visitors should approach the paintings on display. He informed readers that they should look for quality and bring "temperament, . . . education, and experience" to their judging. Praiseworthy works of art, he added, possess "a certain beauty and fitness of expression in paint" that reveal to the "connoisseur a thoughtful analysis of the essentials in the production of the emotion induced by the complex phenomenon of vision." Sickert identified the circuitry that produces a reaction to art: the artist identifies the essentials of the scene (an aim that Clausen also pursued), and the viewer's optical and mental apparatus leads to an emotional response. Second, Sickert addressed the nature of impressionism, concluding definitively that it "is not realism" but rather

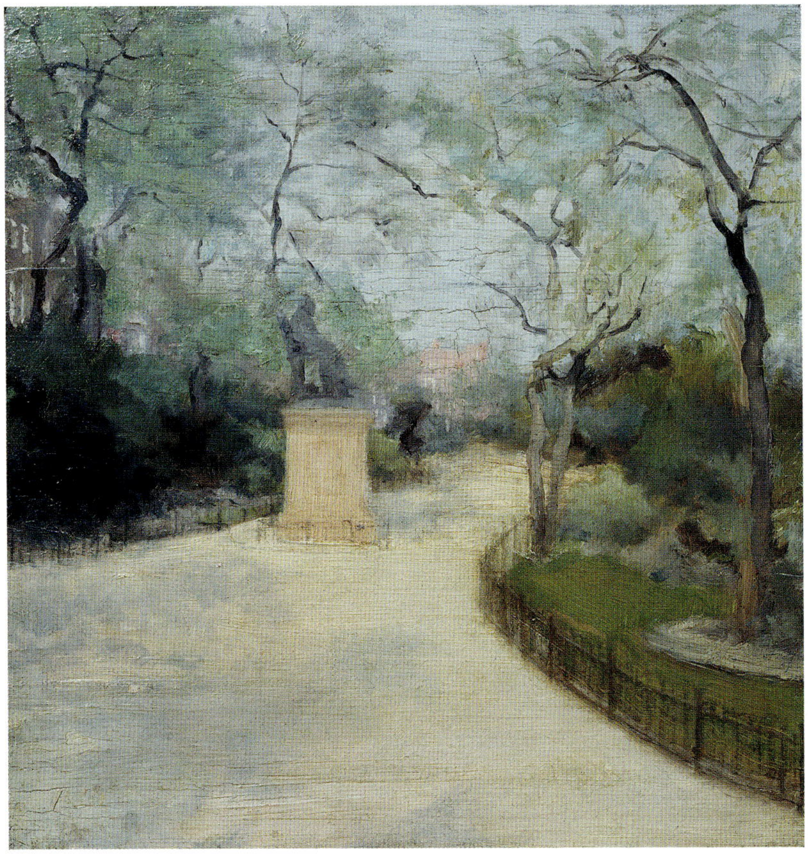

FIG. 48
Paul Maitland (1863–1909), *The Gardens, Chelsea Embankment*, ca. 1889. Oil on wood, support 280 × 265 mm. Presented anonymously in memory of Sir Terence Rattigan, 1983. © Tate, London 2015.

the pursuit of beauty in the urban landscape, the "magic and poetry" that Londoners see daily around them.[116]

So what did Sickert consider "the essentials"? Not factual veracity, apparently, but the judicious selection of both subject matter and mode of delivery. "Real quality," he explained, demands of the artist "complete knowledge of the subject treated, and of simplicity and directness of the treatment."[117] In other words, knowledge is demonstrated not by completeness but by conciseness and brevity—the ability to make deliberate and well-considered choices. Paul Maitland's *The Gardens, Chelsea Embankment* (ca. 1889) (fig. 48), shown in the Goupil exhibition, elucidates this idea in its sketchy indication of forms and elements specific to the embankment garden—the statue on its plinth, the wrought iron railings, the curving walkways and planted beds. All of these forms are registered in a generalizing, suggestive manner without regard to distinctive details that could signify, for example, a botanical specimen or the precise contours and surfaces of a sculpture.

It should be noted that impressionism in British artistic discourses was a highly ambiguous term that could include Whistler's circle and rustic naturalists such as Clausen, as well as progressive French painters like Monet and Degas. Sickert's contribution was to separate impressionism from realism and to reinforce the movement's association with the urban subjects that Degas pioneered. Yet Sickert's project intersects with that of Clausen and Emerson in his emphasis on visual processing. Robins points out that Sickert's position was analogous to that espoused by the art critic George Moore, who used the metaphor of a "net of sensations" to describe the flood of perceptions that should engulf the viewer upon seeing a work of art. Achieving this end depended on many factors, not least the surface of the work of art and attention to painting as painting (rather than as pictorial illusion).[118]

It is worthwhile at this juncture to sketch in a few details about George Moore, whose views on art were quite close to those of Sickert at this time and whose public statements helped to advance the painter's agenda. Upon his death, Moore was best known as a writer, but he had trained as a painter as a young man, studying in Paris in the 1870s at the École des beaux-arts and the

Académie Julian. This period exposed him to the work of progressive French painters, including Degas and Édouard Manet (who became friends with the young Irishman), and to the realist novels of Émile Zola and the symbolist poetry of Stéphane Mallarmé. After ten years in Paris, Moore returned to Britain, taking up residence in London rather than at his family estate in Ireland. While developing his reputation as a novelist, Moore also worked as an art critic, publishing reviews that signaled his allegiance to a younger generation of French and British artists, among them Sickert and Steer. He contributed regular columns to the *Speaker* from 1891 to 1895 and later wrote for the *Saturday Review* and the *Spectator*. He also published *Impressions and Origins* (1891), a collection of his reviews, and *Modern Painting* (1893), which included essays on Whistler, Manet, and Monet. Moore saw his writing as a mission to change public taste, and he urged writers like D. S. MacColl to join him in this cause.[119]

Moore was aided by the willingness of Sickert and even Steer to speak publicly in favor of new painting. Steer, though not a polemicist like Sickert, weighed in on the nature of impressionism in a talk given to the Art Workers Guild in 1891. Although George Clausen did not formally join this group until 1893, it is tempting to consider whether he heard Steer's paper, which was also delivered at the Society of Arts. Steer insisted that true art was "inspired by nature and nature only," and, like Sickert, he charged the artist with finding "his pictures in the scenes around him." Turning to technique and recognizing the possible antipathy of his audience, Steer made an impassioned plea for the new painting. "We are told that Impressionism is a passing craze and nothing but a fashion. Is there any fashion I would ask in painting grass green instead of brown[,] in making a sky recede and hold its right place in the picture instead of hitting one in the eye? Is it a craze that we should recognize the fact that nature is bathed in atmosphere? Is it a fashion to treat a picture so that unity of vision may be achieved by insisting on certain parts more than others? No! it is not fashion it is a <u>law</u>." A successful artist, he added, could raise a subject above the level of recorded facts to poetry, a rhetorical flourish reminiscent of the criticism that greeted Millais's late paintings. Steer insisted that this artistic goal had a long and distinguished genealogy that could be traced back to at least Reynolds and Gainsborough.[120] As Ysanne Holt observes, this argument was not unique to Steer but was also put forward by critic D. S. MacColl, who argued that the British painting tradition, going back to Reynolds, had long recognized the need to eliminate distracting details. Pre-Raphaelitism was an aberration and a departure from this lineage.[121]

Although Clausen was not part of the circle surrounding Steer and Sickert, his practice was shaped by the debates that followed their attempts to consolidate impressionist practice. Clausen's square-brush technique came under attack, criticized as a mechanical response regardless of his subject matter. He eventually abandoned this technique after studying the work of Degas and, in particular, Degas's use of pastel. Clausen's new mode of handling paint—shorter and more staccato strokes—was better suited to registering movement as well as atmospheric effects and reflective passages of color. By rejecting a uniform square brush technique for a handling of paint reminiscent of pastel, Clausen was also forcing himself to rethink each touch of brush to the canvas, which aligns with his comment to Emerson that he was "in touch <u>all the time</u>." In Clausen's new technique one can also detect the influence of Monet, whose

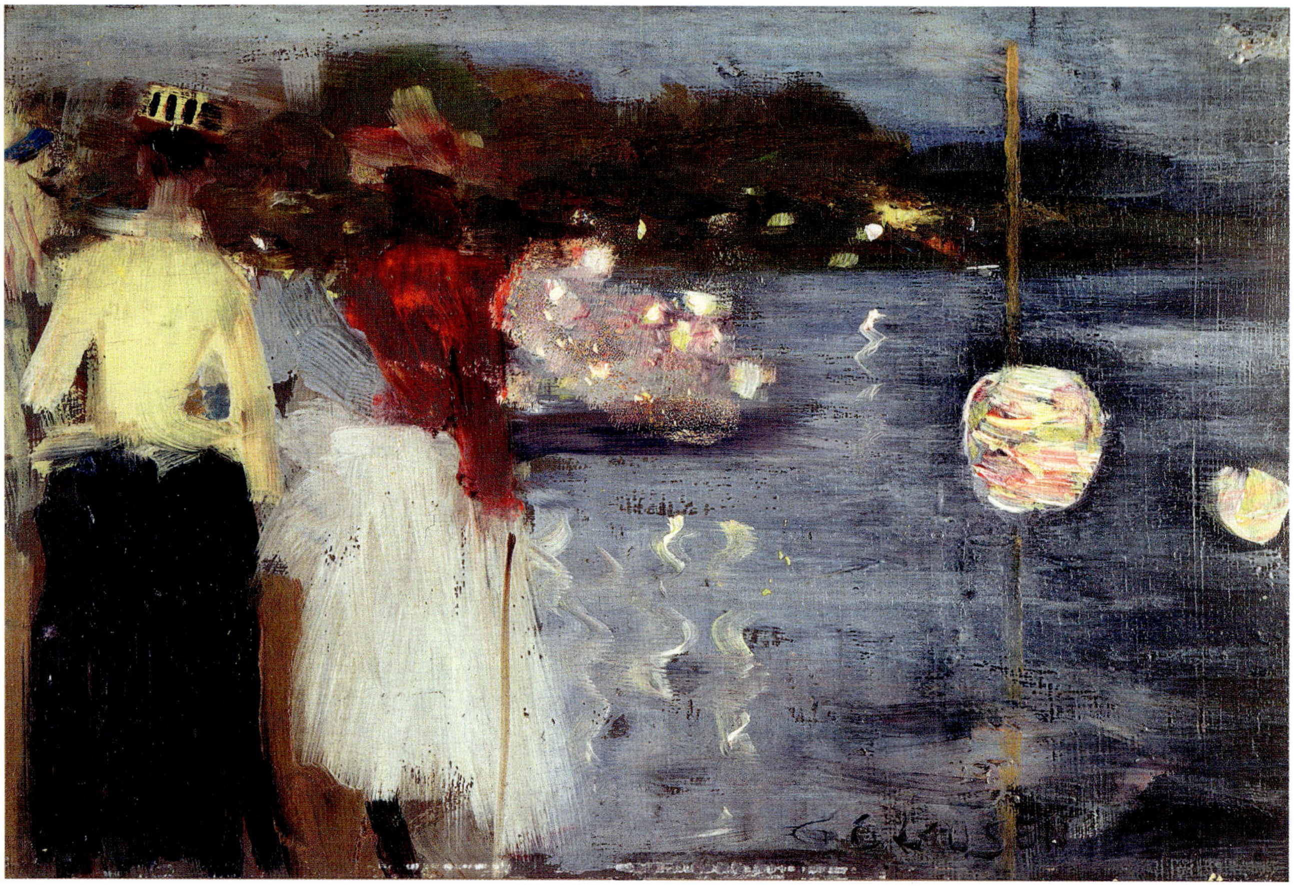

FIG. 49
George Clausen, *Souvenir of Marlow Regatta*, ca. 1889. Oil on panel, 16.5 × 24.6 cm. Leeds Museums and Galleries (Leeds Art Gallery), U.K.

handling of paint became both admired and questioned within British modernist circles at this time, facilitated by his one-man show at the Goupil Gallery in 1889.[122]

Clausen's *Souvenir of Marlow Regatta* (ca. 1889) (fig. 49), a work that the artist referred to as a *pochade*, reveals Clausen's experimental directions at this moment. The subject bears some similarity to Philip Wilson Steer's *The Bridge* of 1887–88 (fig. 50), in the silhouetted figures, sketchy handling of paint, and interest in the play of reflective light on the water, and to Sargent's *Carnation, Lily, Lily, Rose* in the attention to effects of light and, in particular, Japanese lanterns. Clausen's subject was derived from a decidedly more urbane locale than his naturalist subjects: the Marlow Regatta, a spectacle that drew crowds from the metropole. Forms are suggested by visible strokes of paint, which appear to be quickly

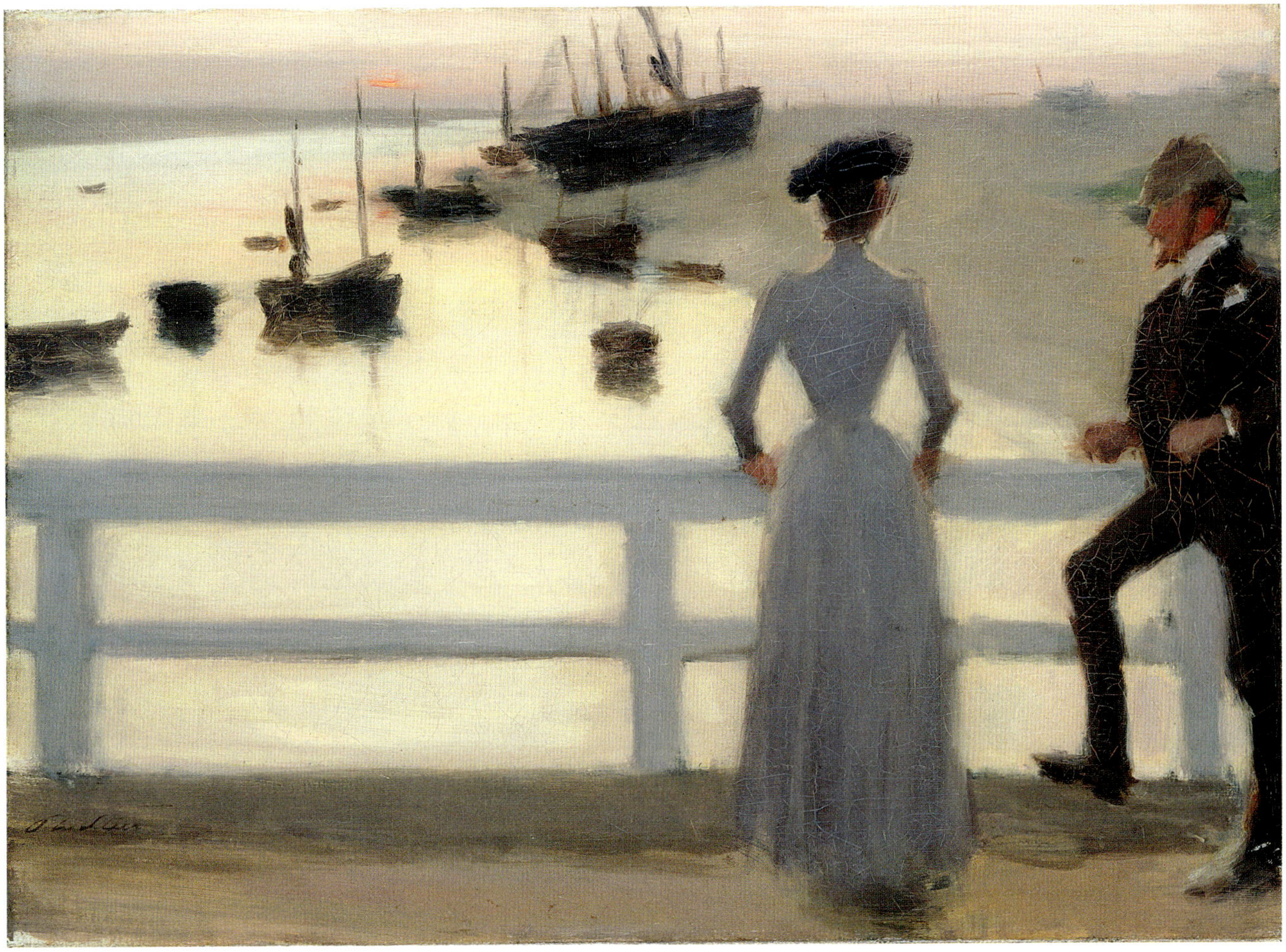

FIG. 50
Philip Wilson Steer (1860–1942), *The Bridge*, 1887–88. Oil on canvas, support 495 × 655 × 20 mm. Purchased 1941. © Tate, London 2015.

laid on in layers; no details are represented. The handling of paint does not differentiate between figure, lantern, water, or land, unifying the work rather than achieving a differential depth of field, as in *The Stone Pickers*.

Critics recognized Clausen's participation in the new movement in painting, albeit as a latecomer. R. A. M. Stevenson's profile of Clausen, published in 1890 and illustrated with reproductions of such paintings as *The Stone Pickers*, makes many of the same points that Steer deployed with respect to impressionism. Stevenson critiqued earlier artists' attempts to achieve "truth" through "glassiness of surface, over-crowded composition, niggled workmanship, and tight handling"; he also referred to their "blind persistent labour"—a dig at Pre-Raphaelitism. Art, he explained, should be "less a blind striving after what is fact, and more as the realising of how we ourselves felt nature." Above all, the painter should strive for unity. In this context, Stevenson praised Clausen's work for its "pursuit of truth" and its acknowledgment of the "need for new, personal, and genuine investigation of one's own feelings about nature."[123]

Stevenson's aim to position Clausen as an emerging leader of a generation that had rejected the practices of the Pre-Raphaelites and absorbed the lessons of French modernism was reinforced by the critical response to Clausen's breakthrough painting *The Mowers* (1891) (fig. 51). By this time, Clausen had moved to Widdington, Essex, to a property on a farm, and he once again used his locale as a subject for his painting. The painting shows three mowers in a field. Two figures in the foreground are shown scything the grass; their stance—heads down as they focus on their task, legs apart to stabilize them as they push their blades through the grass, the sinews standing out in their arms—suggests their physical power. The ripples of their clothing, accentuated by passages of light and shadow, convey a fluid sense of motion. Behind them, to the left, a third figure strides further into the field. The sky overhead is blue and the trees in the background lean from the wind, echoing the movement of the mowers. The paint is built up in layers, the bright highlights in the men's white shirts and the dashed wildflowers applied last so that they stand out.

Clausen endows his laborers with dynamic energy and motion, a sharp contrast to the stillness in *The Stone Pickers*. In his later painting, the viewer, like the mowers, is encouraged to sweep the scene, panning from side to side, rather than to inch upward from foreground to background, as in *The Stone Pickers*. The implied S-curve composition of the mowers speeds visual passage through the work. One can measure the distance Clausen has traveled from his earlier work by comparing *The Mowers* with Bastien-Lepage's *Joan of Arc*, in which the scene unfolds like a series of expanding frames (fig. 45). Bastien-Lepage's composition directs attention to the figure of Joan of Arc first, and then slowly moves outward, detail by accretive detail. Clausen's canvas, by contrast, has none of the plentitude found in the work of Bastien-Lepage, whose details established the "reality effect," to borrow Roland Barthes's phrase. Clausen's flowers are simply dots of color sprinkled over the brown field indicating grasses. Here is Clausen's solution to the quest for "pure impression," the task of paring painting down to the "essentials" of the scene.

Clausen's suppression of detail in favor of a generalizing impression drew praise.[124] The *Saturday Review*, for example, declared that the painting showed that Clausen "has long been a student of nature, now he has discovered beauty."[125] As Kenneth McConkey points out, Clausen's "summary treatment" of the figures helps to convey a sense of "instantaneity" and underscores the distance Clausen had traveled from his origins in naturalism.[126] George Moore made this point as well, observing that the painter had finally "shaken himself free" of the influence of Bastien-Lepage and claimed his own ground. Moore praised the artist's representation of movement—exemplified in Clausen's decision to generalize the handling of the trousers—and of light, which evoked the sultry conditions of a hot summer's day. Moore called attention to the sky, describing it as "deep and lowering—a sultry summer sky, a little unpleasant in colour, but true." This was the only painting of that

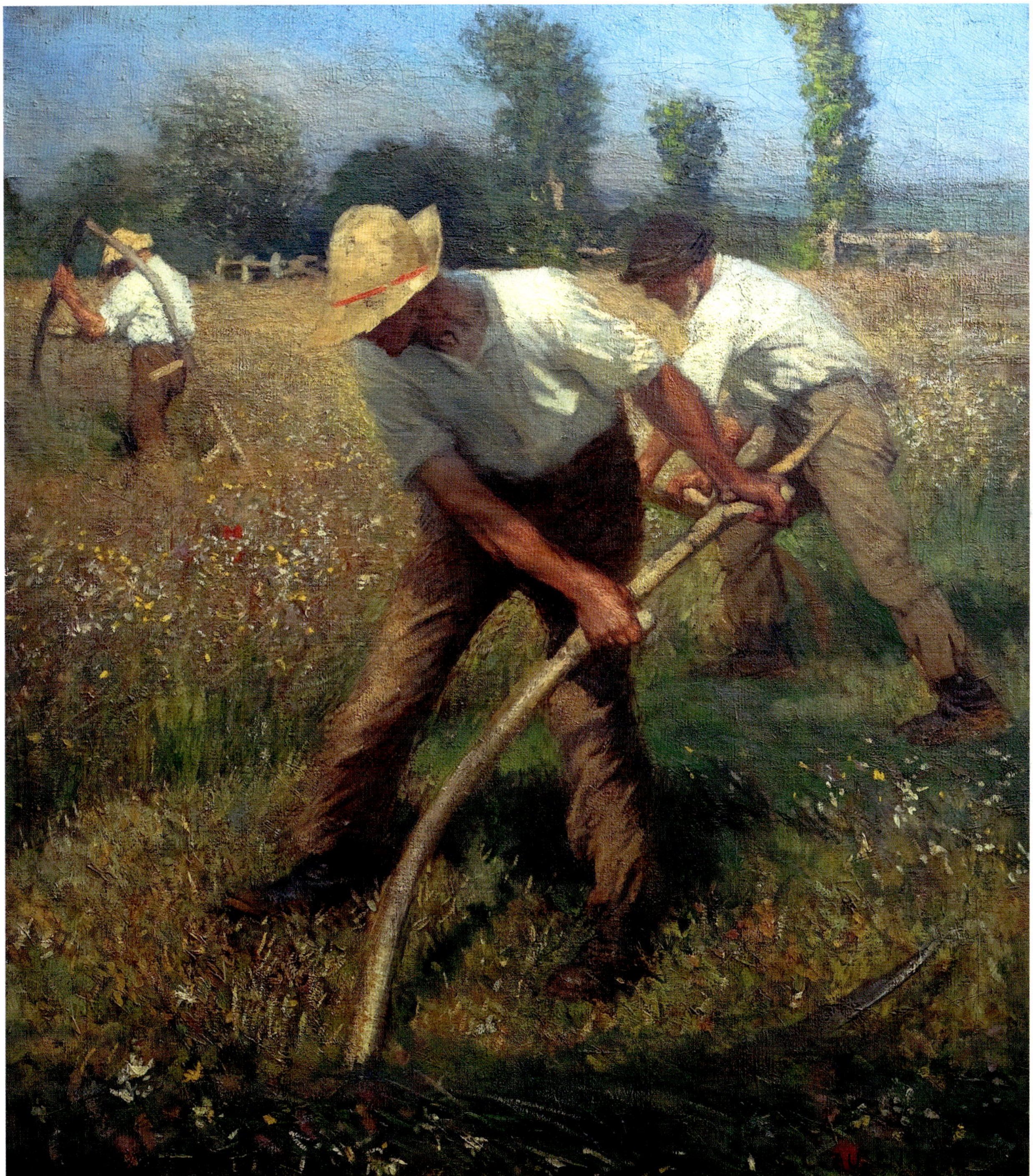

FIG. 51
George Clausen, *The Mowers*, 1891. Oil on canvas, 101.6 × 86.4 cm. The Collection: Art and Archaeology in Lincolnshire (Usher Gallery, Lincoln), U.K.

year's Royal Academy exhibit, Moore proclaimed, worthy of critical analysis.[127]

Around the same time, Clausen again took up the topic of Bastien-Lepage in a tribute volume published in 1892. His important essay reveals that the painter had reached a turning point in his quest to "find out what is essential & vital & to hold on to that and & let the rest go." Clausen's essay reiterated his admiration for Bastien-Lepage for bringing to painting "a new view of nature" in keeping with the modern generation's interest in finding beauty in the quotidian. Clausen praised the artist's ability "to get at the actual optical appearances and to suppress any hint of his own feelings," his devotion to carrying "the finish of the foregrounds in these pictures to the farthest possible point," and his dedication to "the strict literal truth." But these features were precisely the aspects of painting that progressive British artists and critics wished to cast aside, arguing instead for individual expression, unifying effect, and selective truth.

Indeed, Clausen recognized his master's limitations: "I think he found himself hedged in by the conditions necessary to the realization of the qualities he sought." Clausen had experienced this epiphany when confronted at the Exposition Universelle with the juxtaposition of Bastien-Lepage's *Joan of Arc* (1879) and Courbet's *The Stone Breakers*. Clausen found Courbet's work more successful because it was pared down to essentials and thus communicated at a distance, whereas Bastien-Lepage's painting, filled with details based on close observation, failed to do so. Clausen concluded, in his 1892 essay on the painter, that it was "impossible" to have both "truth of impression and of detail." For Clausen, truth of impression now counted most: "the building-up or combining a number of facts, each of which may be true of itself and to the others, does not in its sum total give the general impression of truth. It is but a number of isolated truths."[128]

Like Emerson, Clausen undermined the entire premise on which the truth-to-nature project had originally rested: the inductive, objective, and positivist recording of nature built up from closely observed details. Clausen concluded his essay by describing the endgame to which Bastien-Lepage had brought him. The French painter had "carried literal representation to its extreme limit." He left Clausen pondering "whether it is possible to attain literal truth without leaving on one side much of that which is most beautiful in nature? And further, the question arises, whether literal truth is the highest truth."[129]

Clausen's query, quietly and logically proposed, was nonetheless potentially earth-shattering to the modes of artistic practice that had persisted in British painting from the origins of Pre-Raphaelitism to Brett's late canvases of the 1890s. Brett believed implicitly that "literal truth is the highest truth," and this drove his persistent observation of the scene in nature. But his struggle to adjust a practice rooted in the belief in literal truth to new scientific findings, particularly those related to optics and cognitive processing, hints at the dilemma that Clausen now faced.

Clausen's response was to declare that "realism, as an end in art, leads nowhere; it is an *impasse*."[130] But rather than throw up his hands and end his practice, as Emerson had done, Clausen drew on the lessons provided by his recent ventures into pastel, the growing admiration for Degas among progressive British artists, and the advice of progressive critics to take his work in a new direction.

Clausen's assessment of Bastien-Lepage, in the end, squares with that provided by the more acerbic Walter Sickert in the same 1892 volume of tributes to the French artist. Like Clausen, Sickert found *Joan of Arc* wanting—he criticized its drawing, color, and execution, all of which he found reflective of the deeply flawed realist school. He dismissed the realists for producing nothing more than "a catalogue of facts." Millet, by contrast, possessed "an unerring analysis [that] enabled him to select what were the vital constituents of such a scene." Significantly, Sickert argued that this act of selection required that the moment of executing an artwork be separated from the moment of observation, so that the painter did not become mired in the exigencies of the moment. He thus foreclosed on the possibility of working in plein air. In this passage and elsewhere in his essay, Sickert made clear his belief that Bastien-Lepage's work fell short of the great masters Manet, Millet, Whistler, and Degas.[131] This is perhaps odd in a volume intended as a tribute to the French painter, and it indicates how high the stakes were for those participating in debates about truth to nature and the future of British painting. Sickert could not afford to miss an opportunity to state his case.

George Moore also saw the tribute volume as an opportunity to assert his views about modern realism. It is worth noting that by this date Moore was regarded as a realist novelist in the mode of Zola. He defended the French writer's work when a case was brought against Zola's British publisher under the National Vigilance Act. Moore's own novels had also come under attack: *A Modern Lover* (1883) was banned by circulating libraries because of its explicit sexuality, and the bookseller W. H. Smith refused to stock *A Mummer's Wife* (1885). In assessing the tributes to Bastien-Lepage, Moore reviewed only Sickert's essay, which mirrored his own views. Indeed, he was pleased to see Bastien-Lepage knocked off the pedestal of fame. To be sure that readers grasped the full significance of Sickert's analysis, Moore summarized it: "Observe nature, study nature; but do not copy nature."

Moore also used his review as an opportunity to underscore Sickert's achievement. He explained how Sickert worked from memory (a process with a clear debt to Whistler and also reminiscent of Hamerton's advice on the topic): "For many years Mr. Sickert has practised what he preaches. He had made thousands of drawings in the fields and streets, in the theatres and the music-halls; he has perfected his memory by constant application, and his pictures are painted from his studies."[132] In other words, the motif is not the urtext for a painting; it is the artist's memory and sketches that matter. Working from these sources implies selectivity, as psychology had established that memory was not a mirror of experienced reality, and the practice of sketching involved shorthand references.

Moore reiterated his criticism of painting directly from the motif in his review of the NEAC exhibition of 1893. He declared that a shift had taken place over the previous decade or so: "Fifteen years ago there was but one way to paint, and that was to paint direct from nature. He who did not take a six-foot canvas into the fields and copy all that he saw, and exactly as he saw it, was an old fogey. Now it is the old fogeys who paint direct from nature, the young men know that though the palette can give you the exact value of the earth, it cannot reach the light of the sky—therefore, he [who] would paint from nature ends with a false value."[133]

Moore repeated here what Emerson had learned from Helmholtz: that it is impossible for the artist to replicate the colors found in nature. The new generation, Moore explained, was now looking to the Old Masters, such as Reynolds, for inspiration, despite

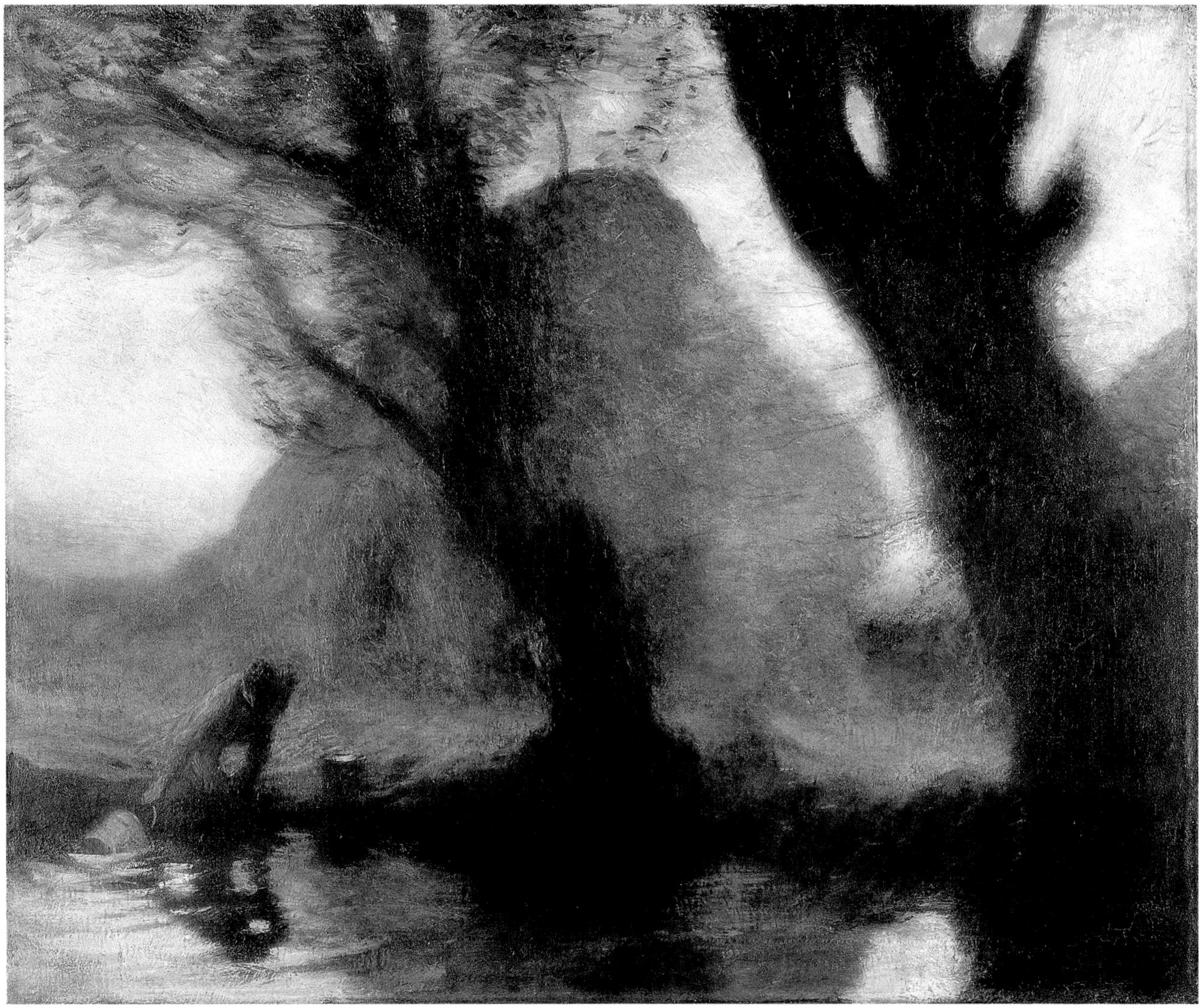

FIG. 52
George Clausen, *Dusk*, 1903. Oil on canvas, 62.2 × 73.7 cm. Laing Art Gallery, Newcastle upon Tyne, U.K.

Ruskin's condemnation of the older painter's generalizing method, a path that Millais had also pursued to solve the quandary in which he found himself. Clausen, too, looked to the past for inspiration, a maneuver that Steer had also recommended in his lecture on impressionism. In 1894, Clausen wrote David Croal Thomson, who had just returned from Italy, "I should much like to have seen the pictures in Italy and propose to go at the next opportunity—for it begins to be apparent to me that an artist should study all that has been done, and as much as he studies nature. The more I look at nature the more I see the reasons for the generalizations and conventions of the older men, and the more convinced I am that 'realism' is a vain thing!"[134]

By 1895, Clausen had pronounced realism, or naturalism, a dead end, according to Moore. In an essay titled "Exteriority," in which Moore recounted a recent conversation with Clausen, he described how Clausen's practice had changed from his early naturalist canvases, such as *Labourers After Dinner* (1884), which conveyed the "impersonal vision of the camera" in the "blank brutality of their faces." Moore had written a highly critical review of *Labourers After Dinner* when it was first exhibited, which sparked an ongoing dialogue with the painter. When the two men met at the Royal Academy exhibition of 1895, to which Clausen had submitted *The Harvest* and *The Farmer's Boy*, Moore complimented the painter for abandoning Bastien-Lepage as a model. Clausen explained, "I pursued the method until I found myself facing a dead wall. I could go no further, and had to retrace my steps." In turn, Clausen observed that Moore had also undergone a profound change in his novels, an assessment with which Moore concurred, explaining that he was striving to get away from "exterior realism." Instead, he aimed "to write a book in which everything shall come from within," meaning the subjective self. Clausen declared that he shared a similar ambition, adding, "I am painting a picture of night: a figure in the sky. I haven't got it right yet, but it will come right one of these days."[135]

It is not evident to which painting Clausen was referring—and perhaps he never completed it. But his paintings of the succeeding decade reveal that he continued to explore how to pare down the scene to its essentials and achieve a unifying composition, as in *Dusk* (1903) (fig. 52). The reduced light and palette fuse the various parts of the scene into a whole. Built-up layers of paint knit together water, laborer, trees, and haystack; subtle brushstrokes with no distinct edges treat figure and landscape alike. Truth is now measured by the evocative effect of the whole rather than the reality effect of the part.

But the succeeding generations of modernists did not take up Clausen's solution to the quandary of how to depict truth to nature. While they agreed that realism was a dead end, Clausen's attempt to balance the study of nature with generalized forms that suggested the softening effects of light and atmosphere was not a path they chose to follow. They fought for a more radical reform of art making, while still wrestling with the relationship of representation to nature.

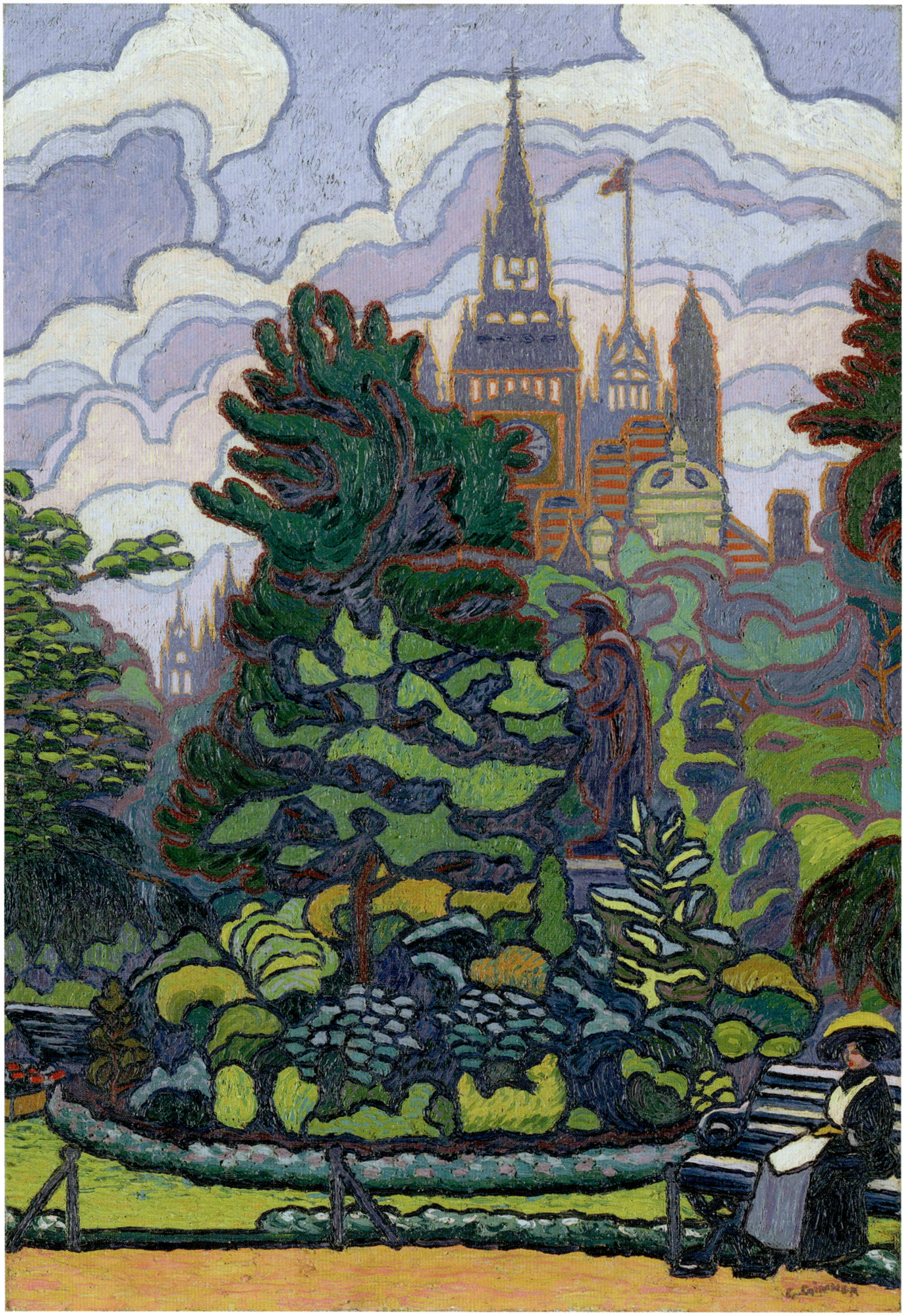

FIG. 53
Charles Ginner (1878–1952), *Victoria Embankment Gardens*, 1912. Oil on canvas, support 664 × 461 mm. Purchased 1984. © Tate, London 2015.

{ CHAPTER FOUR }

# Neorealism

## *Truth to Nature in Modernist Critical Debate*

The paint announces itself. Thickly applied, it stands up in small ridges; brush marks are visible. The paint underscores the highly structured composition, which is cartoonlike in the use of outline to define forms as bounded fields of color, sometimes pure and sometimes heightened by the inclusion of contrasting tones, yellow against green, pink next to orange, and beige against blue. The stylized scene depicts the embankment garden along the Thames near the British Parliament, the clock tower of Big Ben silhouetted against the cloud-filled sky. The middle ground is filled with a conical garden bed, crowded with climbing clumps of color that suggest flowers, shrubs, and trees, and ringed at the ground by dots of color. In the far right foreground is a woman seated on a bench and wearing a large hat crowned with bright yellow; her presence, highlighted by her red lipstick, is balanced by five bright red spots at the left of the image, perhaps connoting geraniums. Red appears again in the outlines of some of the greenery in the upper registers of the image and in the architecture in the far distance. These spots of red animate the composition and establish zones of visual interest, while the canvas as a whole is unified by the rhythmic repetition of curves throughout, from flower bed to the clouds in the sky.

Charles Ginner's *Victoria Embankment Gardens* (fig. 53) was shown at the Goupil Gallery in 1914 on the occasion of an exhibition with fellow painter Harold Gilman. The two artists had participated in the formation of the exhibiting society the Camden Town Group, which held a series of three exhibitions between June 1911 and December 1912 at the Carfax Gallery.[1] Goupil and Carfax, commercial dealers aiming for the high end of the market and specializing in contemporary art, particularly that of France and Britain, competed for many of London's modernists. Goupil's spacious and luxuriously appointed venue on Regent Street signaled Ginner's and Gilman's ambitions, as did Ginner's essay "Neo-Realism," which accompanied the exhibition. Like Emerson before him, Ginner was concerned with demarcating his signature concept and polemicizing his pictorial practice within a densely occupied field. The rise of little magazines and the concomitant expansion of the commercial art market had created multiple venues or platforms from which the modernists could declare their aims, as in the case of Roger Fry's Omega Workshops or the Vorticist journal *Blast* (1914–15).

In his essay, Ginner acknowledged the possible semantic resemblance between neorealism and naturalism but dismissed the latter as an ill-informed movement. He critiqued it along the same lines that Emerson had done in his "Renunciation," for merely copying nature and providing nothing beyond surface mimicry.

> *Naturalism is a poor relation of Realism. It is the production of a Realist with a poor mind. A mind that goes to search out and reveal the secrets of Life and Nature, but has not the power to find. Naturalism is the photography of Nature. The Naturalist, with infinite care, goes out to her and copies the superficial aspect of the object before him. He only sees Nature with a dull and common eye, and has nothing to reveal. He has no personal vision, no individual temperament to express, no power of research. Nature remains a mystery to him in spite of all his work. Plastic Art then ceases, the decorative interpretation and intimate research of Nature,* i.e. *Life, are no more. It is in the R.A. that the last embers of this short-lived Naturalism are burning out.*[2]

Ginner's rant reveals not only a prolonged reaction against naturalism but also the persistence of key questions that plagued Emerson and Clausen: To what degree should the work of art be mimetic of nature? What is the role of the artist's subjective perception, and how should it be recorded in the final image? Should the artist bring to the act of representation an "innocent eye" or vision informed by artistic precedent and/or scientific knowledge? What constitutes research on nature? Ginner's diatribe is particularly striking because he nonetheless holds on to the paradigm of truth to nature—a paradigm that, on the surface, seems unrelated to the issues of modernism that are more commonly associated with a desire to abstract or stylize the motif and to signal the power of the artist's imagination.

The *Spectator*, in its review of the New English Art Club, offered a guide to the generational shift taking place: "After a generation of impressionism, which has largely meant a preoccupation with the external side of nature, a heightening of tone and brightening of colour, a generation has sprung up that is questioning the whole representation side of art and is eagerly hunting back amidst the conventions of ancient art, from Siena to China, to recapture an inner message that they think has escaped in the advance and settling down of modern painting with its elaborate machinery upon the whole province of vision."[3] This critic decisively rejected the optical perception of external nature as the source for modern art and suggested instead the pursuit of an undefined "inner message."

The ground for such a redefinition of modern art had already been well prepared. The question of subjectivity, as articulated by scientists such as Tyndall and Helmholtz and philosophers and psychologists such as Spencer, Sully, Bain, Lewes, and James, had provoked a crisis within Pre-Raphaelitism, naturalism, and the photographic practice of Peter Henry Emerson. Yet none of the artists ever questioned the significance of human vision, despite their recognition of its flaws. The *Spectator*'s critic suggested that advancing painting (while looking back) was not about probing the mechanics of vision but about finding a new form of expression. Yet the province of vision was still an important, if troubling, aspect of modernism.

Ginner's tactics—issuing a manifesto and displaying his work—were in keeping with those of his generation, which took up the mantle of a combative avant-garde in order to define its own ground for intellectual debate. At the same time, we can see Ginner's mode of operation as a continuation of strategies deployed throughout the truth-to-nature debate, reaching back to Talbot and his publications, pronouncements, and exhibitions concerning photogenic drawing and the calotype. For Ginner, the stakes seemed high; modernism in Britain at this moment was not a unified field. It was a contentious movement, highly charged by strong personalities, artists such as Sickert, Wyndham Lewis, and Augustus John, who had learned from Whistler the benefit of cultivating a public persona, even celebrity status, as well as the value of using exhibitions, alliances, and critical reviews to gain public attention and to assert a personal program. Such an agenda fueled factions within the exhibition societies, such as the Camden Town Group, that modernist artists working in London established in order to leverage shared labor and acumen, in addition to mutual intellectual and financial support. Gilman's polemic was aimed at his colleagues in the Camden Town Group as much as at the public sphere. Arguably, he was in particular provoking Augustus John and his colleague J. D. Innes, who had also taken up the question of how to modernize or reform landscape painting.

## J. D. Innes and Augustus John: Out of the Coal Hole

The concept of an innocent eye and the truth-value of plein-air painting were given new life in the early twentieth century by J. D. Innes and Augustus John. Innes and John, like their predecessors, were engaged in a quest for authenticity, but what constituted the authentic for them was shaped less by debates about optical vision than by those concerning mental perception, particularly as articulated by Henri Bergson, a French philosopher-scientist whose works were translated into English in the first decade of the twentieth century.

Both Innes and John were Welsh, and this shared background helped foster their friendship, as did their delight in bohemianism and outré escapades. John, the older of the two artists, began a course of study in 1894 at the Slade School of Fine Art, and Innes matriculated to the school in 1905. The Slade, the first art school located within a liberal arts university, the University College London, had gained a reputation for progressiveness under the tutelage of Slade Professor Fred Brown. This reputation was enhanced by the achievement of its students, who included, in the 1890s, Augustus's sister, Gwen John, and, in the following decade, Wyndham Lewis, Harold Gilman, and Spencer Gore, a close friend of Innes's. In 1893, Philip Wilson Steer was hired as the assistant teacher of painting, and as Innes's instructor had a great influence on the student, who took up his mentor's methodology. By the time of Innes's studies, this concerned distilling formal design from the studied motif in nature and assimilating his personal vision to the British landscape tradition, as in *Nutting* (1905) and *Chepstow Castle* (1905). In 1900, University College London accepted the bequest of Henry Vaughan, which included important landscape prints, among them Turner's *Liber Studiorum* and rare mezzotints from Turner's *Little Liber* and Constable's *English Landscape Scenery.* This combination of resources meant that students had access to both the newest directions in art and the finest models of English artistry, according to contemporary standards.

John left the Slade in 1898 and Innes in 1908; by 1907, the two had struck up a friendship. The years when these two artists emerged professionally witnessed a profusion of important art exhibitions in London that stirred up debates about modernism and fueled the pluralism that characterized the London art world at this moment. These included *Manet and the Post-Impressionists*, organized by Roger Fry and held at the Grafton Gallery (November 1910–January 1911), which highlighted Paul Cézanne, Maurice Denis, André Derain, Paul Gauguin, Vincent van Gogh, Édouard Manet, Henri Matisse, and Georges Seurat, among others; and *The Second Post-Impressionist Exhibition* (1912), featuring art by progressive French, British, and Russian artists. To this list can be added *Cézanne and Gauguin* (1911), sponsored by the collector Michael Sadler, and an exhibition at the Stafford Gallery in October 1912 showcasing artists associated with the British modernist journal *Rhythm*.[4] These exhibitions were widely reviewed in the press, both favorably and unfavorably, and they encouraged other art dealers to venture into modernism, negative commentary notwithstanding. Innes and John began to exhibit professionally in this period, sponsored largely by the Carfax and Chenil galleries, relatively small, simple spaces that contrasted with the luxurious accommodations of the galleries lining New Bond and Regent Streets.[5]

John and Innes, despite their rootedness in urban exhibition culture and their obvious debts to the previous generation of Steer and Sickert, broke decisively with the urban paradigm and, as Lisa Tickner has pointed out, thus garnered a much more equivocal place in the history of modernism than artists engaged with metropolitan spaces.[6] In 1901, through his new teaching post at an art school linked to Liverpool University College, John met John Sampson, the university librarian, who had taken on the monumental task of studying and preserving the Romany language and way of life. John was attracted to the gypsy lifestyle—or, more accurately, to how the gypsy life appeared to an educated urbanite seeking an alternative source of identity. Rather than strike the pose of an urbane dandy—as Whistler and Oscar Wilde had done—John became a primitive bohemian (fig. 54). In his autobiography, John cited Walt Whitman's *Leaves of Grass* as a catalyst for his decision to cast off the trappings of city life and the code of his father's generation: "I found Walt Whitman intoxicating! Here, I thought, was my Poet! Henceforth I was to live for Freedom and the Open Road! No more urbanity for me, no more punctilio, and, as for clothing, according to my Golden Rule, the 'importance of appearances' insisted on at home, already made my neglect of them a duty. But now a new kind of exhibitionism was born; in its way, as exact and conscientious as my father's cult of the clothes-brush: a kind of inverted Dandyism."[7]

Instead of a top hat, John adopted a wide-brimmed gaucho-style hat commonly associated with the working classes and, most strikingly, a beard—as *Colour* noted in 1917: "The fact that a modern Briton should grow a beard in defiance of fashion is alone sufficient to prove that he belongs to the minority, and the rumoured attempt of his to viridify [*sic*] the sign of his virility ought assuredly to convince doubters that he is a perfect representative of this minority."[8] Innes, whom John described as an original and a "naïf," likewise adopted a bohemian persona. As John recalled, "Innes was a remarkable looking young man, and dressed with a distinction which owed nothing to the rules of fashion. His pale and ashen cadaverous face with lank black hair, black beady eyes and large flexible mouth would arrest attention anywhere; attired in a long black over-coat and a broad-brimmed Quaker hat, a voluminous silk scarf took the place of a collar, and he held a silver-mounted malacca cane in his gloved hand."[9]

John's career was structured around a powerful set of paradoxes: the pursuit of bohemianism, enabled by commissions from wealthy elites; the search for inspiration in nature and the gypsy life while maintaining a studio in town; and the rejection of social proprieties while using extensive social networks to fulfill his ambitions.[10] In 1917, *Colour*'s critic referred to him as "a sort of inverted Thoreau."[11]

Such rejection of bourgeois values in both art and life was not new; the paradigm, arguably, had been invented in the earlier romantic age by such figures as the poet Lord Byron and Thomas Chatterton, an archetype of the mistreated and misunderstood artist. Moreover, John's fascination with gypsies can be read as a reinvigoration of the Orientalist paradigm.[12] Yet there was a distinct note to early twentieth-century versions of the romantic—a roughness, frankness, and sexual aggressiveness not found in earlier incarnations.[13] While Innes reveled in his dangerous lifestyle, by December 1912, when he met up with Augustus John and artist Derwent Lees in Marseilles, it had caught up with him. At this juncture, John and Innes's friendship collapsed; John reported to Quinn that Innes's "arrival somehow spoiled the trip as he got really ill— He had had a very dissipated time at Perignan. . . . We took him back to Paris & sent him off to London to see a doctor. . . . The company of a sick man gets on one's nerves in the end."[14] John's vision of health and vigor in the open air did not allow for a consumptive individual, and although he continued to express admiration for his colleague's work, he soon despaired of Innes's ability to recover fully.

Nonetheless, particularly around the years 1910 to 1912, John and Innes were engaged in a shared project of shedding urbanism and the persona of the dandy while simultaneously charting a new course for painting independent of the paradigms of naturalism and impressionism, a quest that guided their choice of painting sites. In deciding where to travel, they chose places that provided them with the requisite escape from the metropolis while also allowing them to insert themselves into the art-historical narrative. Thus, in 1909, Innes traveled the

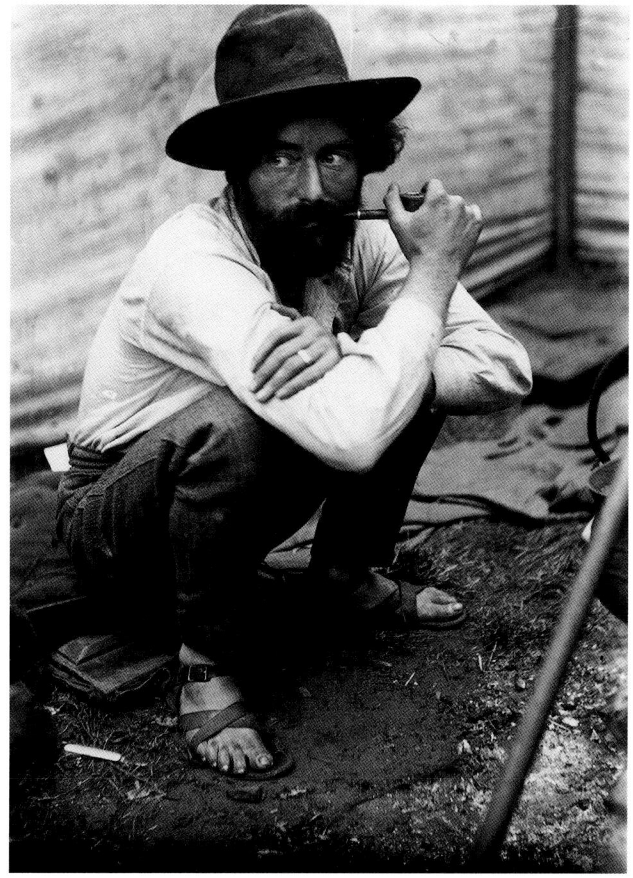

FIG. 54
Charles F. Slade, *Augustus John*, September 1909. Snapshot print enlargement, 155 mm × 111 mm. © Reserved. Collection National Portrait Gallery, London. Given by Ewan Phillips, 1982.

course of the River Wye, visiting, among other destinations, Chepstow, where Turner and later Steer painted. But he reported to his friend John Fothergill, who helped to found the Carfax Gallery, that "the word Chepstow has become almoste [*sic*] repulsive. We would not stay there."[15] To escape the crowds of tourists, John and Innes eventually returned to Wales and traveled together in the mountain district in 1911 and 1912. Although this landscape was long associated with the eighteenth-century British painter Richard Wilson, the two artists never engaged directly with Wilson's legacy; instead, they used the opportunity to reinforce their virile bravado—drinking with innkeepers and consorting with gypsies—and to hone their painterly responses to nature. John recalled in his autobiography, "At this time Innes's activity was prodigious; he rarely returned of an evening without a couple of panels completed. These were, it is true, rapidly done, but usually meant long rambles over the moors in search of the magical moment. Perhaps he felt he must hasten while there was time to make these votive offerings to the mountains he loved with religious fervour."[16]

While searching out unpopulated and rural venues in the United Kingdom, the two artists also traveled to the Continent and alighted in the south of France. In 1908, Innes and Fothergill traveled to Bozouls, in the Aveyron department of southern France. Their travels also took them to the fishing village of Collioure on the Mediterranean coast and the medieval town of Perpignan in the Pyrenees, the southern light and the sea and mountains supporting their search for essential motifs and a more intense palette. The artists were seemingly unaware of the Fauves' previous forays into this region, but they were nonetheless drawn by the region's associations with a "golden age"—utopian, authentic, primitive, and yet classicist, as Ysanne Holt has explained.[17]

Indeed, when recounting his decision to travel to the south of France in the winter of 1910, John mythically recalled that he was drawn there "as if in answer to the insistent call of far-off Roman trumpets."[18] He settled first at Arles, then traveled to Marseilles, where he could observe and mingle with gypsies, and then went on to Italy. There, he continued to pursue his obsession with gypsies but also sought to reinvent his painting practice by engaging with great examples of art from the past. He wrote to Quinn, upon his return to Martigues, of the power of his visit to Italy. "My imagination and sense of reality seems to me just twice as strong as it was before—and no exaggeration," he told Quinn. "I wouldn't dream of 'putting it on' to you (as we say in England) but I tell you frankly & sincerely, I feel nobody dead or alive is so near the guts of things as I am at present—and its devilish hot, and what is surprising—together with this infallible realism my sense of beauty seems to, has grown simultaneously."[19]

John's rhetorical pairings—imagination and reality, realism and beauty—is revealing. On the one hand, these phrases underscore his penchant for the dialectical; on the other, they indicate his approach to the problem of rendering nature. Realism, for him, could not be separated from idealism, as achieved through the artist's imagination, the knowledge of past art, and the sensibility for beauty.

John continued his search for answers to this problem in the south of France during the summer of 1910. While there, he visited the home of Cézanne, whom he had grown to admire deeply, like many moderns of his generation. Shortly after his visit, John wrote to Quinn of his admiration not only for Cézanne but also for Pierre Puvis de Chavannes, whose attempt to marry a sense of painting tradition with his own distinct sensibility set a strong example for the British painter.[20] In addition to artistic models, the south of France provided John with a transformative physical environment. In his autobiography, he described how, "after a necessary interval spent in observation and experiment, I attuned my eye at last to the marvelous light of this region, and a series of panels attest, I think, to the beneficial effect of it on my palette."[21]

In 1911, John wrote Quinn about the new directions in painting he hoped to pursue. The letter was a response to one in which Quinn praised the recent work of the American modernist painter Alfred Maurer, who had adopted "a high, bright key"; indeed, Quinn was hopeful that "artists are beginning to learn more and more that their pictures should not look as though they were painted in a coal cellar or in a prison cell with very little light coming through a small hole in the wall."[22] This dig was perhaps targeted at Whistler and his circle; Whistler's pupil Mortimer Menpes had claimed publicly that he and other followers of Whistler had adopted a "'nature from a coal-cellar' theory," referring to Whistler's metaphor of looking through a "chink" in a "coal-cellar" to clarify and purify the perception of nature to the absolute essential of "five tones, and no more." Robins observes that this exercise was designed to visualize flatness—to see the scene afresh without acquired knowledge—and derived from Helmholtz's theory of binocular vision.[23]

John responded to Quinn that in light of his friend's concerns regarding the coal-cellar effect, his current body of work, small panels done directly from the motif in Provence (fig. 55), would surely find favor. "I am awfully pleased with your remarks on colour," John wrote Quinn, "which I entirely sympathise with and all the more delighted as they make me sure you will be pleased with my recent stuff and the Provence things which are all in pure colour. No shit or smuts on them I hope. I am on my way I think to get back (or forward) to a purely delightful way of decorating which shall in no sense compete with the camera or the coal hole. But one has alot to unlearn before the instinct or soul or what ye call it can shine out unobstructed. Have you read Bergson, the Paris philosopher?"[24]

John was clearly eager to clear a space for his project distinct from both the Whistlerian inquiry into visual perception and the analogous inquiry conducted by Emerson. Like Whistler, he eschewed a priori knowledge of nature and such principles of organizing vision as complementary color contrast. By arguing for an innocent eye, to repurpose Ruskin's term, John distinguished his project from that of Sickert as described in his essay on the London impressionists. John declared that he would produce pure color, and therefore, he hinted, pure nature. Furthermore, untutored vision, according to John, would allow the "instinct or soul," a concept privileged by Bergson, to become the dominant motif of a work of art. Truth to nature, through John and Innes, was being recalibrated yet again, this time according to a whole new set of value judgments about the aim and ambition of painting.

John's reference to Bergson in his letter to Quinn—the same letter in which he praised Innes as "an entirely original chap . . . and a virgin intellectually" and extolled the "beauty of the Maoris at the White City"[25]—indicates that John, while attuned to his physical surroundings, had found a way out of the problem of realism that differed from the solutions proposed by Emerson, Clausen, Sickert, and Steer. In order to achieve the desired purity of light and color that he saw in Provence, John proposed to reject the paradigms of photography,

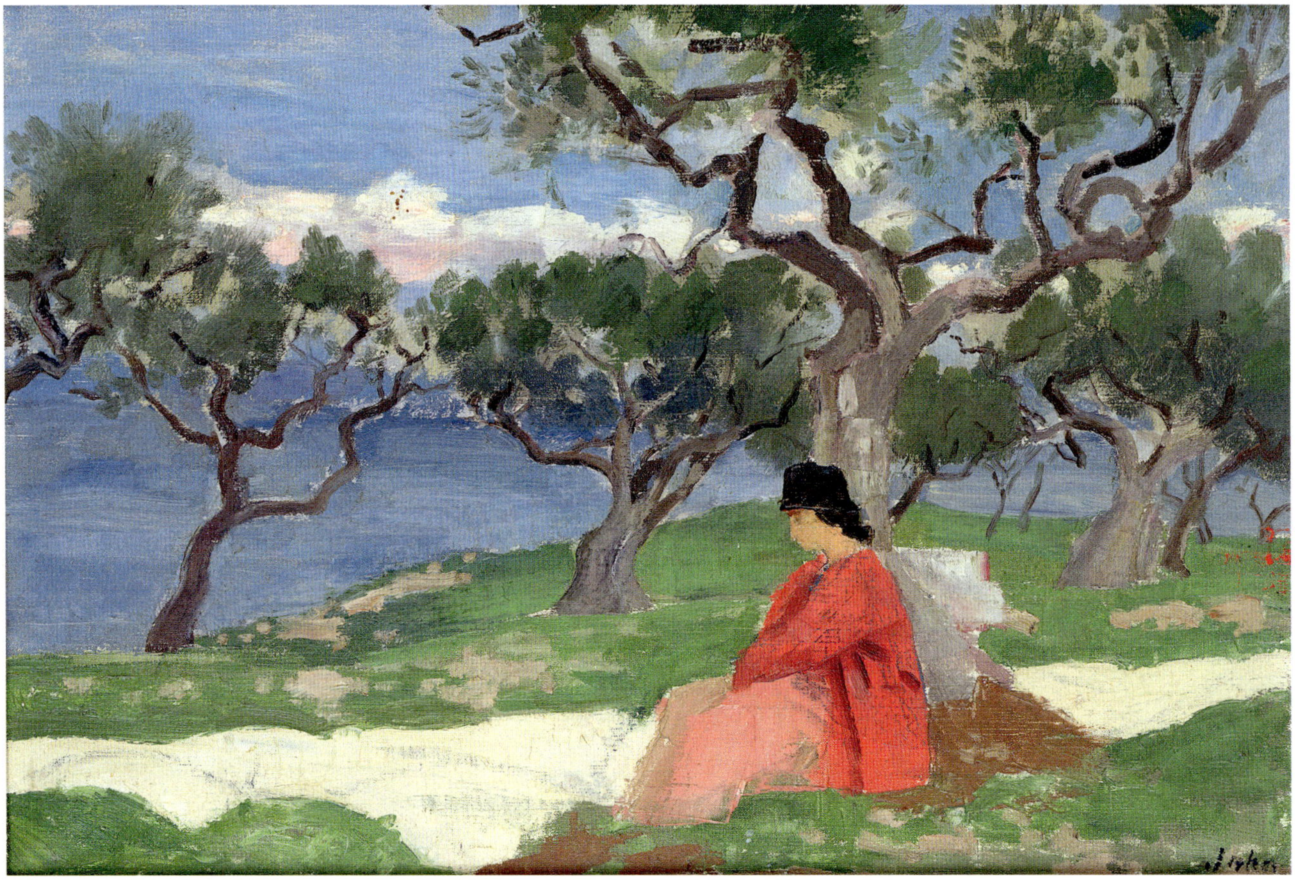

FIG. 55
Augustus John (1878–1961), *Blue Lake, Martigues*, ca. 1910–12. Oil on canvas, 38.5 × 55 cm. Private collection. © Estate of Augustus John / Bridgeman Artist Copyright Service.

naturalism, and impressionism; indeed, he wanted to "unlearn" in general in order to allow "the instinct or soul or what ye call it . . . [to] shine out unobstructed," a quasi-primitive gesture.

Quinn responded to John's question about Bergson by listing the books by Bergson that he owned. "William James," he continued, "all of whose writings I have read with interest, for years, was a great admirer of Bergson and thought him 'infernally clever.'"[26] This train of thought in their correspondence was drawn to an abrupt close as they became preoccupied with securing a painting by John that had been acquired by Frida Strindberg, who was initially reluctant to part with it.

Bergson was one of several turn-of-the-century writers, including William James in the United States, Edmund Husserl in Germany, and James Ward in Britain, who questioned the "mechanistic and positivistic analysis of nature," undermining scientific naturalism.[27] Intuition, for example, one of Bergson's key areas of investigation, was a troubling category for the naturalists. After showing early promise in mathematics, Bergson demonstrated an interest in philosophy and psychology. Around the turn of the century, he secured a prestigious post at the Collège de France, in recognition of the growing significance of his publications, which included *Time and Free Will* (1889, translated into English in 1910) and *Matter and Memory* (1896, translated into English in 1911). In 1907 he published *Creative Evolution* (translated into English in 1911), which caused considerable uproar and led to his growing reputation not only in France but also in Britain and the United States. By the outbreak of the Great War, Bergson had achieved notable influence in both intellectual circles and popular culture—in 1914, the *Times* of London reported that young girls fainted at his overcrowded lecture hall at the Collège de France.[28] He likewise lectured to crowded halls in Britain. In October 1911, he gave a series of four lectures on "the nature of the soul" at the University of London. In May 1913, he gave the keynote address in London as president of the Society for Psychical Research.

Like his intellectual forebears, among them James Sully, whom he cited in *Matter and Memory*, Bergson integrated aesthetics into his arguments, which helps to explain his significance for modernist literary and artistic practice in Britain and Europe just before the Great War. As Mark Antliff points out, "the cultural phenomenon known as Bergsonism" in France cannot be separated from the complex political situation of the republic; in Britain, his popularity was differently configured. To explain this phenomenon, Mary Ann Gillies describes the accessibility of Bergson's works in translation, his celebrity as a public lecturer, dating back to his first trip to England in 1911, and, most significantly, his engagement with highly charged issues then central to the British intelligentsia, including evolution, psychology, and metaphysics.[29]

Even without having read Bergson's books, Augustus John was likely to have known Bergson's work through one his most significant interpreters in London, the writer T. E. Hulme, who published extensively on Bergson in the pages of the *New Age* and other publications and translated the philosopher's *Introduction to Metaphysics* into English in 1912. The *New Age*, under the editorship of A. R. Orage and Holbrook Jackson, first appeared in 1907 and quickly established a reputation for engaging with issues of modernism. John, even when traveling, had the journal sent to him and also recommended it to Quinn.[30]

Hulme took the occasion of a review of William James's "Pluralistic Universe," published in the *New Age* in 1909, to expound on Bergson, whom James had recognized as an important influence.[31] James and Bergson were both engaged in investigating the Cartesian mind-body problem; more specifically, their writings teased out issues of perception and the role of the mind in discerning and constructing both external reality and internal selfhood. The latter bifurcated, Bergson argued in *Time and Free Will*, between a self understood through "deep introspection" and "constantly *becoming*," and the other lived "outside ourselves" and unfolded "in space rather than in time."[32]

In his article on James, Hulme categorized Bergson as the "exact antithesis of the Platonic metaphysics, where the changing flux is dismissed as appearance and reality is found in the stable concepts of the intellect."[33] One could add, the exact antithesis of positivism as exemplified by British inductive philosophy. Bergson argued that although science—biology, chemistry, psychology, and physics—was an expression of the intellect, it failed to understand life truly, staying only on the surface, offering only a "mechanistic conception of all nature."[34] As Hulme clarified, "rationalism," according to Bergson, was "the abuse of the power of translating the flux of immediate experience into a conceptual order."[35] Here, Bergson picked up on the same theme found in many of the Royal Academy banquet speeches by scientists: Science is associated with thought and intellect, but it fails to grasp the full complexity or emotional timbre of life. To understand and engage with life required intuition, according to Bergson, which he defined as "instinct that has become disinterested, self-conscious, capable of reflecting upon its object and of enlarging it indefinitely."[36]

This emphasis on intuition is of a piece with Bergson's analysis of differing modes of perception. In *Creative Evolution*, Bergson explained how flux—the trafficking of events and matter in everyday life—was typically explained as occurring outside the perceiving subject. The observer dissected fluid experience into small fixed parts in order to digest it, according to common understanding. But, Bergson proposed, what if perception was understood as taking place not with "the eyes of the intellect alone" but "with the spirit"?[37] This mode of perception—the intelligence installing "itself in the flux of reality by means of that sympathy that one calls intuition," as Hulme approvingly described it—was unencumbered by intellect or memory.[38] It could be fluid and holistic; the perceiving subject was part of, inseparable from, the flux.

In *Matter and Memory*, Bergson elaborated upon this point, explaining that "pure intuition" required giving up "certain habits of thinking and even of perceiving," including empiricism, which failed to fully grasp the inner life because "it can see in the self only a succession of juxtaposed facts."[39] What Bergson desired most of all was a fluid continuity between the inner and the outer worlds, between the perceiver and external reality.

Art was a crucial means by which this continuity might be achieved. In *Creative Evolution*, Bergson argued that the artist was uniquely able "to regain, in placing himself back within the object by a kind of sympathy, in breaking down, by an effort of intuition, the barrier that space puts up between him and his model."[40] The artwork was an act of both perception and articulation. Such a description of art making would have been attractive to John—for what drove him to pursue the life of the gypsy but a desire to break down the distance between himself and his models? It also validated his pursuit of light and color without resort to the types of scientific analysis that Emerson and Clausen had reveled in, with their diagrams and other means of investigating vision. Indeed, critics recognized John's desire to put the perceiving subject in the scene itself. The journal *Colour*, in 1917, expounded on the success of John's landscapes: "He is a *poseur* in a new sense: he poses nature by disposing her, and us inside her, according to his pleasure."[41]

The vibrancy of John's palette was one means of breaking down the "the barrier that space puts up between him and his model," as in *Llyn Treweryn* (1911–12) (fig. 56). In this scene of the Welsh mountains, the intense blue of the lake, echoing that of the sky above, creates immediate visual interest. The contrasting pinkish-white of the clouds helps to underscore the brilliancy of the blue tones throughout the painting. While John did not follow the rules of linear perspective, he nonetheless adopted compositional devices that point to the

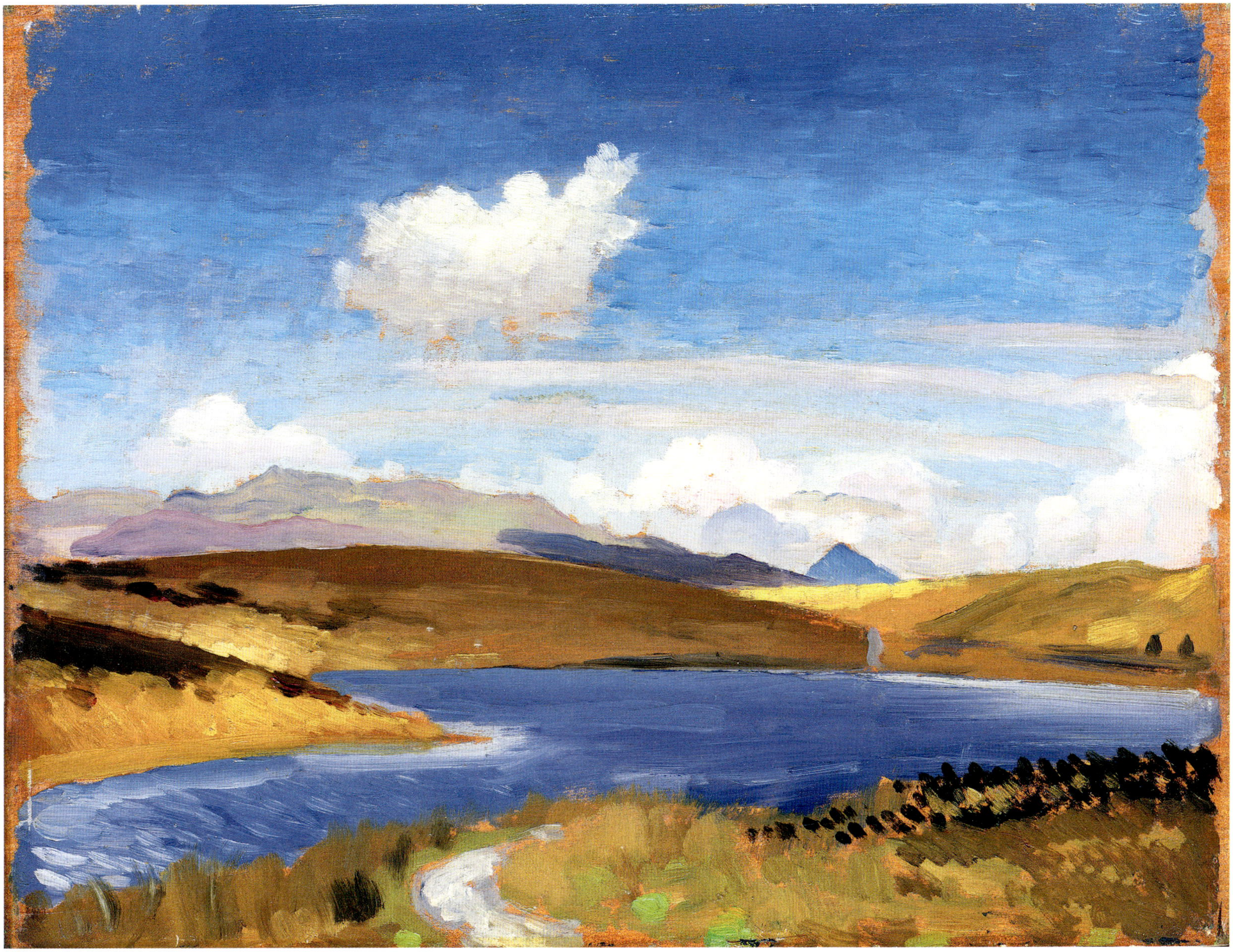

FIG. 56
Augustus John, *Llyn Treweryn*, 1911–12. Oil on wood, support 316 × 407 mm. Purchased 1932. © Tate, London 2015. © Estate of Augustus John / Bridgeman Artist Copyright Service.

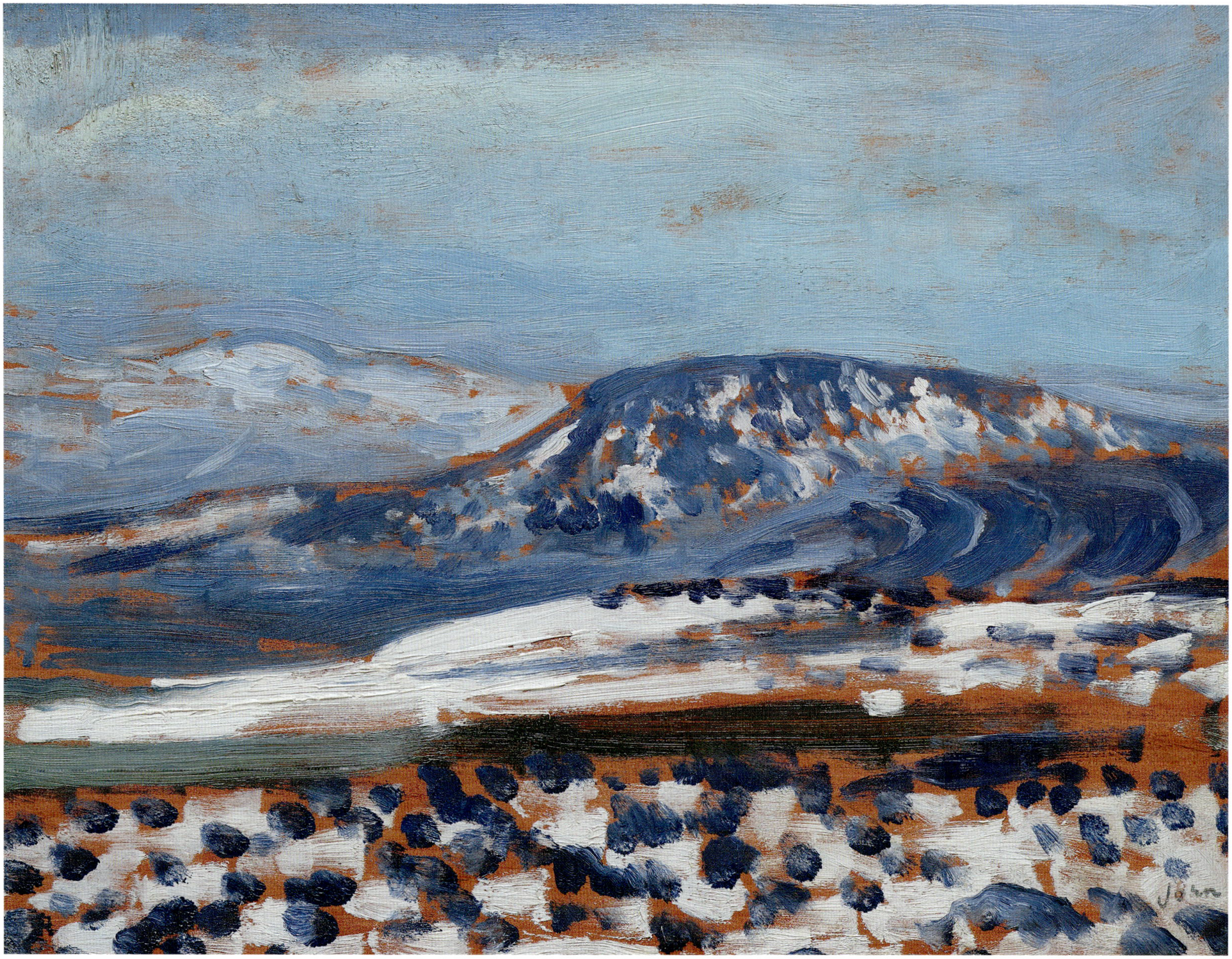


FIG. 57
Augustus John, *Welsh Mountains in Snow*, 1911? Oil on wood, 30.4 × 39.6 cm. Art Gallery of New South Wales, Gift of the Contemporary Art Society, London, 1944. © Estate of Augustus John / Bridgeman Artist Copyright Service.

heart of the scene, such as the small path curving from the immediate foreground to the lake and the bands of clouds that lead to the bluish triangular peak seen protruding above the green-brown hills surrounding the lake and the river that feeds it. At the same time, bold brushwork, by which forms are suggested rather than transcribed, unifies the picture plane and directs attention across the whole scene. These strokes, as in the black ovoids along the bank of the lake at right, make visible the artist's touch.

Generalized treatment rather than specific details was in keeping with Bergson's conception of intuition, which he contrasted to positivistic analysis.[42] The scientist, according to Bergson, understood "that nature contains only facts, and that, the facts once posited, the relations are simply the lines running between the facts." The intuitionist, by contrast, believed that the perceiving subject could not be separated from that which is perceived. Thus the self understood, according to Bergson, that "I can as well claim that it is the form of thought that has determined the shape of the facts perceived, and consequently their relations among themselves." That is, the fact does not exist outside the intuitive mind; the material world, therefore, is "a simple flux, a continuity of flowing, a becoming."[43]

The notion that a successful work of art was the result of merging the inner and outer worlds would have suited John, whose writings are distinctly unencumbered by the question of how to represent external nature accurately, free of concern with fact or detail, and more interested in living life than strenuously observing it. The question for him was how to let his soul shine out while also representing the natural world. In short, he was untroubled by "truth to nature" as understood by the Pre-Raphaelites, recognizing that nature or external reality could not be separated from the perceiving self—both were part of a continuous flow.

That sense of flow—of an intuitive response to the scene—and of painting released from habitual ways of seeing and representing is conveyed by another of John's Welsh landscapes from this period, *Welsh Mountains in Snow* (1911?) (fig. 57). The foreground is composed of a band of loosely articulated and overlapping dots of white and blue, achieved by loading the brush with paint and touching the panel briskly with it, then quickly lifting it up—signifiers of the artist's presence as well as a rocky topography. This band does not recede into space according to an orthogonal system of linear perspective, although a slight recession is indicated in the middle ground by sweeping curves of paint at right that assist in the transition from foreground to background and also suggest rising hills. The bright palette also resists traditional conventions of aerial perspective—the blues of the foreground echo those of the background and thus flatten space. The tannish-brown underpaint visible throughout the panel, from the foreground band, through the hills in the middle ground, to the sky strewn with clouds, also insists upon the two-dimensionality of the painted scene and gives the lie to painting as a three-dimensional illusion of external nature. The panel proclaims that it is both a painted decoration and the product of intuition. The play of tan and blue in relation to a dynamic rendering of space that is both illusionistic and resists that illusion also recalls the work of Cézanne, who likewise sought to privilege his own sensations of nature in his painting.[44]

John's panel, with its echoing curves of paint to demarcate the hills and the scattered touches of paint in the foreground, also evokes a strong sense of rhythm associated with the Bergsonian sensibility. In 1911, around the time that John was corresponding with Quinn, the art critic Huntly Carter explored the link between new directions in art and Bergson's philosophy in a review of the Société des Artistes Indépendants exhibition for the *New Age*. "The new form of art is the perception and expression of continuity," wrote Carter. "The new intuition is the apprehension of Reality underlying forms of life, of things living and evolving."[45] Modern art, according to Carter, should be concerned with evoking the knowledge of energy and life pulsing through nature, sensed or intuited rather than optically acquired. It is particularly striking that Carter did not mention Bergson by name in his review. Nonetheless, it is evident that readers understood precisely to what framework Carter referred; the writer and critic John Middleton Murry submitted a letter to the editor in the following issue in which he confirmed Carter's assessment and explicitly linked the "Conscious-Intuitional Movement" then discernible in Parisian culture to Bergson. Moreover, Murry hoped that British critics would come to grips with "the implications of Bergsonian theories of time and change," so that "creative art would receive a new lease on life."[46]

Carter's review also drew attention to the artists' concern with rhythm: "On all sides artists were engaged in processes involving the production of inspired pieces of music, putting together new pictorial material, composing lyrics in colour, lyrics in line, lyrics in light to the new deity, rhythm."[47] This emphasis on rhythm reflects Bergson's theories of aesthetic feeling, as explained in *Time and Free Will*, in which Bergson argued that rhythm was the best way for an audience to enter into sympathy with a work of art. Rhythm, because of its repetition, "establishes a kind of communication between him [the performer] and us," and allows us to "suspend the normal flow of our sensations and ideas by causing our attention to swing to and fro between fixed points." In short, such art inserted the spectator directly into the perception of flux and suspended habitual ways of receiving and perceiving sensations. This art should not be mimetic because, Bergson argued, "imitation of nature" was not the most "efficacious" means of "impressing feelings on us." Moreover, nature did not, in and of itself, possess rhythm. The artist's task, according to Bergson, was not to imitate nature but to give viewers "a share in this emotion, so rich, so personal, so novel," thus breaking down "the barrier interposed by time and space between his consciousness and ours."[48]

Bergson's emphasis on rhythm as a direct route to intuition inspired the founding of the British journal *Rhythm*, established in the summer of 1911 by the writer Murry and the painter J. D. Fergusson. The journal clearly staked its intellectual agenda on Bergson, who was the focus of the opening essay, written by Murry. Like Hulme, Murry understood Bergson's philosophy as a validation of the significance of the subjective self. "Human personality," Murry explained, "by the consciousness of reason drags reality into its knowledge, into itself, the rational self asserting its ultimate triumph over an externality, which is, while merely external, meaningless." External reality possessed significance only when it was internalized by the self, and it could be known only through the self; the detached observer did not exist. Nature, in this system, had no truth-value in itself because it could be truly known only through the apprehending self. Nature, in short, was not the final or universal arbiter of truth—art was: "Art is the true and only expression of reality."[49]

Art, according to Murry, should be arrived at through intuition; the artist should "return to the moment of pure perception to see the essential forms, the essential harmonies of line and colour, the essential music of the world." Through this process, the "unessential dissolves away—memories and that false knowledge." The artist's perception "penetrates beneath the outward surface of the world, and disengages the rhythms that lie at the heart of things, rhythms strange to the eye, unaccustomed to the ear, primitive harmonies of the world that is and lives."[50] Uncannily, Murry zeroed in on the same problem that had plagued Clausen—how to arrive at what is essential—but he proposed a solution that was not yet available to the painter.

John's letter to Quinn of June 1911, in which he yearned for the instinct or soul to shine out unobstructed in his paintings, suggests that he was driven by a desire to arrive at the moment of pure perception that Murry described. The treatment of the foreground in *Welsh Mountains in Snow* perhaps registers the moment at which John came closest to the Bergsonian project. His patterned touch and flattened space help us perceptually merge with the scene and experience the artist's delight in the open air and rugged scenery. Disengaging from the urban scene and traversing the mountains of Wales or the rocky coasts of southern France enabled John to rethink processes of perception.

Through such approaches to painting landscapes, John drew close to the understanding of Bergson expressed in the work of the British circle around the journal *Rhythm*—Jessica Dismorr, J. D. Fergusson, S. P. Peploe, and Anne Estelle Rice, who worked together in a quasi-colony in the Mediterranean—and the Paris-based Puteaux cubists.[51] Mark Antliff describes how the Puteaux cubists rejected "Albertian perspective." They "*distorted*, rather than mirrored, a qualitative, durational reality that could only be grasped through intuition" and "substituted rhythmic *passage* for the quantitative measurement of space."[52]

In addition, John would have sympathized with the rhythmists' position vis-à-vis notions of masculinity and femininity. As Antliff has discerned, the rhythmists perceived women and "female fecundity [as] an example of the élan vital's productivity. Women were the product of this élan, but men, by virtue of their intuitive powers, were both product and producer of the élan vital."[53] This philosophy translated into a liberated sexuality for Fergusson and his compatriots in Paris, as it did for John and Innes. It can also be seen in landscape scenes in which women are shown as intimately connected to nature, as in John's *Blue Lake, Martigues* (ca. 1910–12) (fig. 55). The female figure is figured as an organic part of the landscape, just as rooted to the earth as the trees around her. The paint rendering the female form is undifferentiated from that of the landscape; the strokes of red and pink that form her jacket and skirt are akin in breadth and handling to those composing the bright tan path and gray rocks protruding from the green grass.

A critic for the *Saturday Review*, writing in 1913, read such handling of paint, which emphasized overall pattern and effect rather than exacting details, as a mark of spontaneity and truth. It enabled John to reconcile "even the most unlikely figures with their background. Indeed, it seems indivisible from the very nature of art, implying unself-consciousness and truth, and manifesting itself as ceaseless newness."[54] Truth to nature is achieved here not through the carefully calibrated and visibly laborious application of paint to canvas, as in Brett's paintings, but through the illusion of a quickly captured scene that suggests the artist's bond with the flux of the external world.

That impression of quickness, to the point of disregard of artistic conventions, was also significant for Innes's practice. In a letter to John Fothergill in 1909, Innes supplied a short anecdote that underscored his seemingly cursory technique: "A man came in to see my work the other day and said of one of my sketches 'That's a good slap at nature.' Slap in the face for nature said I. at this he went red all over and was quite embarrassed."[55] Fothergill's recollections of the painter further entrench Innes in the Bergsonian paradigm, as when Fothergill recalled that "nature had no fears for him—he made what he liked of it, he used it freely for expressing his many and marked moods."[56] The natural scene was a screen onto which Innes projected himself.

Innes likewise rejected scientific analysis; induction held no charm for him. As far as he was concerned, the new disciplines of geology and astronomy and their related modalities of vision were vehicles for romanticism—means to envision worlds other than his own, not to see his own more clearly. He wrote to Fothergill about his father, who was "sometimes getting a headache . . . at other times going of into dreamlands of geology and astronomy (which seemes to be fosselised poetry) where it is wonderfull to follow him. He tells me that with present telescopes they can see the moon as if we were only 30 miles from it and that there are great canals across Mars as if the life on it were making a last effort to collect water to keep itself alive as it is all dry except the poles."[57]

Innes's romantic projections were perhaps strongest with respect to Mount Arenig, in Wales, where he reportedly buried a silver casket containing the effects of his relationship with Euphemia Lamb, then married to the painter Henry Lamb.[58] Two scenes of the mountain reveal how Innes was able to convey different moods, albeit relying on different points of view and conditions to do so. In the version from 1911 (fig. 58), the handling of the scene is particularly childlike in its simplicity. This sense of naïveté suits the energetic sensibility of the work, suggesting that the artist spontaneously reacted with joy to the scene before him. The rocks that dot the left foreground and compose the low stone wall at right are registered with only a few strokes of paint each, rather than through carefully modulated chiaroscuro, with darker hues laid boldly alongside lighter ones to suggest volume. The clouds that cling to the low hills in the middle ground and the foot of the peaks in the background, and those that streak throughout the sky, are rendered with obvious rounded strokes or, in a few places, dabs of the brush loaded with paint. This technique of creating the illusion of forms with relatively unmodulated passages of paint is seen most starkly at far right, in the touches that suggest shrubs along the wall and the topography of the distant hills. The color equally endows the scene with a sense of exuberance. The mountain peaks are cloaked in red, set off by the dark blue folds of ground just below. The pink clouds, stretching out behind the darker blue-gray cover, provide a note of dynamism through the contrasting colors. The green of the immediate foreground is set off by the tawny brown middle ground. Each band of landscape plays off the adjacent field.

The 1911 painting, with its assertive brushwork, radiating rays, vivid color contrasts, and mood evocative of the dawn of a new day, creates a sense of energy, whereas Innes's scene of Arenig from 1913 has a decidedly more somber tone (fig. 59). The overall blue cast of the painting helps to project this mood, although again there are touches of pink in the distant clouds, and the mountain caps contrast in reddish tone against the darker flanks of the slopes. The clouds, rather than

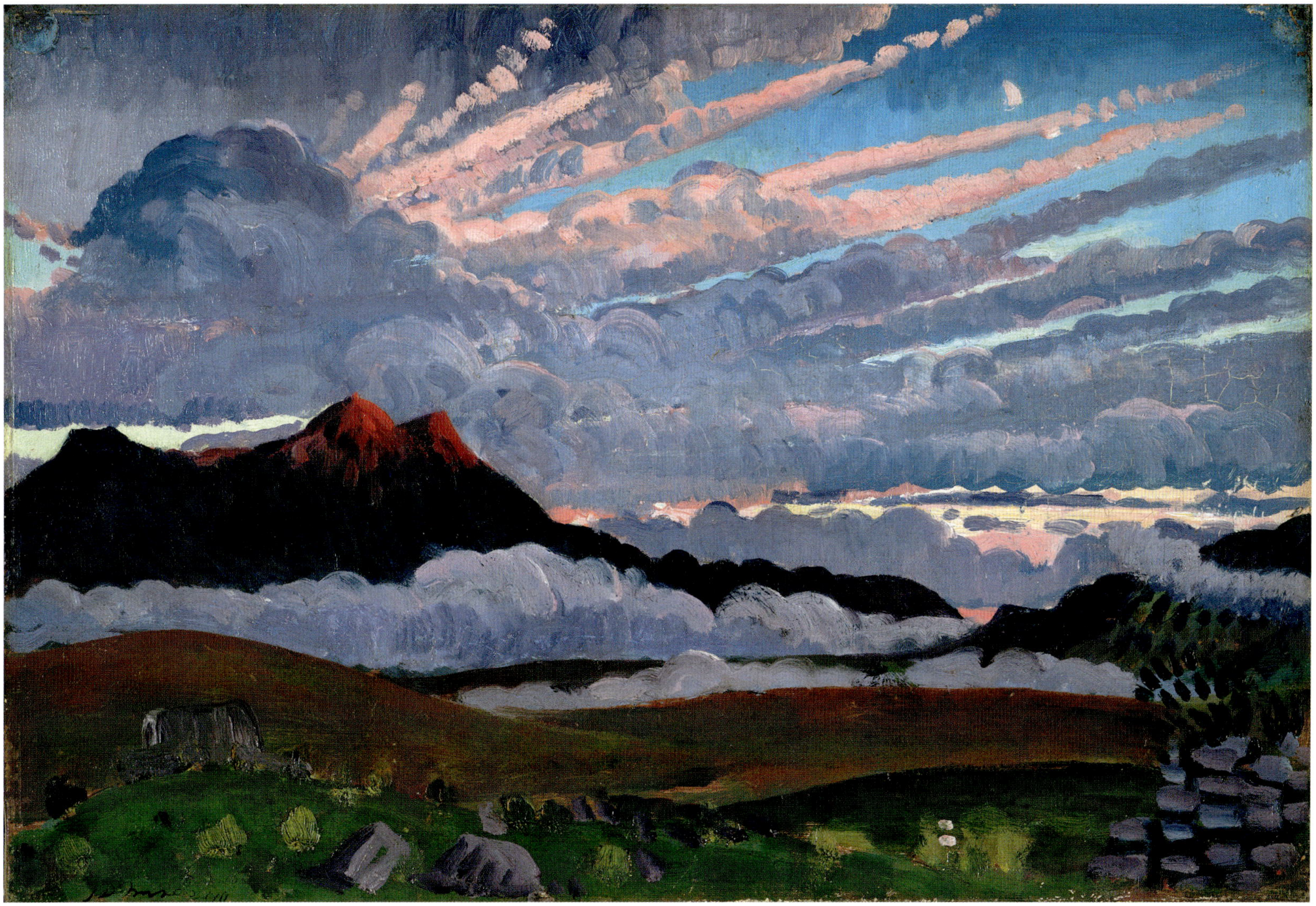

FIG. 58
James Dickson Innes (1887–1914), *Arenig*, 1911. Oil on canvas, 36 × 51 cm. Gift of the Massey Collection of English Painting, 1946, National Gallery of Canada (acc. no. 4774).

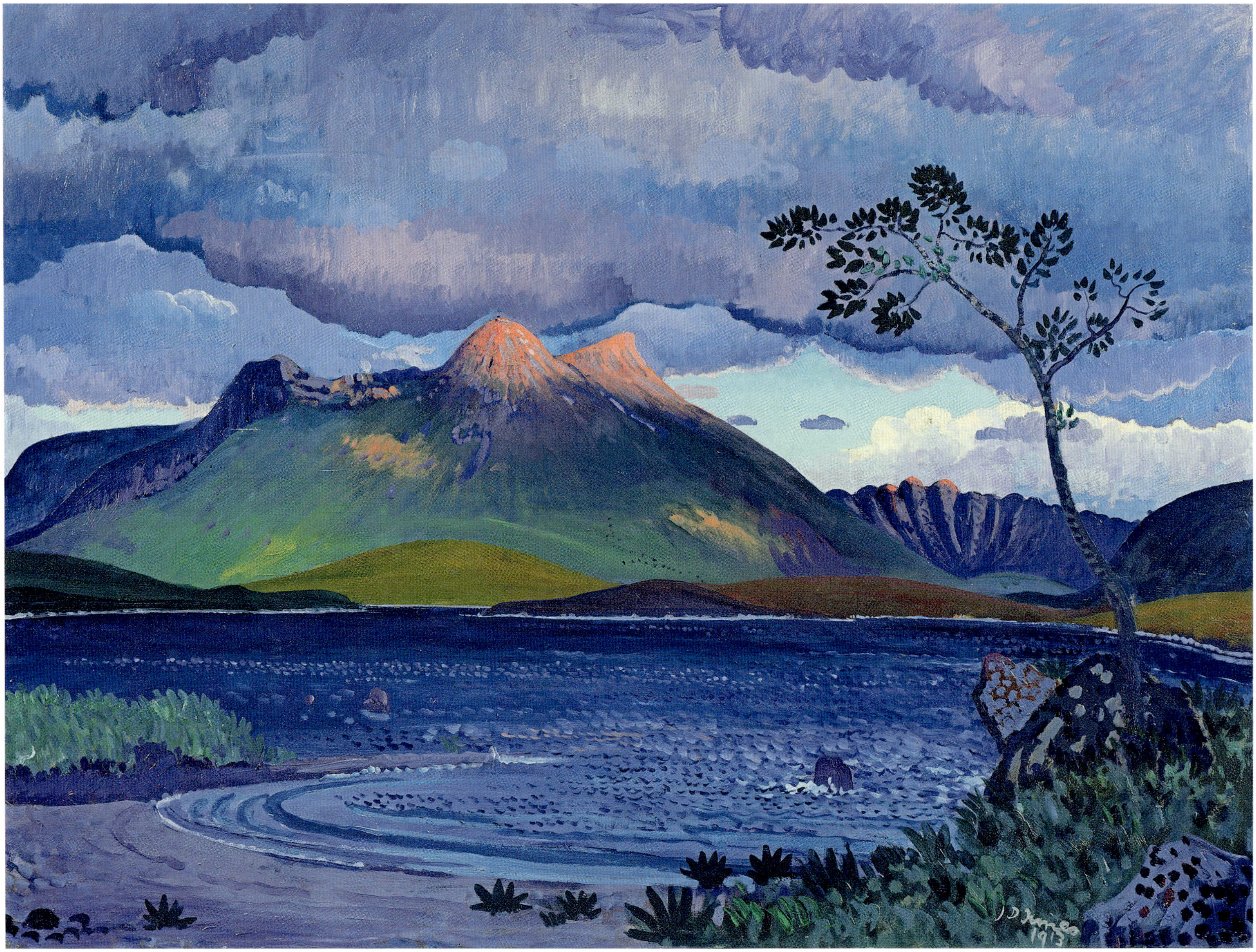

FIG. 59
James Dickson Innes, *Arenig, North Wales*, 1913. Oil on plywood, support 857 × 1137 mm. Presented by Rowland Burdon-Muller, 1928.

suggesting outward-projecting energy, hover above the mountain, nearly echoing the ridgeline. The mood of isolation is underscored by the absence of signs of humanity. The sparseness of the scene, with its boldly simplified forms, helps to direct attention to the slender tree at right; its fragility is reinforced by its juxtaposition against the more massive mountains.

Like John, Innes tried to create the impression of an untutored vision, an effect achieved most successfully via his brushwork, composition, and color choices. The reception of his work, however, created a paradox that dampened its possible Bergsonian overtones. Laurence Binyon, who worked as assistant keeper of prints and drawings at the British Museum while pursuing a literary career and maintaining friendships with progressive artists such as Sickert and William Rothenstein, recognized Innes's connection to the current trend of post-impressionism in a review of Innes's exhibition at the Chenil Gallery in 1911. But Binyon, who was developing his expertise in Asian art at the time, also complained that Innes's act of simplification was "imposed on nature from without rather than an extraction of the nude and strong essentials." Binyon understood that this process of extraction or distillation should proceed from an intuitive response to the scene, and his description of this methodology was clearly informed by Bergson. "It is the eye of the seer piercing through to the energies behind appearances," he wrote, "which we crave for our enlightening; we want realities; and we measure the artist's mind by his capacity to ignore the insignificant things, and to seize and understand the big things, the things that matter."[59] Binyon, like Carter, believed that the ultimate reality was the underlying life force or energy. He clearly understood the system in which Innes hoped to matter but found that he fell short. And, provocatively, like Clausen before him, Binyon wanted the artist to identify the essential—eschewing the Ruskinian dream of rejecting nothing, selecting nothing, and scorning nothing.

Critics often used Innes's work as an opportunity to search for artistic influences, a strategy that countered notions of spontaneity and personal authenticity because it implied that Innes's action of seeing was informed by prior models or systems of art making that he imposed upon the scene. John Quinn, for example, complained to John that he did not care for two of the four Innes watercolors that John had selected for him because they seemed to be based on Japanese prints, a point that John admitted, though he countered that the influence might have been unconscious, given the ubiquity of Japanese prints at the time.[60] O. Raymond Drey, in a review of the New English Art Club exhibition of 1913, in which Innes showed several works, described them as "pleasantly coloured pictures in his pleasant Chelsea-Japanese manner." Furthermore, Drey charged that "until he has carried his experiments in this means of expression much further than he has done, it will be impossible to know whether anything profound underlies it," a critique all the more biting for its appearance in the pages of *Rhythm*.[61]

Most frequently, critics attempted to assimilate Innes within the British landscape tradition. Innes's friend John Fothergill repeated in numerous texts the anecdote that "the Turner watercolours, then in the National Gallery, were his ideal and almost daily visitation, and, so long as I knew him, he always carried a book of Constable's coloured reproductions and the *Liber Studiorum*."[62] This associationist strategy was verified by the National Gallery's decision, in 1921, to couple its retrospective exhibition of Innes's work with a reinstallation of works relating to J. M. W. Turner's *Liber*

*Studiorum*. The catalogue accompanying the exhibition situated Innes within both postimpressionism and the genealogy of the landscape tradition in Britain. In its notice of the exhibition, the journal *Colour* observed of Innes, "His importance is that he was one of the leaders in England in the break-away from Impressionism towards a formalized treatment of landscape. The tendency, though 'modern' in its reappearance, is not without tradition, since it is found in the works of both Cotman and Wilson."[63] This notion of a "formalized treatment" revisits the concept of a system for producing landscape, such as those developed by Cozens and Cotman, which is antithetical to a highly personalized reaction to nature.

Innes's conundrum was not unlike that faced by Emerson, who wanted to claim direct, unmediated vision in his photography and simultaneously to validate his project by associating it with recognized landmarks in the history of art. But such quotations, as Binyon pointed out, potentially undermined claims of intuitive response. Innes's quandary also indicates how strong a hold the British landscape tradition had on the contemporary imagination, a purchase reinforced by the easily accessible displays in the National Gallery, the British Museum, and the South Kensington Museum. The task that the Pre-Raphaelites and the New English Art Club had set themselves—of ridding their art of convention in order to claim authenticity and truth to nature—was still proving difficult for succeeding generations to achieve.

## Charles Ginner: Redefining Realism

In 1914, Charles Ginner entered into the debate about truth to nature in art with his polemical essay "Neo-Realism," published in his exhibition catalogue and in the *New Age*. This essay provoked a controversy fueled by the whirlwind of new styles and modes of expression that swept through London exhibition halls in the years just before the Great War. Ginner had been born in France and had initially studied engineering and architecture before turning to art, attending Académie Vitti and, briefly, the École des beaux-arts. In 1909, he settled in London. The opportunity to see the work of Van Gogh was decisive for his painting practice, encouraging his use of bright colors and complementary contrasts and an expressive application of paint. Unlike Van Gogh, who turned his back on Paris and located his utopian studio in the rural south of France, Ginner chose to root his practice in the urban sphere. This direction was confirmed by the members of the Camden Town Group, who situated their paintings in the northern, downtrodden district of London so as to offer an authentic vision of quotidian city life. Even potentially shocking murder scenes were treated with deadpan dispassion in the hands of Walter Sickert.

Ginner, however, often avoided much of the grime and noxiousness of the city in favor of pursuing Whistlerian nocturnal transformations, as in *Evening, Dieppe* (1911), and spaces of *rus in urbe*, as in *Battersea Park, No. 2* (1911), *Neuville Lane* (1911), *Victoria Embankment Gardens* (1912), and *Leicester Square* (1912). When he took up more straightforwardly urban scenes, he delighted in the tightly knit fabric of the city rather than its sordid passages: the forms and lines created by buildings, telegraph and telephone wires, flag and light poles, and iron railings, and the shifting patterns of light and dark created by the movement of the sun through the density of the built environment. Such features are at the heart of *Leicester Square* and *Victoria Station, London, the Sunlit Square* (1913) (fig. 60).

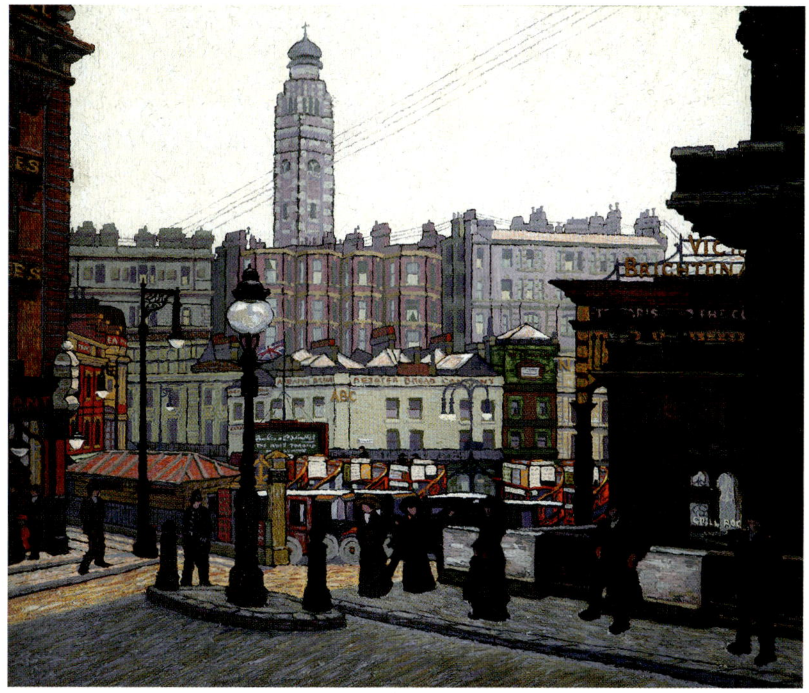

FIG. 60
Charles Ginner, *Victoria Station, London, the Sunlit Square*, 1913. Oil on canvas, 75 × 87 cm. Reproduced by kind permission of The Atkinson, Lord Street, Southport, U.K.

In 1913, Ginner first visited Leeds, home to Michael Sadler, a significant patron of modern art. Ginner portrayed the city from a low point of view, looking up at the factories and houses that lined the hills, rendering the urban fabric as patterns of geometrical forms that extend from the small scale of painted windows and contrasting bricks to the larger scale of the built environment. The manufacturing center thus appears as a well-oiled machine in such paintings as *Leeds Canal* (1914) (fig. 61). No matter the subject, a strict sense of order governs his work, the product of his technique of developing his final oils from squared-up sketches (fig. 62). These sketches were created, Robert Upstone explains, by using "a homemade viewfinder with a grid, transferring each element to paper."[64]

Ginner sketched directly from the motif, taking his inspiration from the observed scene. The viewfinder was not intended to improve his vision in the sense of seeing what might not be visible to the naked eye, as in the case of Brett's use of telescopes. It was intended to help him organize his vision, to select and to frame a scene and then to find its underlying pattern or rhythm. The tool also endowed the composition with a sense of clarity, as the attention Ginner devoted to each square of the grid lends a material concreteness to every depicted object. The weightiness of his painted scenes, even in the clarity and structure of the shadows, makes his work wholly distinct from both impressionism and naturalism, a point he underscored in his short essay "Neo-Realism."

In 1910, Ginner met Harold Gilman through their joint participation in the Allied Artists' Association, an exhibiting society founded on democratic principles (no jury) by the art critic Frank Rutter in 1908. The association's hanging committee was organized alphabetically, bringing together Ginner and Gilman, along with Spencer Gore. Gilman had trained at the Slade (around the same time as John), studying with Steer, among others. He exhibited early in his career with the New English Art Club and then participated in the founding of the Fitzroy Street Group, an informal artists' group formed around 1907 at the urging of Walter Sickert. Assembling a group of younger painters, including Gore and Ginner, for weekly gatherings in his studio, Sickert aimed to reinforce new directions in painting and to bring his colleagues into contact with amenable buyers.

The Camden Town Group, formed in 1911, was in many ways an outgrowth of the Fitzroy Street Group,

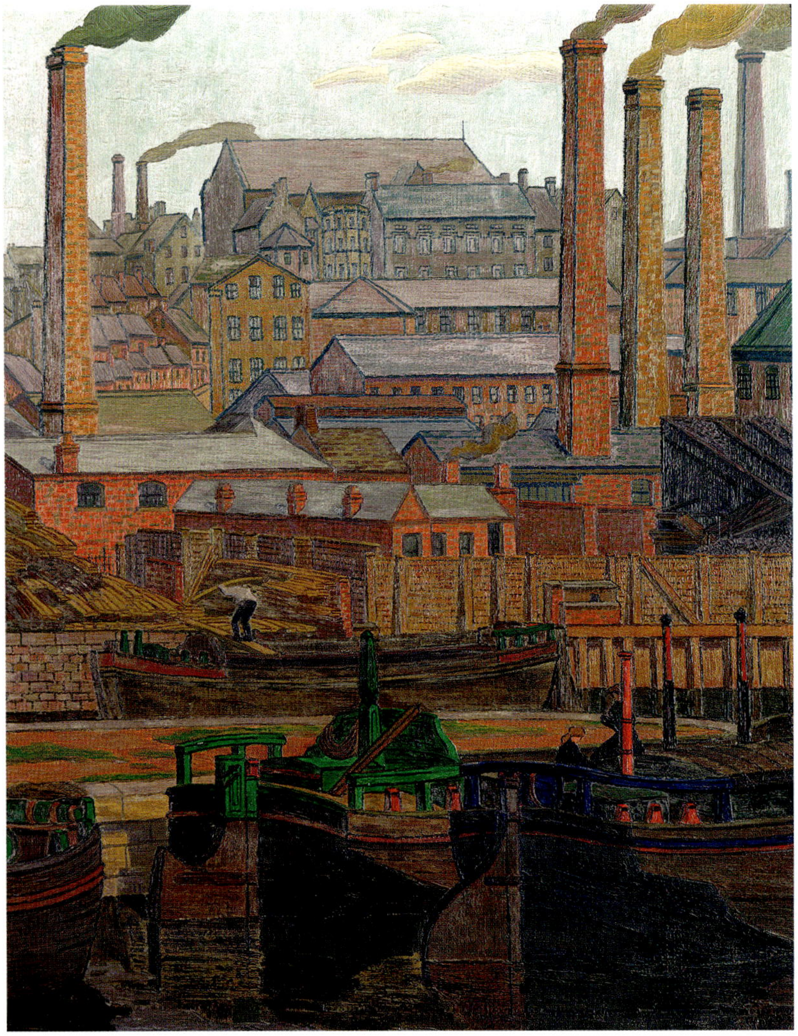

FIG. 61
Charles Ginner, *Leeds Canal*, 1914. Oil on canvas, 73.6 × 58.4 cm. Leeds Museums and Galleries (Leeds Art Gallery), U.K.

although the Camden Town membership was formalized, in contrast to the informal sensibility of the Fitzroy Street Group. The Camden Town Group came to include, in addition to Sickert, Ginner, Gilman, Robert Bevan, Spencer Gore, Walter Bayes, Malcolm Drummond, J. D. Innes, Augustus John, Henry Lamb, Wyndham Lewis, James Bolivar Manson, Lucien Pissarro, William Ratcliffe, Doman Turner, and Herbert Lightfoot (who was replaced after his death, albeit temporarily, by Duncan Grant).[65] The group organized a series of exhibitions at the Carfax Gallery, where William Rothenstein served as artistic advisor, in June and December 1911 and December 1912. As Ysanne Holt notes, critics perceived new trends in the group's work, but no single theory dominated.[66]

Gilman's oeuvre is harder to characterize than Ginner's; Gilman produced figural interiors early in his career and continued to do so throughout his life, adding scenes of everyday life, such as urban markets, to his repertoire. Gilman's choice of subject and composition was initially shaped by his admiration of Whistler and later Van Gogh, a preference that he shared with Ginner and that was encouraged by Roger Fry's *Manet and the Post-Impressionists* and a trip to Paris with Ginner around 1910. Gilman's *Canal Bridge, Flekkefjord* (ca. 1913) (fig. 63), for example, echoes Van Gogh's renderings of the Arles drawbridge, the Pont de Langlois.[67] Gilman's drawing for this painting reveals the use of a heavy line to define forms, a practice reminiscent of drawings by both Ginner and Van Gogh (fig. 64). The annotations to the drawing not only indicate that it was rendered from the motif but also underscore Gilman's interest in color. Although the drawing was executed on the spot, the final painting was not intended as a transcription of the original scene. The alphabetical labeling of portions of the drawing and the corresponding color code (found in the upper right quadrant of the drawing) suggest that Gilman was concerned with using repeated colors to establish a strong unifying pattern.

*Eating House* (ca. 1914), a subject also derived from Van Gogh, reveals that Gilman, like Ginner, used a framing device. To transfer the initial drawing to canvas, Gilman adopted a perspective frame—a string grid—seen in the faintly rendered squares repeated across the canvas (fig. 65). These squares create a sense of order sympathetic to the inherent linear qualities of the scene established by the patterned wallpaper and the horizontals and verticals that make up the booths of the restaurant.[68] This order endowed Gilman's paintings, even those of bustling city squares, with a sense of "static order" that David Peters Corbett finds indicative of the Camden Town painters' concern with "reducing social action to the function of elements in the overall design" and with translating the social experience of modernity into "the application of paint."[69]

The perspective frame suggests a methodology of rigor and self-discipline akin to that associated with Pre-Raphaelitism and can be read as translating nature and landscape into an object clearly separate from the observing subject. But in fact this method was understood as a means of unifying perceiver and subject. This can be seen in the series of landscapes Gilman produced in Gloucestershire. In the painting *In Gloucestershire* (ca. 1916), Gilman attempted to rethink the figure-ground relationship by using broad, intermingled touches of paint to render leaves and sky so that passages of gray-blue, suggestive of sky, lie on top of areas of green, associated with leaves (fig. 66). This maneuver is reminiscent of Cézanne in flattening out the space of the scene. The sense of flattened patterning is underscored by underpassages of black that set off the luminosity of the shrubbery around the base of the tree and the dark outline used to render the tree, a cloisonné-derived technique that references Gauguin.[70] Louis Fergusson later recalled that in these paintings Gilman had been able to achieve a deeply felt connection with the scene: "You felt that Gilman planted himself in the heart of the woods and apprehended them passionately. He searched out the secrets of their being, and, however objective the

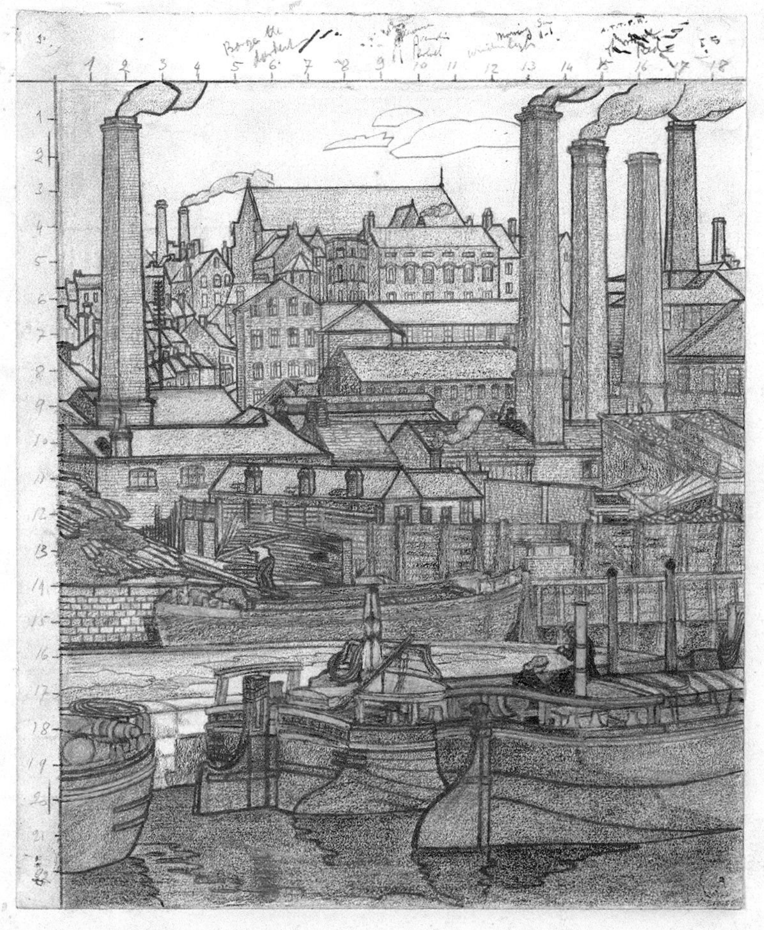

FIG. 62
Charles Ginner, *Leeds*, ca. 1914. Pen and brown ink and colored chalk mounted on cardboard, 31.6 × 25.3 cm. (Registration no. 1959-0011-3.) Purchased 1959 with Mary Buick Bequest Funds. Courtesy of the Museum of New Zealand Te Papa Tongarewa.

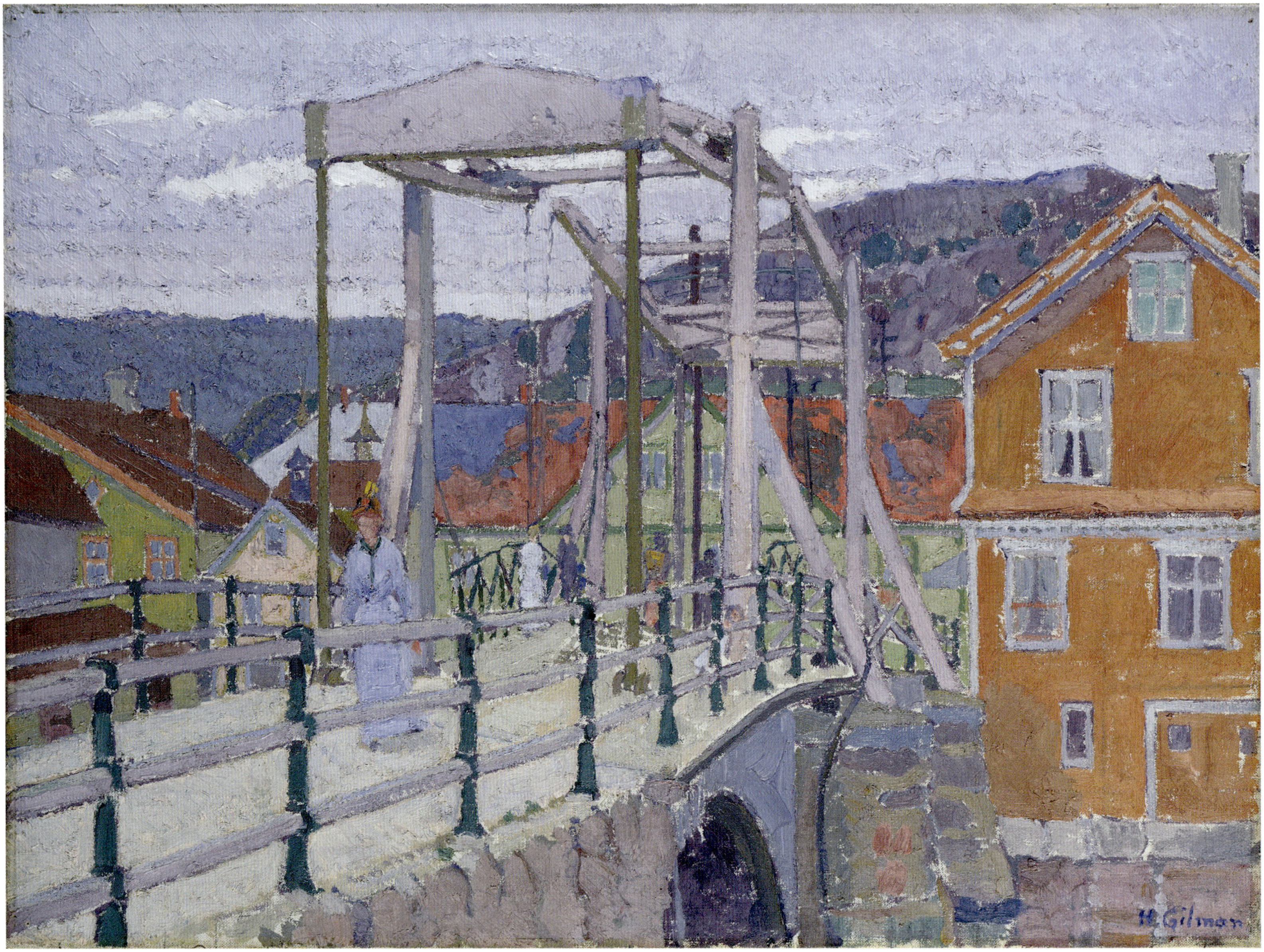


FIG. 63
Harold Gilman (1876–1919), *Canal Bridge, Flekkefjord*, ca. 1913. Oil on canvas, support 464 × 615 mm. Purchased 1922. © Tate, London 2015.

transposition, design never lost the whip-hand of representation."[71] The concept of transposition was explicated in Ginner's essay on neorealism.

Ginner's essay "Neo-Realism" appeared in the *New Age* on 1 January 1914, just when the affiliations of the younger generation of modernists were again being reconfigured. In the fall of 1913, the Camden Town Group met to reassess its strategies, frustrated by the limited space available at the Carfax Gallery.[72] From this meeting and others held that fall there emerged a new exhibiting society, the London Group, with Gilman as president, Robert Bevan as treasurer, and James Bolivar Manson as secretary. As in the case of the New English Art Club, artistic factions vied for control of the group. Duncan Grant, John, and Innes bowed out even before the new group was formalized. Pissarro, Sickert, and Lamb resigned before the first exhibition in March 1914.[73] Sickert felt increasingly embattled. He was at odds with the two emerging factions discernible in the exhibition *English Post-Impressionists, Cubists, and Others* (December 1913–January 1914), which presented the Fitzroy Street Group and Camden Town Group as one constellation and the cubists as another.[74]

Ginner's essay "Neo-Realism" was his attempt to wield the weapon of publicity that the older generations—exemplified by Whistler and Sickert—had used so cunningly and that his contemporaries were adroitly employing. It was the vehicle by which the painter defined his version of realism: "the plastic interpretation of Life through the intimate research into Nature." The artist should bring to this task "his personal ideal, his personal vision of nature and attitude towards life." Ginner's concept of neorealism is saturated with Bergsonian "*intuition*" or "*intellectual sympathy*," whereby "one places oneself within an object in order to coincide with what is

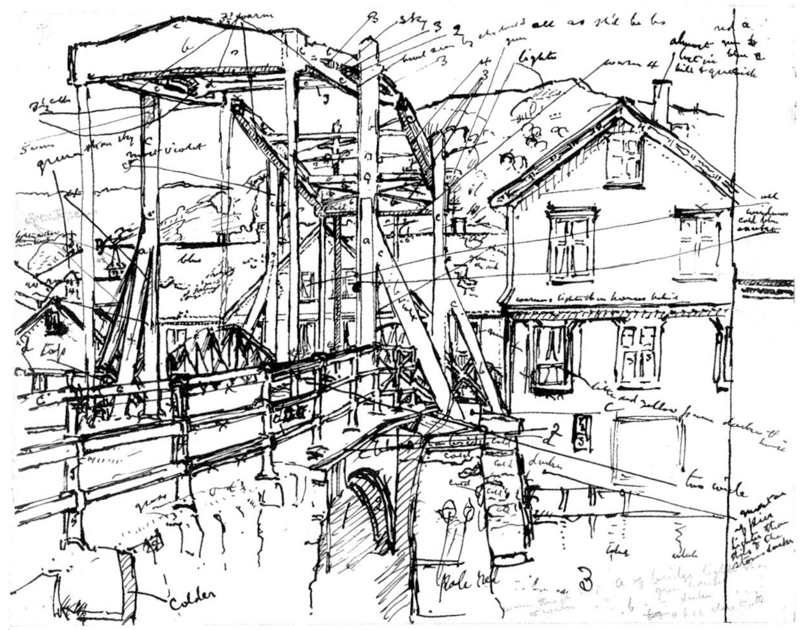

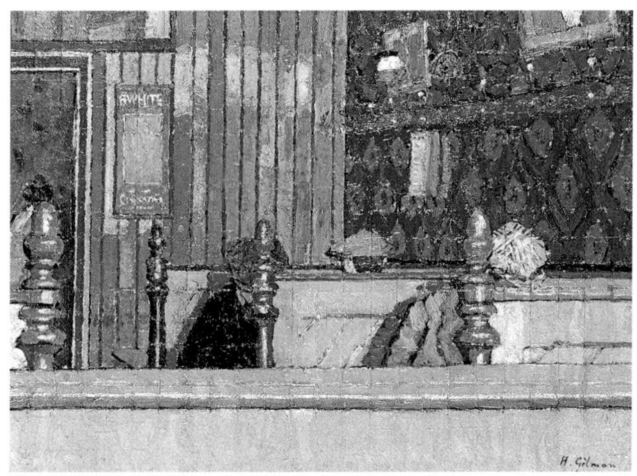

FIG. 64
Harold Gilman, *Study for Canal Bridge, Flekkefjord*, ca. 1913. Ink on paper, support 230 × 290 mm. Purchased 1955. © Tate, London 2015.

FIG. 65
Harold Gilman, *Eating House*, ca. 1914. Oil on canvas, 57.2 × 74.8 cm. Sheffield Galleries and Museums Trust, U.K.

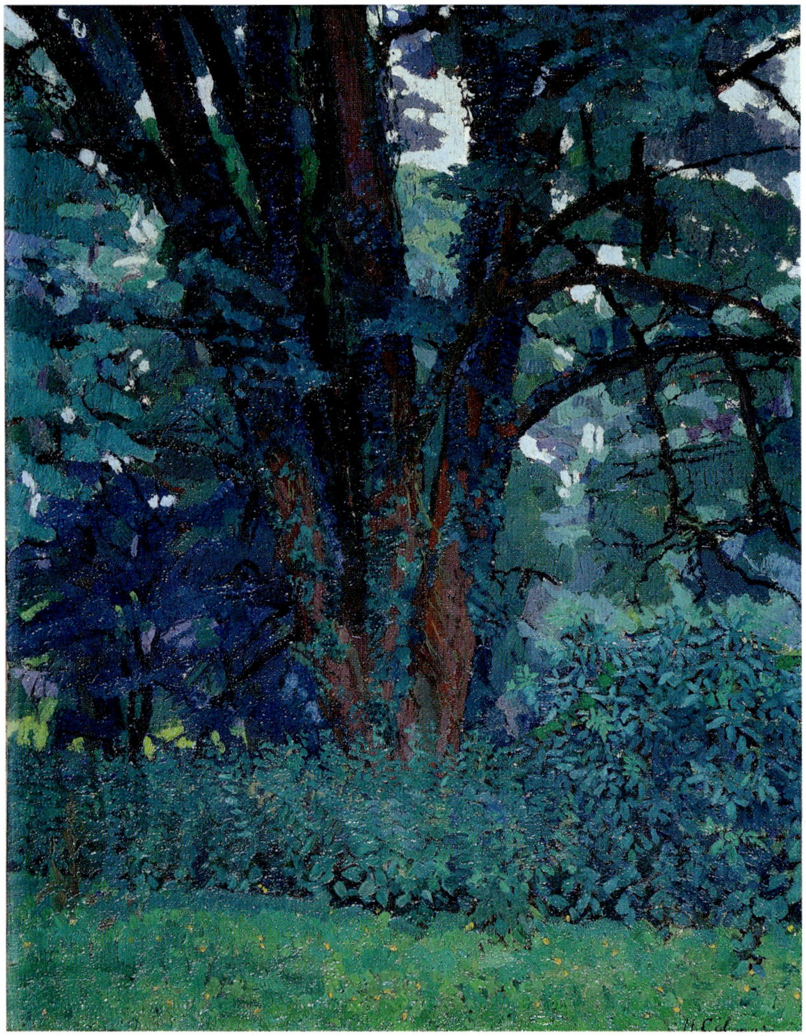

FIG. 66
Harold Gilman, *In Gloucestershire*, ca. 1916. Oil on canvas, 59.7 × 46.3 cm. Leeds Museums and Galleries (Leeds Art Gallery), U.K.

unique in it and consequently inexpressible." Specifically, Ginner described "Neo-Realism" "as a deliberate and objective transposition of the object (man, woman, tree, apple, light, shade, movement, etc.) under observation, which has for certain specific reasons appealed to the artist's ideal or mood, for self-expression. . . . It is only this intimate relation between the artist and the object which can produce original and great work. Away from this we fall into unoriginal and monotonous Formula."[75]

Ginner was likewise informed by the Bergsonian notion of reaching self-understanding by penetrating the outer crust of self, passing through what Bergson described as sequential zones of perceived objects, memories, and "tendencies and motor habits," before arriving, "beneath these sharply cut crystals and this frozen surface, [at] a continuous flux." There one would find a "self in duration" that is unrepresentable by either image or concept and is understood or, more accurately, experienced through "metaphysical intuition." Ginner envisioned an art that would rely on this sort of intuitive connectivity between self and the external world. His ideal work of art would be "an unconscious creation produced by the collaboration of Nature and the Artist Mind."[76] The artist's eye, so privileged in Pre-Raphaelite discourse, had disappeared, and nature, given agency by Talbot, was once again active in the act of representation.

Like P. H. Emerson and George Moore, Ginner wanted an artistic genealogy for his practice. Looking to the past, he could find a form of realism—what he described as the method by which "all great painters by direct intercourse with Nature have extracted from her facts which others have not observed before, and interpreted them by methods which are personal and expressive of themselves"—that extended back to the early Egyptians. He also included "the early Greeks, the early Italians, the early French, the early Flemish," and the "most intense of modern realists, Vincent van Gogh." Significantly, he omitted the British landscape tradition—Cotman, Cozens, Turner, Constable, and the Pre-Raphaelites—from his family tree.

While Ginner clearly admired the legacy of realism—he proposed that neorealism shared with realism the aim to "interpret . . . life in all its effects, moods and development"—he also distinguished neorealism from its forebears. The neorealist, he explained, first selected a scene that "appealed to the artist's ideal or mood, for self expression," and then undertook "deliberate research, concise study and transposition."[77] Rather than select a scene in nature for its scientific significance, historic ties, or literary links, Ginner proposed that the artist choose a topic of purely personal appeal, because the intent of the final work of art was to express the artist's subjective self, not nature and its associations.

While research, study, and even transposition are familiar terms in the pursuit of truth to nature, Ginner's use of them was distinct in the context of Bergson. Although Ginner did not define transposition, he used the word in a way that suggests "rewriting." It is helpful to consider that in music transposition means shifting to another key, keeping a trace of the original while changing the tone. The squared transfer process that Ginner and Gilman used is key to this process of transposition—of extracting a subject from nature and recasting the scene for the final painting. The artist's direct observation of the external world is mediated through the grid, which becomes the meeting point between the artist's individualism and the inherent rhythms of nature. Ginner referred to his practice as a form of partnership between the artist and the landscape: "this deliberate research by the artist into Nature, this collaboration." Furthermore, he added, "this objective transposition must necessarily bring with it good and sound craftsmanship . . . and must result in a strong and solidly built up work." Ginner went on to criticize sketchlike or "slapdash, careless, and slick painting which has been and is still so much in vogue," a critique seemingly aimed at John and Innes. He concluded that these methods of transposition and craftsmanship, combined "with a continued renewal with Life, i.e., collaboration of the Artist and Nature, must result in a strong, individual, and interesting interpretation of Life."[78]

The communal practice of pursuing a verifiable and repeatable universal truth about nature and its laws held no appeal for Ginner. His practice was highly individualistic and drove toward knowledge of the perceiving self rather than external nature—or, more accurately, the meeting point between external reality and the perceiving self.

Yet it is also important to recognize continuity. While Ginner and the art critic Philip Hamerton made strange bedfellows, they shared a similar understanding of the limits of mimicking nature. Hamerton's recourse was to memory, which would force the artist to recall the essence of a scene, whereas Ginner had recourse to an intuitive bond between landscape (as life) and artist. In both cases, it was a project of perception explained through the lenses of psychology and philosophy—in the case of Hamerton, via William Carpenter, and in the case of Ginner, via Henri Bergson. It was a project also shared with Emerson, who learned from Hermann von Helmholtz that the artist could not "transcribe" nature but should instead aim to "translate," in this case, in response to the physiological limitations of the eye and in recognition of the role of experience in shaping how the mind responds to visual stimuli. In other words, while Ginner may have positioned his essay "Neo-Realism" as a critical interjection into heated debates of the contemporary avant-garde, he was participating in a longer-term investigation into the nature of representation.

Ginner's essay on neorealism was reprinted in the catalogue for his two-person show with Gilman, held at the Goupil Gallery in April–May 1914. The exhibit included such works as Gilman's *Eating House*, Ginner's *Victoria Embankment Gardens*, and drawings for a number of oils that allowed viewers to learn more about the artists' process. The exhibition, following on the heels of the publication of Ginner's essay in the *New Age*, sparked a critical debate with Sickert and Hulme.

Sickert took the position of elder statesman and issued his general approval of Ginner's essay. He understood the younger man's main thesis to be that "art that is based on other art tends to become atrophied, while art that springs from direct contact of the artist with nature tends at least to be alive." But he objected to creating yet another label and also criticized Ginner for regarding Poussin's art as derivative and for misrepresenting academic art. Despite the ostensible mildness of Sickert's review, Ginner took the bait, writing a letter to the editor defending himself; it was published in the same issue of the *New Age* in which Italian poet Filippo Marinetti's "Geometric and Mechnical Splendour in Words at Liberty: Futurist Manifesto" appeared, along with an essay on Bergson's popularity. Sickert fired back with a review titled "The Thickest Painters in London," wittily describing Ginner's and Gilman's techniques while also suggesting their stupidity. The argument did not stop there—Gilman responded with a letter to the editor in which he charged that Sickert had misrepresented him.[79] Ginner attempted to continue the debate by submitting another letter to the editor on 9 July 1914, but he received no response. Instead, the next letter to the editor concerning modern art jabbed at the arguments of T. E. Hulme.[80]

This exchange between Ginner and Sickert never directly confronted one of the key problems concerning Ginner's polemic—to what degree could it adequately explain his paintings and those of Gilman? While the textual account of transposition suggested a strong alignment with Bergson's philosophy, the discernible traces of the perspective frame in the paintings suggest that the landscape has been pinned down and controlled by the artist. This sense of mastery could be construed as separating and distinguishing the landscape as object from the observing subject. Moreover, the static order of the compositions pushed against a sensibility of flux and the mutability of nature.

T. E. Hulme offered the most decisive critique of Ginner's essay on neorealism in a series of articles on modern art in Britain that appeared in the *New Age* in the spring of 1914, just when Italian futurism and British vorticism were breaking onto the London scene. Both movements strove to create art for the modern age that incorporated admiration for machines, technology, the city, dynamic movement, and speed. They took visual cues from French cubism and the fragmentation of forms into facets and planes that suggested multiple points of view simultaneously. Hulme's purpose was "to define the characteristics of a new constructive geometric art which seems to me to be emerging at the present moment."[81] By this date, Hulme had fostered friendships with many of the young artists in London through his Tuesday night salon, which he held weekly between 1911 and 1914 and which Ginner and Gilman attended, along with Spencer Gore, Wyndham Lewis, and C. R. W. Nevinson, among others.

As the scholar Rebecca Beasley explains, Hulme had recently taken up the cause of abstraction and realized that he needed to reconsider his descriptive language accordingly.[82] To capture the full scope of modern art in Britain, he divided it into two broad groups: "geometrical or abstract, and vital and realistic art." Each type, he argued, had emerged to "satisfy a different desire of the mind." Realistic or naturalistic art "corresponded to a certain rational humanistic attitude towards the universe," whereas geometric art "may be the precursor of the . . . final break up of the Renaissance," that is, the dissolution of that rational humanism that had produced realism.[83] Hulme's conclusions were provoked in part by Wilhelm Worringer, whom he had met in 1913 and whose *Abstraktion und Einfühlung* (1908, published in English in 1910) he read closely. Worringer argued that an "urge to abstraction" could be traced to "primitive culture," and that "the less mankind has succeeded, by virtue of its spiritual cognition, in entering into a relation of friendly confidence with the appearance of the outer world, the more forceful is the dynamic that leads to the striving after this highest abstract beauty."[84]

While Hulme made clear that he preferred the London Group—Spencer Gore, Harold Gilman, Charles Ginner, and Robert Bevan—over Roger Fry's Bloomsbury set (Vanessa Bell and Duncan Grant), his most extensive praise was reserved for Wyndham Lewis, C. R. W. Nevinson, Henri Gaudier-Brzeska, and especially David Bomberg, who were associated with vorticism.[85] Thus he felt it necessary to challenge Ginner's theories using logic rather than wit, as Sickert had done. Hulme clearly understood that with neorealism Ginner laid claim to two essential lines of argument that the critic wanted to harness to his desired goal of abstraction: the validation of this art form by earlier, archaic art, and the unique connection to life itself that this art promised.

Hulme first attacked Ginner's claim that "all good art is realistic." This fallacy, he charged, had led Ginner to perceive abstraction negatively as a formulaic response, whereas it was in fact "a method of expression." To Ginner's observation that decorative abstraction was a dead end, Hulme retorted that this again reflected a fundamental misconception of abstraction, which was in fact a mode of creative expression. To shore up his point, he turned to Cézanne, whom Ginner had treated as a realist, and reinscribed the French artist within the genealogy of abstraction, explaining that Cézanne had "initiated a break-up of realism and provided the material for an abstract art" that Picasso then developed. The third point that Hulme challenged lay closest to Bergson: that, according to Ginner, only an "intimate relation" between the artist and nature could produce original art.[86] This belief, Hulme charged, derived from a naïve Rousseauism, or romanticism, that failed to distinguish between nature and life and, moreover, endowed nature with an undue ethical morality that negated the practice of any artist who strayed from using nature as a source.

Here, Hulme identified a key assumption behind much of the truth-to-nature discourse: a reverential regard for nature as the source of all truth and knowledge. Hulme declared that art no longer needed to be tied to nature as a source of creativity and validation. He acknowledged that an artist, in order to "keep his work alive," must conduct research into nature. But this did not mean that an artist should produce realist art.[87] Nature should simply provide the creative artist with elements to be reconfigured, distorted, or distilled in the process of creative expression. In discussing David Bomberg's art, Hulme's ideal exemplar, he explained further that "*form* becomes the *porter* or *carrier* of internal emotions." Viewers, he acknowledged, then had to learn how to read such artworks, forgoing their

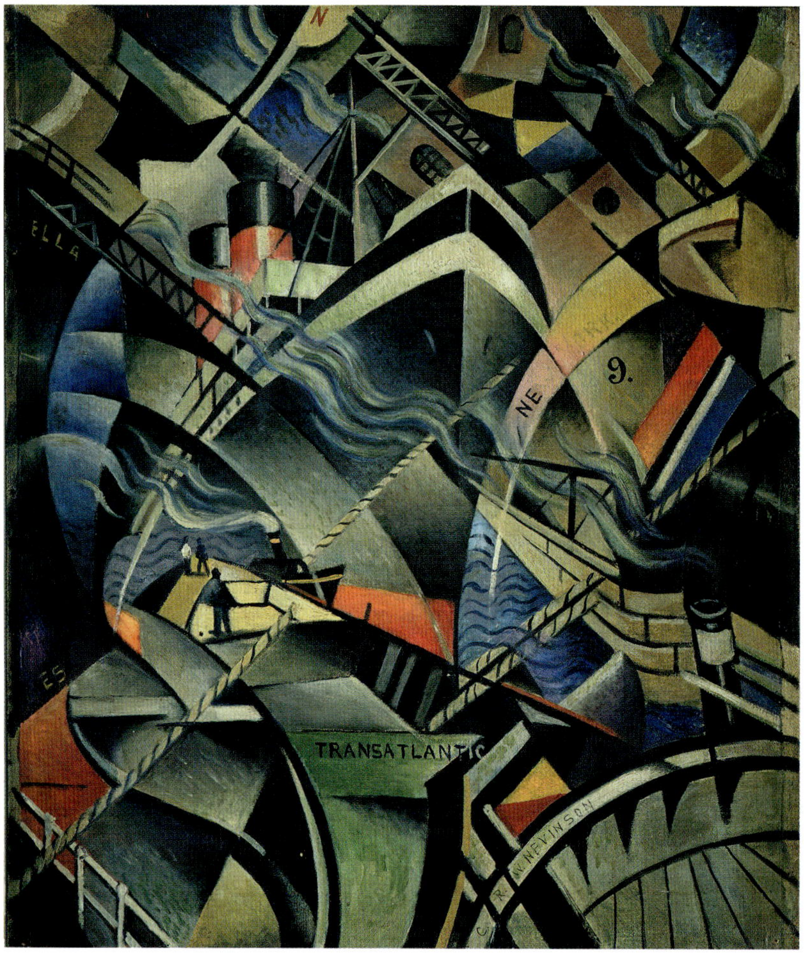


FIG. 67
C. R. W. Nevinson (1889–1946), *The Arrival*, ca. 1913. Oil on canvas, support 762 × 635 mm. Presented by the artist's widow, 1956. © Tate, London 2015.

accustomed practice of "living our own emotions *into* . . . atmosphere, colour, and landscape" in favor of "outside shapes and colours" and "abstract form."[88]

Hulme concluded his refutation of Ginner's neorealism by stating that "both realism and abstraction, then, can only be *engendered* out of nature, but while the first's only idea of living seems to be that of hanging on to its progenitor, the second cuts its umbilical cord."[89] With this statement, Hulme attempted to put an end to the quest for truth to nature, releasing art from the task of representing the external world. He robbed the pursuit of truth to nature of its moral efficacy as described by John Ruskin and F. G. Stephens, an efficacy that still lingered in the writings of Clausen, Steer, Sickert, and even Augustus John, who shifted the focus of truth to nature from honoring God's handiwork to honoring self.

Moreover, Hulme turned his back on the idea that science could rejuvenate art. Talbot, Ruskin, and Stephens had believed that science, with its emphasis on facts and disciplined inquiry, would cleanse art of its dependence on the past and would purify and modernize artistic practice. But by 1914, the notion that science could bring the artist closer to understanding nature or provide access to truths otherwise inaccessible was in doubt in artistic circles. That art and science were bound together in the pursuit of truth rested on a rational empiricism that Bergson had demolished. Moreover, a new generation of young artists, tied closely to the urban sphere, denigrated the appreciation of nature; the vorticists, in their 1914 manifesto, blasted "bowing the knee to wild Mother Nature." They took their cues instead from the machine, from blessed "PORTS, RESTLESS MACHINES . . . ENGLAND, Industrial Island machine."[90]

C. R. W. Nevinson, in his essay "Vital English Art," published in the *New Age* on 18 June 1914, laid waste to the truth-to-nature paradigm while taking up the Bergsonian mantle. Nevinson's arguments had originally been aired at the Doré Gallery (then hosting the *Post-Impressionist and Futurist Exhibition*, organized by Frank Rutter), and his published version appeared in the same issue as Sickert's critique of Ginner's neorealism.

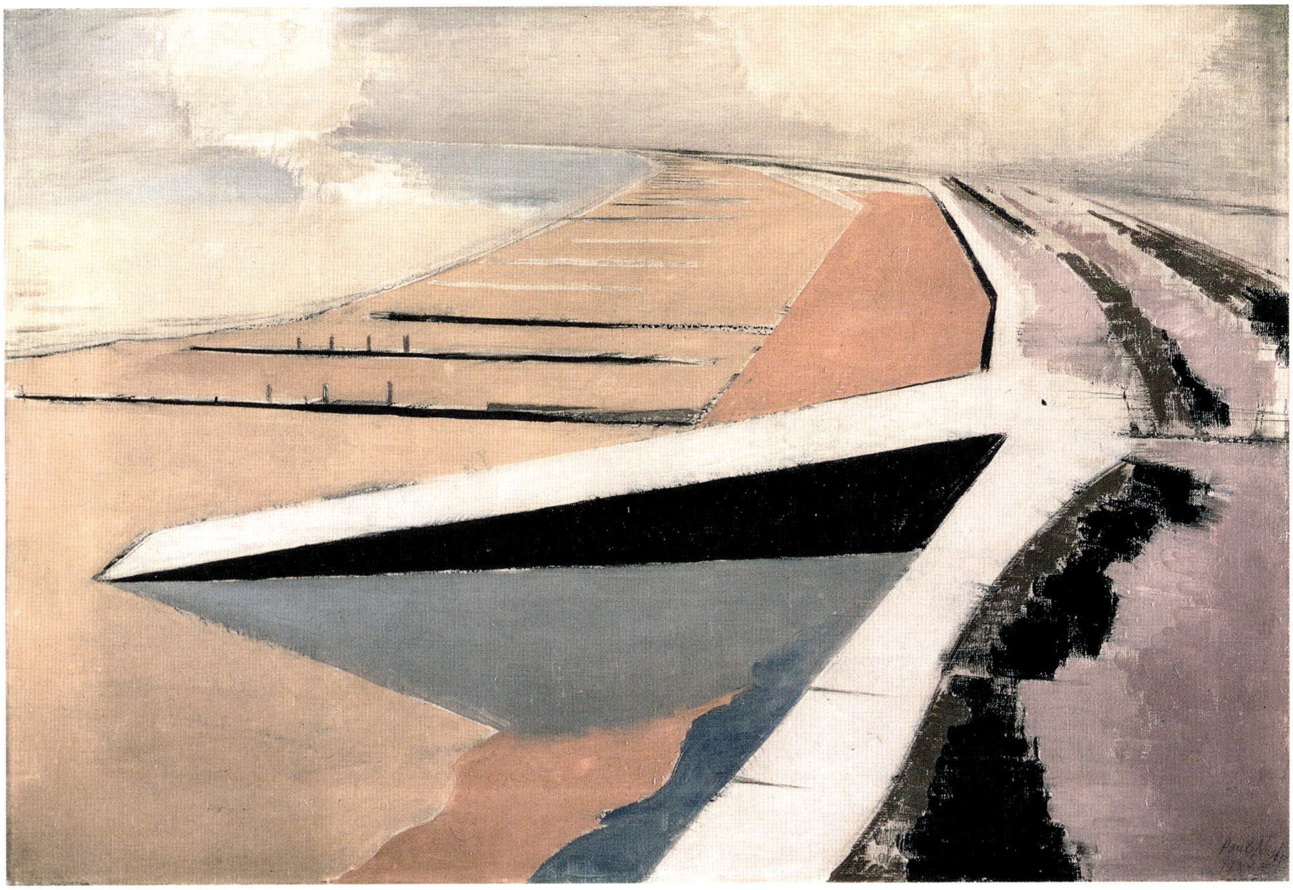

FIG. 68
Paul Nash (1889–1946), *The Shore*, 1923. Oil on canvas, 62.2 × 94 cm. Leeds Museums and Galleries (Leeds Art Gallery), U.K.

Nevinson, who had attended the Slade and then the Académie Julian and Matisse's Cercle Russe, was at this moment most closely affiliated with Filippo Marinetti, the two having co-signed "Vital English Art: A Futurist Manifesto," which appeared in the *Observer* on 7 June 1914. His admiration for the technology of the modern age was on display in the painting he submitted to the London Group exhibition in March of that year—*The Arrival* (ca. 1913) (fig. 67), which depicts a large steamer. The ship's route is announced by the letters that march across the lower third of the canvas. The image of the steamer is intercut with port scenes—signaled by references to docks, tugboats, ropes, and wharves—along with fragments of shapes that suggest girders and cogs, and lines that break up planes of color into facets reinforced by contrasting colors and differentiated shades.

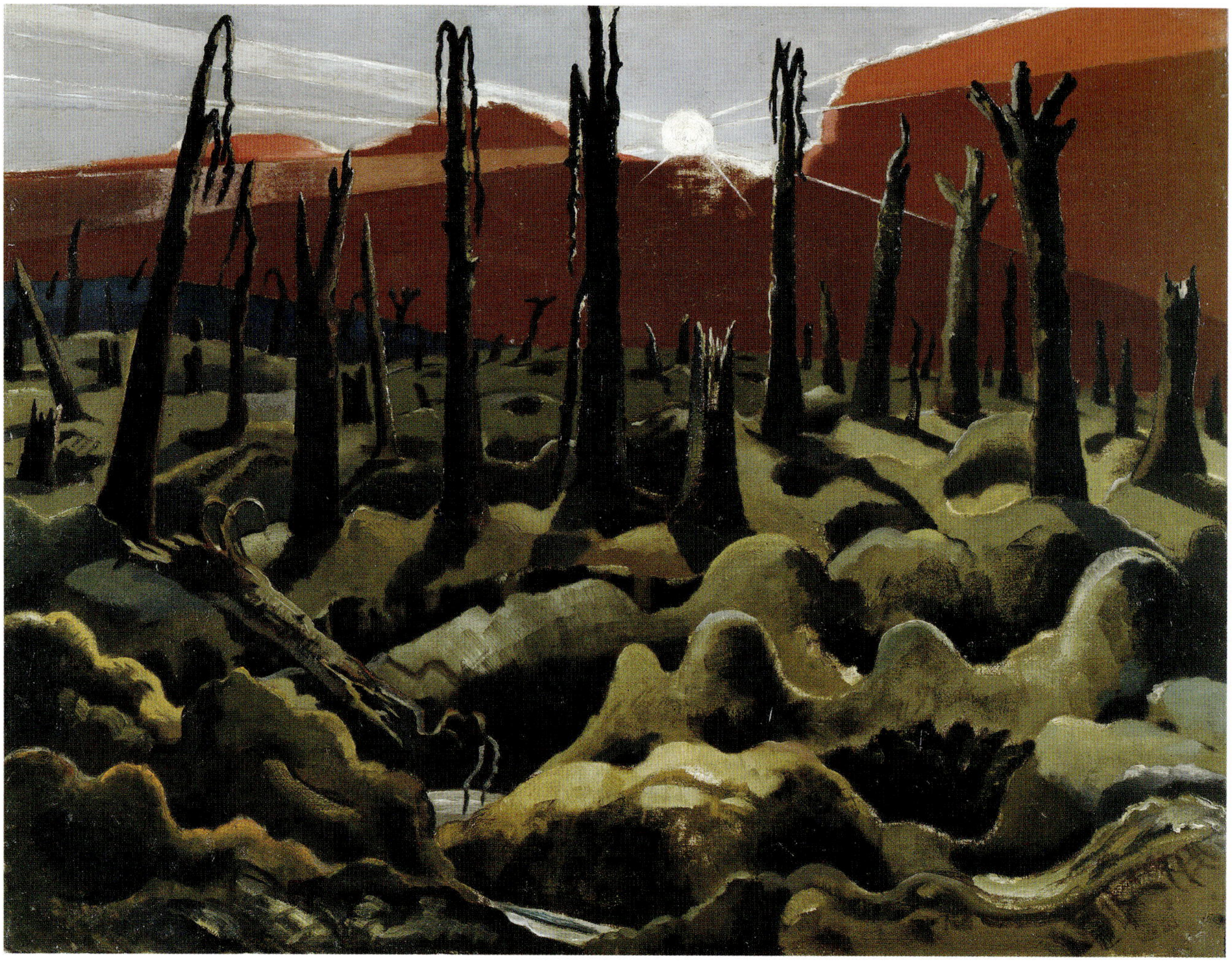

FIG. 69
Paul Nash, *We Are Making a New World*, 1918. Oil on canvas, 711 × 914 mm. Imperial War Museums.

Nevinson was clearly engaged in the project of radical modernism, which included formal innovation, adopting a public persona as the rebel, and engaging with the conditions of modern life.[91]

In "Vital English Art," Nevinson set forth four principles. First, "no picture should be a mere representation"; instead, "a picture must be a plastic abstraction of an emotion, remembered, seen, smelt or heard, not visually, but mentally felt." Second, "art must be the expression, intensification and concentration of life." In other words, the picture was about the artist's subjective sensation rather than optical impression. These two principles completed the shift from optical perception to mental perception as the source of art that had been ongoing since the 1870s. The third principle dictated that if art was to represent modern life, it should be preoccupied with speed and simultaneity, since these were the major characteristics of modernity. In order to represent the states of mind produced by speed and simultaneity, artists should use "contrasted colours, lines and dimensions"—Nevinson's fourth principle. Indeed, modernity was the inspiration for the entire structure of Nevinson's argument, which he believed many critics would find "too scientific" in its clarity and logic. But, he retorted, he lived in a "mechanical age."[92] Nevinson reclaimed science as a partner in making art, but now with a focus on its contributions to technology rather than to psychology or inductive philosophy.

Ginner explicitly rejected the new terms of modern art proposed by Hulme and Nevinson in his essay "The Cézanne Stunt" (1919), published in *Art and Letters*, a journal established in 1917 by Frank Rutter and Herbert Read. In this essay, Ginner sought to reassert the primacy of optical perception by insisting that Cézanne's mode of representation was the product of the landscape around him; since the landscape of Provence was already "simplified" and elemental, the painter derived a corresponding art form by observing it. Cézanne, he stated unequivocally, was "a strong and sincere realist." He cautioned fellow artists that Cézanne's emphasis on solidity and elemental structure was ill suited to England because of the differing nature of the English landscape, where the ground was covered with "thick vegetation . . . and we have a much more complex and detailed surface to work with."[93] Ginner firmly believed that sound art derived from the artist's physical environment.

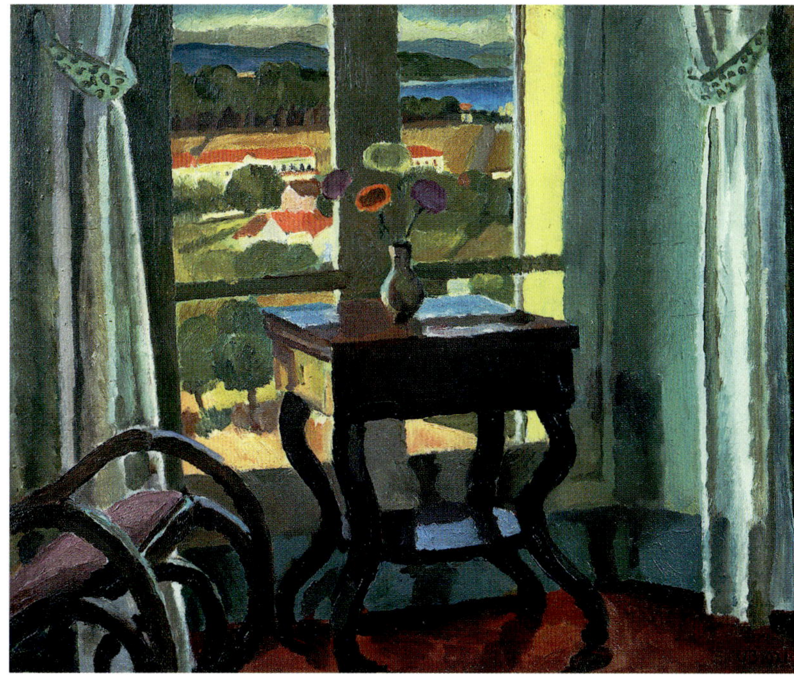

FIG. 70
Vanessa Bell (1879–1961), *Interior with a Table*, 1921. Oil on canvas, support 540 × 641 mm. Bequeathed by Frank Hindley Smith, 1940. © Tate, London 2015.

But this argument that British landscape painting should respond to the uniqueness of the British landscape created an inescapable trap, for it privileged motif over mode of expression and made nature the ultimate judge. Indeed, to close his essay and to point the way forward, Ginner returned to the discourse of truth to nature, indicating its extraordinarily strong attractive pull: "John Nash and Lucien Pissarro—each with a quite different outlook—are giving us English landscapes seen with an extraordinary truthful vision. I pick out these artists as examples, because they are striving after truth each according to his own light. In that endeavour British art will find its salvation."[94] Ginner was already eagerly building an artistic genealogy for his practice, looking to the younger artists John Nash, brother of Paul Nash, and Lucien Pissarro, son of Camille Pissarro. But the postwar debate over truth to nature looked profoundly different from its prewar predecessor.

In the postwar environment, scientists headed to university laboratories and away from public lectures and spectacular entertainments, retreating from their former public ambition for cultural authority in favor of establishing technical expertise.[95] The representation of nature continued to engage British artists, but their art suggests a twin preoccupation with questions of modern artistic vocabulary—as in Paul Nash's *The Shore* (1923) (fig. 68)—and the expressive possibilities and symbolic associations of landscape. Nash, for example, explored a fusion between technology—in the form of the sea wall—and nature. David Peters Corbett argues that the artist aimed to repress modernity while allowing its presence to inform the structures of his landscapes.[96] Further examples can be cited, including Paul Nash's ironic *We Are Making a New World* (1918) (fig. 69), which depicts the war-torn earth of trench warfare lit by a rising sun, or John Nash's *The Cornfield* (1918), a richly fertile landscape that evokes a postwar sensibility of return to order and fecundity through the pronounced line of carefully aligned stacks of grain that march down the hillside, or Vanessa Bell's *Interior with a Table* (1921) (fig. 70), which speaks of the calm pleasure of contemplating nature.

Artistic discourses were similarly reconfigured; in the 1920s, the journal *Colour* repeatedly praised landscape paintings that established a strong sense of decorative pattern built upon the underlying structure of the scene.[97] Truth to the material world and its processes was no longer paramount; critics had come to expect that the artist should offer an original viewpoint rather than universal truisms. And while this point might seem to go without saying in today's pluralistic, postmodern art world, it is worth remembering that this was not always the case, and that modernism's project, as it unfolded in nineteenth-century Britain, was once very differently configured by a dynamic exchange between scientists and artists.

# { *conclusion* }

The belief that art and science could unite to discover the truth of nature was born of a distinctive historical moment. In the 1830s, the inductionist methodology, which prized close observation of nature, held sway. It fueled the birth of photography and also added impetus to calls to reform fine art. To modernize art, the Pre-Raphaelites were urged to turn to scientific models of investigation, to study nature closely. A new aesthetic emerged, one that valued the finely wrought detail and factuality. This aesthetic, particularly as described by John Ruskin, conveyed a new moralism associated with the perseverance and dedication of working directly from nature. It required setting aside the self, paring away emotional responses to the scene, in favor of an objective, rational assessment. Thus Ruskin distinguished between the Pre-Raphaelite John Everett Millais and the romantic J. M. W. Turner.

But this dedicated pursuit of truth based on the close observation of nature proved to be difficult to achieve, a fraught enterprise. The challenging tasks the artists set for themselves, such as capturing changing colors and patterns of the setting sun, proved their undoing, an unraveling reinforced by a changing framework for scientific investigation. By the 1860s, John Tyndall had begun to rework the inductionist paradigm to allow a greater role for the imagination. His efforts were reinforced by new explorations of cognition, psychology, and physiology led by philosophers and scientists like Hermann von Helmholtz, James Sully, Alexander Bain, and George Henry Lewes. Their findings also challenged the notion of the "innocent eye," the ability to see things unmediated by preexisting knowledge, which Ruskin had regarded as fundamental to the reform of painting.

The debates fueled by the lectures and writings of these scientists and philosophers shaped the production and reception of the mid- to late-career oeuvres of Millais and John Brett. These two artists, whose early work was closely tied to the Pre-Raphaelite aesthetic, chose divergent paths. Brett stayed faithful to his roots, refining his theories of representation in light of the new challenges posed by changing scientific paradigms. Millais, by contrast, explored new modes of expression, eschewing dispassionate objectivity in favor of poetic and emotive responses to the landscape. His new painterly approach and subjects also reflected a newfound appreciation of older modes of painting, particularly those associated with Constable. With the rise of Darwinism and challenges to inductionism, nature was no longer purely a repository of God's handiwork or an encyclopedia waiting to be read but a potential battleground for the survival of the fittest.

This changing scientific context also challenged fine art photography, as revealed in the career of P. H. Emerson. Emerson moved from supreme confidence in his skill to capture nature fully and clearly to utter dejection. By the end of his career, he despaired that he could not overcome photography's purely mechanical nature, concluding that it could not rival painting, after all, as a fully expressive mode of representation. Throughout his career, Emerson had struggled to find a compass for photography: should it align itself with science or with art? He finally chose the latter, a choice embedded in his repeated attempts to create an artistic genealogy for photography. Rather than look only forward, Emerson, like Millais, began to look backward to recover potential models for artistic responses to the natural scene. He thus blunted the progressive claims of photography and painting while also challenging the earlier partnership of art and science.

Following generations inherited this legacy of both doubt and confidence and the lingering remains of the paradigm of art and science conjoined in the search for truth to nature. They sought to reinvigorate the modernist paradigm of looking forward, cutting ties to the past. The painter George Clausen found to his disappointment that the styles of naturalism and realism that he and his colleagues in the New English Art Club had forged from a confluence of the remains of Pre-Raphaelitism and French models of modern painting, including impressionism, were no longer pathways forward but dead ends. In dialogue with Emerson, Clausen pursued a new way to observe and record nature, focusing on the effect of the whole instead of the earlier dedication to the rendering of detail.

Augustus John and J. D. Innes, like their Pre-Raphaelite predecessors, sought to experience nature directly, and they traveled extensively in search of the perfect motif. But their interactions with nature, informed by their understanding of Henri Bergson, were more intuitive than inductionist. They abandoned the attempt to capture the surface of nature in minute detail, as Brett had done in *The Glacier of Rosenlaui*, in favor of a sensation of breadth, as revealed in John's *Welsh Mountains in Snow*. While Clausen, too, had sought the impression of the whole, Innes and John rejected Clausen's method of brushwork and rendering of space in favor of new approaches that they hoped could convey a sensation of the internal harmonies or rhythms of nature. Rather than function as detached observers, driven by moralistic compunction and rational objectivity, they sought to merge self with external reality and to recover the notion of an "innocent eye." Their efforts met with mixed results; some reviewers easily grasped and affirmed their agenda, while others resorted to the genealogy of art making to make sense of their practice.

Harold Gilman and Charles Ginner, taking up the crisis of realism that Clausen had articulated, argued for a new solution that they dubbed "Neo-Realism." Like Clausen, Innes, and John, Gilman and Ginner argued that truth to nature lay in deriving the essential elements of a scene. And, like Innes and John, Gilman and Ginner pursued an intuitive and sympathetic response to the scene and envisioned an art form that would achieve the unification of the artist's mind with nature. But the neorealists' push for order, like their methodology of transposition, seemed at odds with the notion of an open-ended, intuitive expression. Moreover, as Gilman made clear, the two painters still gave nature ultimate authority.

This did not sit well with T. E. Hulme, who argued that the path forward in modernism lay in cutting the umbilical cord to nature. Likewise, Hulme no longer saw science as art's partner in reforming art and expressing the modern experience. British modernism, as Lisa Tickner has observed, by 1914 had charged itself with cleansing English art of its ties to the rural landscape and was instead engaged with "the urban, the virile, the industrial and the 'primitive.'"[1] The truth-to-nature paradigm had no value in this new economy. The dialogue with science in which this paradigm was embedded no longer offered any promise.

Despite the perhaps disappointing end of the dynamic relationship between art and science, that relationship was of indelible and lasting significance for the course of modern art in Great Britain. It explains how and why the truth-to-nature paradigm had such lasting purchase on the British artistic imagination. It also helps to explain the particular configuration of British modernism. As scholars such as David Peters Corbett and Lisa Tickner have reminded us, to understand effectively the course of British modernism, we must analyze it on its own terms—in light of the specific social and historical conditions in which it was embedded.

Leading artists, scientists, and philosophers of the Victorian age believed that they would forge a new future together and create new ways of knowing and visual expression that embodied the experiences of the modern age. Their attention focused upon pursuing truth to nature. But the truth-to-nature paradigm lost its potency and certainty for the artistic community over the course of the Victorian era, leaving artists to renegotiate their former partnership with science.

In investigating these sociohistorical conditions, art historians have perhaps unconsciously replicated the "two cultures" paradigm separating the humanities from the sciences that C. P. Snow so deplored. Science has typically been called upon to help answer specific questions of practice—for example, how complementary colors were understood—but not to investigate larger directions in art making. Tackling such questions is challenging, given the training most art historians receive, which reflects the increasing specialization of higher education. But just as art historians choose to learn new languages to pursue new areas of study, we must venture into the realm of science and its histories to do full justice to the art we study, art that was never subjected to the disciplinary boundaries that divide the arts and sciences today.

# { notes }

## Introduction

1. Talbot, "Photogenic Drawing," 115.
2. [Stephens] [Seward, pseud.], "Purpose and Tendency," 61.
3. Little, "Art, Nature, and Photography," 85.
4. Quoted in "Banquet at the Royal Academy," *Times* (London), 1 May 1871, 6.
5. Wiseman, *Points of Contact*, 11, 3.
6. For an overview of this topic, see Lightman, "Science and Culture."
7. Wiseman, *Points of Contact*, 4, 27–30.
8. David Peters Corbett, in *The World in Paint*, offers an insightful definition and historiographical overview of the concept of visuality (15–16).
9. Gates, "Ordering Nature," 181. See also, for example, Christ, *Finer Optic*.
10. See Smith, *Fact and Feeling*. Smith's historical framework for shifts in scientific thinking in relation to literature was extraordinarily useful for this study, as these changes apply equally well to art making.
11. Flint, *Victorians and the Visual Imagination,* 37. Reading both text and images closely and focusing upon such tropes as dust, blindness, and the mind's eye, Flint foregrounds the tension between the seen and unseen and debates about the perceiving individual and his or her place in the world that shaped Victorian culture. Her work is an important starting point for this project, but my focus here is more specifically art-historical and concerned with the dynamic between art and science.
12. See Levine, *Darwin and the Novelists*; Beer, *Darwin's Plots* and *Open Fields*; Dale, *In Pursuit of a Scientific Culture.*
13. Tickner, "English Modernism," 23.
14. As the historian of science Timothy Lenoir observes in *Instituting Science*, "knowledge is always situated, local, and partial." He amplifies this point by noting that "the object of knowledge and the interpreter do not exist independently of one another; knowledge is a form of interpretation, involving temporal, bodily engagement with the world rather than the detached, disembodied, contemplative stance favored by the sociology of science" (5). Thus Lenoir envisions knowledge as "interested in the twofold sense, encompassing active interpretative engagement with the world as well as the social and economic interests of the actors involved in the construction of knowledge" (7–8).
15. Smith, *Body of the Artisan*, 8.
16. Hanson, *English Virtuoso*, 3, 15. See also Gibson-Wood, *Jonathan Richardson*. Gibson-Wood observes that the Royal Society fellows, particularly those engaged in experimental philosophy, valued image making and painting as modes of investigation (as in the camera obscura) and detailed records of nature (as in botanical specimens), stressing the value of art for "empirical, not humanistic knowledge; according to this set of values, history painting did not rank above still life" (13).
17. Meyers, *Knowing Nature*, 5–6.
18. Yeo, *Science in the Public Sphere*, ix. See also Yeo, *Defining Science*.
19. Thomas Sprat, in his *History of the Royal-Society*, recorded that it was founded to serve a twofold purpose: "to make faithful *Records*, of all the Works of *Nature*, or *Art*, which can come within their reach: that so the present Age, and posterity, may be able to put a mark on the Errors, which have been strengthened by long prescription: to restore the Truths, that have lain neglected: to push on those, which are already known, to more various uses: and to make the way more passable, to what remains unreveal'd" (61).
20. [Whewell], "On *The Connexion of the Physical Sciences*," 59.
21. Whewell, *Philosophy of the Inductive Sciences* (1840), 1:cxiii.
22. Morrell and Thackray, *Gentlemen of Science*, 96.
23. Shapin and Schaffer, *Leviathan and the Air-Pump*, 3–4, 7, 22 (quotation).
24. Harrison, *Fall of Man*, 1–7 (quotations on 1 and 5).
25. Porter, "Medical Science and Human Science," 57.

26. Harrison, *Fall of Man*, 237. At the same time, leading figures such as William Hutton and Joseph Priestly developed alternative theories of the behavior of matter. As Simon Schaffer has argued in his essay "Natural Philosophy," it is inaccurate to regard eighteenth-century natural philosophy as a unified field centered around Newton. See also Heilbron, "Experimental Natural Philosophy."
27. Locke quoted in Harrison, *Fall of Man*, 223; Whewell, *Bridgewater Treatises*, 287–88.
28. As Bernard Lightman notes, writing from the perspective of the Victorians, "The emphasis on knowledge acquired by observation or experiment into the pantheon of the sciences had taken place relatively recently in the eighteenth century. Previously, the sciences were seen as specialized branches of philosophy and included such bodies of knowledge as grammar, logic, rhetoric, math, music, metaphysics, geometry, and astronomy." Lightman, "Science, Scientists, and the Public: The Contested Meanings of Science in Victorian Britain," in Lightman, *Evolutionary Naturalism in Victorian Britain*, 5.
29. Chambers, *Cyclopaedia*, vii.
30. Hall, *All Scientists Now*, 23, 26–28, 41–44.
31. Ibid., 45, 47.
32. Babbage, *Reflections on the Decline*, 23, 60, 77.
33. Morrell and Thackray, *Gentlemen of Science*, 29, 32, 33 (see also 23–26).
34. Yeo, *Defining Science*; in particular, see the chapter "Reviewing Science," 77–115.
35. Yeo, "Scientific Method and the Image of Science, 1831–1890," in Yeo, *Science in the Public Sphere*, 66.
36. Morrell and Thackray, *Gentlemen of Science*, 157–63; Lightman, *Victorian Popularizers of Science*; and Kember, Plunkett, and Sullivan, *Popular Exhibitions*.
37. Morrell and Thackray, *Gentlemen of Science*, 147.
38. Lightman, *Victorian Popularizers of Science*, 361.
39. Ibid., 32.
40. For more on Victorian art serials, see Helmreich, "Death of the Victorian Art Periodical." For more on the *Art Journal*, see Haskins, *Art Journal and Fine Art Publishing*.
41. Roberts, "Exhibition and Review," 82. See also Prettejohn, "Aesthetic Value and the Professionalization."
42. Solkin, *Painting for Money* and *Art on the Line*.
43. Roberts, "Exhibition and Review," 80.
44. Altick, *English Common Reader*, 357–58; see also 394.
45. For more about specialized societies, see Codell, "Artists' Professional Societies."
46. For more on the art market in this period, see Fletcher and Helmreich, *Rise of the Modern Art Market*.
47. Hanson, *English Virtuoso*, 17.
48. Ibid., 3, 94, 125, 196–97.
49. Dulac, "William Hunter," 19, 27.
50. Kase, "No. 46 Bull Moose," 182.
51. Kase, "Anatomy of the Horse," 106–8.
52. Kase, "'Make the Knife Go with the Pencil,'" 48. For more on Stubbs and science, see Cormack, "Stubbs and Science."
53. For more on this point, see Fraser, "Joseph Wright of Derby."
54. Ibid., 19.
55. Barker, "New Light on 'The Orrery,'" 34.
56. Laura Baudot has proposed that the air pump is intentionally anachronistic, closer in appearance to that devised by Robert Boyle in the mid-seventeenth century than to devices in use during Wright's lifetime. See Baudot, "Air of History."
57. Egerton, *Wright of Derby*, 91.
58. Ritter, *History of Ancient Philosophy*, 565.
59. Yeo, *Defining Science*, 139, 142.
60. Ruskin, *Modern Painters*, in *Works of John Ruskin*, 3:624.
61. Quoted in Turner, *Between Science and Religion*, 8. As Turner explains, "No single word or phrase will exactly describe the cluster of men and ideas associated with Huxley's New Nature. . . . *Victorian scientific naturalism* appears more inclusive than any other phrase. *Positivism* cannot serve because of Huxley's vehement antipathy to the sect and because in England Positivism was simply one part of the larger effort to advance science in the public forum" (11). Huxley himself used the phrase "scientific naturalism," as Bernard Lightman points out, describing it as the culmination of a line of thought traceable to the Renaissance and denoting the study of "observable causes and effects in nature" (rather than of the search for invisible evidence of a divine being). Lightman, *Evolutionary Naturalism in Victorian Britain*, vii.
62. Dawson and Lightman, "Introduction," 4, 5, 7, 8, 10.
63. See Lightman, "Introduction," *Evolutionary Naturalism in Victorian Britain*, xi.
64. Lightman, "Science and Culture," 20.
65. Turner, *Between Science and Religion*, 3.
66. Myers, "Introduction," liv–lv. See also Turner, *Between Science and Religion*, 2.
67. Talbot's letters have been digitized thanks to the efforts of Larry Schaaf, project director, and his team, and are available at http://foxtalbot.dmu.ac.uk/.
68. Michel Foucault reflects upon the notion of discourse and discursive formation in *The Archaeology of Knowledge*, where he proposes to take as his starting point "whatever unities are already given (such as psychopathology, medicine, or political economy); . . . I shall make use of them just long enough to ask myself what unities they form. . . . This field is made up of the totality of all effective statements (whether spoken or written), in their dispersion as events and in the occurrence that is proper to them" (26–27). He continues, "whenever one

can describe, between a number of statements, such a system of dispersion, whenever, between objects, types of statement, concepts, or thematic choices, one can define a regularity (an order, correlations, positions and functionings, transformations), we will say, for the sake of convenience, that we are dealing with a *discursive formation*" (38). Foucault delimits the field, as when he explains that "discursive relations . . . determine the group of relations that discourse must establish in order to speak of this or that object, in order to deal with them, name them, analyse them, classify them, explain them, etc. These relations characterize not the language (*langue*) used by discourse, nor the circumstances in which it is deployed, but discourse itself as a practice" (46). He asks that investigators "discover the law operating behind all these diverse statements, and the place from which they come" (50), by asking who speaks and by what authority, and from what institutional site such statements are issued, as well as by exploring the position of the subject. He insists that discourses should not be treated "as groups of signs . . . but as practices that systematically form the objects of which they speak" (49).

69. Brake, "Writing, Cultural Production, and the Periodical Press," 55.
70. Bourdieu, *Field of Cultural Production,* 30.
71. Fyfe and Lightman, "Science in the Marketplace," 9.
72. Baxandall, *Patterns of Intention*, 11, 42, 43.
73. Ibid., 76.

### Chapter 1

1. Talbot, "Plate X. The Haystack," in *Pencil of Nature*, unpaginated (55). (The fascicles are unnumbered; I have used the penciled number inscribed on the copy in the collections of the Huntington Library, 35021.)
2. [Stephens] [Seward, pseud.], "Purpose and Tendency," 58.
3. Arnold, *William Henry Fox Talbot*, 34, 40.
4. For more on this history, see Brusius and Ramalingam, "Beyond Photography," 15–16.
5. Arnold, *William Henry Fox Talbot*, 65. For the Royal Society medal, see Barrow-Green, "'Merely a Speculation of the Mind'?"
6. Brusius and Ramalingam, "Beyond Photography," 6.
7. Hall, *All Scientists Now*, 57.
8. Arnold, *William Henry Fox Talbot*, 80.
9. "We see, comparatively speaking, nothing we cannot, after a fashion, touch and handle"; the "minds of men . . . thus have become open to tangible, physical things alone." "Art. I-1: First and Second Reports," 8.
10. For more on spectacle at the meetings of the British Association, see Morrell and Thackray, *Gentlemen of Science*, 157–63.
11. "Art. I-1: First and Second Reports," 48.
12. Schaaf, *Out of the Shadows*, xi.
13. Yeo, "Introduction," in Yeo, *Science in the Public Sphere*, ix.
14. Yeo, "Scientific Method and the Image of Science, 1831–1891," in ibid., 70.
15. Yeo, "An Idol of the Market-Place," in ibid., 251.
16. Herschel, *Preliminary Discourse*, 59.
17. Ibid., 102. For Herschel and Baconian induction as it relates to photography, see Schaaf, *Out of the Shadows*, esp. 16–17.
18. Herschel, *Preliminary Discourse*, 118, 169.
19. Schaaf, *Out of the Shadows*, esp. 33–39.
20. Talbot to Jerdan, 30 January 1839, doc. no. 03782, in Schaaf, *Correspondence of William Henry Fox Talbot*. Jerdan was the editor of the *Literary Gazette* and this notice was published in that journal on 2 February 1839.
21. Talbot to Herschel, 25 January 1839, doc. no. 3774, ibid.
22. In his paper, Talbot also situated his findings with respect to the previous literature, namely, an 1802 article that had appeared in the *Journal of the Royal Institution* regarding Thomas Wedgwood's and Humphry Davy's failed attempts to capture the passage of light on chemically prepared paper.
23. Schaaf, *Out of the Shadows*, 47.
24. [Talbot], "Scientific and Literary, Royal Society"; and Talbot, "Photogenic Drawing." For more on the dynamics of Talbot's initial publication, see Schaaf, *Out of the Shadows*, 51–55.
25. See, for example, Talbot to Herschel, 27 January 1839, doc. no. 3774; Herschel to Talbot, 30 January 1839, doc. no. 3780; Herschel to Talbot, 10 February 1839, doc. no. 3801; Herschel to Talbot, 12 February 1839, doc. no. 3805; Herschel to Talbot, 22 April 1839, doc. no. 3864; and Herschel to Talbot, 27 April 1839, doc. no. 3870, all in Schaaf, *Correspondence of William Henry Fox Talbot.*
26. Talbot to Herschel, 27 April 1839, doc. no. 3872, ibid.
27. Talbot, "Photogenic Drawing," 115.
28. Wolf, "Nature as Drawing Mistress," 137.
29. Talbot, "Photogenic Drawing," 115.
30. "Fine Arts: Photogenic Drawing," quoted in Schaaf, *Out of the Shadows*, 100.
31. Parsey, *Science of Vision*, xvi.
32. Constable, *John Constable's Discourses*, 10.
33. While Constable was giving his lectures at the Royal Institution, his son was studying chemistry with Michael Faraday at the Royal Institution. Ibid., 33.
34. Ibid., 69.
35. For more on Constable's sky studies, see Thornes, *John Constable's Skies*.
36. Constable, *John Constable's Discourses*, 67–68.

37. Sloan, *Alexander and John Robert Cozens*, 31.
38. Ibid., 53, 56.
39. Burke, *Sublime and Beautiful*, 13.
40. Gilpin, *Three Essays* (1792), 3, 6, 7, 19; see also 20, 21.
41. Klonk, *Science and the Perception of Nature*, 2–3. Klonk's account maps convincingly onto the account of modern vision offered by Jonathan Crary, who argues that by the 1840s vision had been radically reconfigured to accommodate a "a more adaptable, autonomous, and productive observer" who could "conform to new functions of the body and to a vast proliferation of indifferent and incontrovertible signs and images" produced by modernization's "deterritorialization and a revaluation of vision." Crary, *Techniques of the Observer*, 149.
42. Klonk, *Science and the Perception of Nature*, 3, 5.
43. Ibid., 78, 86.
44. Ibid., 149.
45. Constable, *John Constable's Discourses*, 11.
46. Rosenthal, *Constable*, 124, 129, 132, 171–90.
47. For more on Turner's work on perspective, see Fredericksen, *Vanishing Point*.
48. For more on *Liber Studiorum*, see Forrester, *Turner's "Drawing Book."*
49. Hamilton, *Turner and the Scientists*, 17, 19, 65.
50. Ibid., 128.
51. Kriz, *Idea of the English Landscape Painter*, 129, 137, 141.
52. Klonk, *Science and the Perception of Nature*, 151.
53. Geoffrey Batchen concludes, "In this convoluted formulation, nature is simultaneously active and passive, just as photography itself is simultaneously natural and cultural. Taken as a whole, this description neatly embodies the conceptual paradox that pervades the discourse of all proto-photographers." *Burning with Desire*, 68–69.
54. Babbage, *Reflections on the Decline*, 169. Catherine Rogers discusses how Talbot's investigations into recording light, beginning with the camera lucida, were of a piece with his interest in scientific instruments. "W. H. F. Talbot," 143.
55. See Herschel, *Preliminary Discourse*, 80–82.
56. See Daston and Galison, *Objectivity*, in particular, the discussion of self-surveillance, 174–80, and "the scientific self," 191–251.
57. Worsley to Talbot, 29 January 1840, doc. no. 04009, in Schaaf, *Correspondence of William Henry Fox Talbot*. Whewell claimed that "of the doctrines promulgated by Bacon, none has more completely remained with us, as a stable and valuable truth, than his declaration that true knowledge is to be obtained from Facts by *Induction*." *Philosophy of the Inductive Sciences* (1847), 1:vi.
58. Yeo, *Defining Science*, 52; Todhunter, *William Whewell*, 28–29.
59. Whewell reviewed Herschel's treatise favorably; for more, see Yeo, *Defining Science*, 92–99.
60. Whewell, *Philosophy of the Inductive Sciences* (1840), 2:170–71.
61. Ibid., 2:403.
62. Whewell, *Philosophy of the Inductive Sciences* (1847), 1:37, 41–42.
63. Ibid., 1:634.
64. [Herschel], "Art. VI-1: History of the Inductive Sciences," 219–20. For example, Herschel asserted that "the mind cannot give to arbitrary combinations of its own that impress of reality and unity which it acknowledges when it contemplates realities" (221).
65. [Brewster], "Philosophy of the Inductive Sciences," 266.
66. See Yeo, "An Idol of the Market-Place," in Yeo, *Science in the Public Sphere*, 278–79.
67. [Brewster], "Philosophy of the Inductive Sciences," 302.
68. Snyder, *Reforming Philosophy*, 100; see also 98.
69. Mill, *System of Logic*, 1:8, 11.
70. Snyder, *Reforming Philosophy*, 126, 99.
71. For Herschel's coining of the terms *positive* and *negative* in photography, see Schaaf, *Out of the Shadows*, 95, 96; for more on the printing process, 103–9, 112.
72. Talbot, "Fine Arts: Calotype" (13 February 1841), 108; Talbot, "Fine Arts: Calotype" (27 February 1841), 140.
73. Schaaf, *Out of the Shadows*, 121. Talbot issued his first license in August 1841 to Henry Collen, who practiced in London.
74. This album, compiled by David Brewster and now in the J. Paul Getty Museum, contains 190 photographs representing the work of at least fifteen different photographers from ca. 1849 to 1850.
75. Vered Maimon, for example, focuses on Talbot's account of his voyage of discovery, which positioned him as the "sole originator of the idea," and on the photographer's links to romantic aesthetic theories, such as those espoused by Samuel Coleridge. Maimon, "Talbot's Art of Discovery," 151. My methodological preference has been to locate Talbot explicitly within the scientific community with which he had direct contact. Carol Armstrong has called *The Pencil of Nature* "a new discursive entity"—an "experiment in putting together the photograph with text" driven by a narrative about experiments. *Scenes in a Library*, 112, 117. Armstrong's close reading of this text has been very helpful for my own work.
76. Talbot, "Brief Historical Sketch of the Invention of the Art," in *Pencil of Nature*, unpaginated (21, 12).
77. Talbot, "Introductory Remarks," ibid. (4); "Plate III, Articles of China," ibid. (35).
78. Talbot, "Plate VIII. A Scene in the Library," ibid. (52); "Plate X. The Haystack," ibid. (55). Talbot reiterated this concept with Plate XIII, which he recommended viewing through a magnifying glass, which, he said, "often discloses a multitude

of details, which were previously unobserved and unsuspected. It frequently happens, moreover—and this is one of the charms of photography—that the operator himself discovers upon examination, perhaps long afterwards, that he has depicted many things he had no notion of at the time." "Plate XIII. Queen's College, Oxford, Entrance Gateway" (65).

79. "Plate X. The Haystack," ibid. (55–56).
80. Quoted in Schaaf, *Out of the Shadows*, 147.
81. Green-Lewis, *Framing the Victorians*, 26. For an exploration of scientific photography and the concept of photography as evidence in the Victorian period, see Tucker, *Nature Exposed*; Daston and Galison, *Objectivity*.
82. See Snyder and Allen, "Photography, Vision, and Representation."
83. Talbot, "Fine Arts: Calotype" (13 February 1841), 108.
84. See, for example, Hunt's *Popular Treatise on the Art of Photography*; a fifth edition was published in 1857.
85. "Photography and the Elder Fine Arts," 30, 32.
86. Scott, *Autobiographical Notes*, 251.
87. Smith, *Victorian Photography, Painting, and Poetry*, 99. For an exploration of the links between Pre-Raphaelite painting and photography of the same decades, see also Waggoner, *Pre-Raphaelite Lens*.
88. For more on Rossetti's career as an art critic, see L'Enfant, *William Rossetti's Art Criticism*.
89. Hunt to Chesneau, 16 March 1882, HM 12973, Huntington Library, Art Collections, and Botanical Gardens.
90. For more on Hunt's self-fashioning, see Jacobi, *William Holman Hunt*, chap. 1, 18–40.
91. Prettejohn, *Art of the Pre-Raphaelites*, 35.
92. One critic noted that "the absence of perspective and aerial distance is of a piece with the original school which is imitated." Parris, *Pre-Raphaelites*, 69.
93. Ruskin, "Enquiries on the Causes of the Colour of the Water of the Rhine," 191–92; see also "Note on the Perforation of a Leaden Pipe," 193; and "Facts and Consideration on the Strata of Mont Blanc and on Some Instances of Twisted Strata," 194–96, all in *Works of John Ruskin*, vol. 1.
94. E. T. Cook, "Introduction to Vol. VIII," ibid., 8:xxv.
95. Ruskin, *Deucalion and Other Studies in Rocks and Stones*, ibid., 26:97.
96. Ruskin, "Remarks on the Present State of Meteorological Science," ibid., 1:210.
97. E. T. Cook, "Introduction to Vol. XXXV," ibid., 35:lxv.
98. Ruskin, "Praeterita: Outlines of Scenes and Thoughts, Perhaps Worth of Memory, in My Past Life, Volume 1," ibid., 35:205.
99. Cook, "Introduction to Vol. VIII," ibid., 8:xl.
100. Ruskin, "Pre-Raffaellites," 8.
101. Millais, *Life and Letters*, 1:97.
102. Tyndall addressed his letter "My dear Millais," suggesting that the two men were already well acquainted. Tyndall to Millais, n.d., unbound Millais Papers, accession no. 1485 K741. This letter was originally associated with the Mary Millais album (accession no. 7601), but loose items were removed and rehoused in the unbound Millais Papers, Morgan Library, New York.
103. Combe hosted Millais and Hunt on several occasions and also acquired several of their paintings, including Millais's *The Return of the Dove to the Ark* and Hunt's *A Converted British Family Sheltered a Christian Missionary from the Persecution of the Druids* (1849–50); he also purchased Collins's *Convent Thoughts*.
104. Powell, *Essays on the Spirit*, 15.
105. Hunt, *Pre-Raphaelitism*, 1:147, 137, 150.
106. Ruskin, *Modern Painters*, in *Works of John Ruskin,* 3:4.
107. In *Reading the Pre-Raphaelites*, Tim Barringer points out that the artists of the Pre-Raphaelite circle, particularly William Holman Hunt, began to adopt Ruskin's advice by reading *Modern Painters* even before they met him (60).
108. Ruskin, *Modern Painters*, 3:154, 152, 157.
109. Ibid., 3:622, 32.
110. Reynolds, *Discourses on Art*, 192.
111. Ruskin, *Modern Painters*, 3:33–34.
112. Ibid., 3:162. "They [young artists] have no business to ape the execution of masters; to utter weak and disjointed repetitions of other men's words, and mimic the gestures of the preacher, without understanding his meaning or sharing in his emotions" (3:623).
113. Ibid., 3:624. In the 1851 edition, Ruskin reinforced the connection to the Pre-Raphaelites, adding a footnote citing the work of Millais and Hunt, whom he described as "endeavouring to paint, with the highest possible degree of completion, what they see in nature, without reference to conventional or established rules; but by no means to imitate the style of any past epoch. Their works are, in finish of drawing, and in splendour of colour, the best in the Royal Academy; and I have great hope that they may become the foundation of a more earnest and able school of art than we have seen for centuries" (621).
114. Ruskin, "Pre-Raphaelitism," in *Works of John Ruskin*, 12:359.
115. Ibid., 12:360, 379.
116. "No one mind is like another, either in its powers or perceptions; and while the main principles of training must be the same for all, the result in each will be as various as the kinds of truth which each will apprehend; therefore, also, the modes of effort." Ibid., 12:359.
117. "Fine Arts: Royal Academy," 22 May 1852, 582.
118. Ruskin, "Pre-Raphaelitism," 12:358.

119. Ruskin, *Elements of Drawing*, ibid., 15:120, 116.
120. Bermingham, *Learning to Draw*, 244.
121. Ruskin, *Elements of Drawing*, 15:120, 121.
122. Ibid., 15:27.
123. Ruskin, *Modern Painters*, 3:36.
124. Rossetti, *Fine Art, Chiefly Contemporary*, 169, 176.
125. Ruskin to Dr. Furnivall, Glenfinlas, 16 October 1853, quoted in E. T. Cook, "Introduction to Vol. XII," in *Works of John Ruskin*, 12:xxiv.
126. See Malcolm Warner, entry on John Ruskin, in Parris, *Pre-Raphaelites*, 115–17.
127. For more on the symbolic associations of gneiss for Ruskin, see Malcolm Warner, entry on John Ruskin, in Funnell, *Millais: Portraits*, 90.
128. Quoted in Warner, "John Everett Millais's 'Autumn Leaves,'" 115–16.
129. Macleod, "F. G. Stephens."
130. [Stephens] [Seward, pseud.], "Purpose and Tendency," 58.
131. Ibid., 61.
132. Ibid., 59. For more on the value of work as expressed in Victorian art, see Barringer, *Men at Work*.
133. For an overview of the historiography of these disciplines in the eighteenth century, see Porter, "Terraqueous Globe"; and Crosland, "Chemistry and the Chemical Revolution."
134. Brande, *Dictionary of Science*, 491.
135. Babbage, *Reflections on the Decline*, 45.
136. For the formation of geology in the seventeenth and eighteenth centuries, see Porter, *Making of Geology*.
137. Barber, *Heyday of Natural History*, 216.
138. For more on these events and the larger subject of the relationship between museum architecture and the production of natural knowledge, see Yanni, *Nature's Museums*, esp. 51–61; and Alberti, "Status of Museums."
139. Ruskin, "Notes on Some of the Principal Pictures Exhibited in the Rooms of the Royal Academy (1855)," in *Works of John Ruskin*, 14:21–22.
140. See entry of 26 December 1851, Brett, "John Brett Diary," 7. In a diary entry from 23 April 1852, Brett mused about the relationship between art and science, both of which, he believed, were devoted to the external world, and he reiterated his commitment to astronomy, describing his visit with astronomer William Rutter Dawes (23 April 1852, 12). Several days later, he recounted his experiments in improving his own telescope (27 April 1852, 13).
141. Ibid., entries of 20 May 1852, 15; 18 May 1853, 37; 12 July 1853, 40.
142. "Goodman & Michael Foster took me to the University to hear the discussion on PreRaphaelitism & then persuaded me to get on to the platform." Entry of 26 May 1854, ibid., 55.
143. Entry of 9 December 1856, ibid., 64, 63.
144. Indeed, Alison Smith has suggested that the neat grouping of rocks in the foreground would have been highly improbable in nature and has proposed that the artist may have worked up these rocks in the final painting from individual sketches, a practice adopted in other works. Smith, "*Glacier of Rosenlaui*," 152.
145. Staley, *Pre-Raphaelite Landscape*, 170; see also Smith, "*Glacier of Rosenlaui*," 153.
146. For more on the relationship between the visible and invisible in the context of glaciers, see Flint, *Victorians and the Visual Imagination*, 117–38.
147. Brande, *Dictionary of Science*, 515.
148. Lyell, *Principles of Geology*, 3:149.
149. Forbes, "Illustrations of the Viscous Theory," 205.
150. Forbes, "Third Letter on Glaciers," 18.
151. For more on this painting, see David Cordingly's entry for *The Val d'Aosta*, in Parris, *Pre-Raphaelites*, 175; Barringer, "'Tracing the Finger of God,'" 140.
152. Ruskin, "Notes on Some of the Principal Pictures Exhibited in the Rooms of the Royal Academy (1859)," 14:234.
153. Ruskin nonetheless purchased the work for £200, thus helping the artist out of financial difficulties. Entry of New Year's Day, 1860, Brett, "John Brett Diary," 66.
154. Ruskin, "Notes on Some of the Principal Pictures Exhibited in the Rooms of the Royal Academy (1859)," 14:236–37.
155. Smith, *Victorian Photography, Painting, and Poetry*, 19.
156. Roberta J. M. Olson and Jay M. Pasachoff, in their study *Fire in the Sky*, dispute the suggestion that Dyce relied on photographs for this painting (238). By contrast, Clare Willsdon has analyzed the possible links between the work of Dyce, David Hill, and George Washington Wilson, in her article "Dyce 'in Camera.'"
157. See Pointon, "Geology and Landscape Painting."
158. Warner, *Victorians*, 108.
159. As Chris Brooks points out, "the little boy has the same visual weight as the rocks that dot the seashore." *Signs for the Times*, 136.
160. "Mr. Dyce's Pegwell Bay (141)—one of the most wonderfully elaborated landscapes in the exhibition, with its pallid chalk-cliffs, its dark olive low tide rocks and shallows that reflect the crimson of the sunset, and above all the great comet of 1858. Here all is modern, from the artificially scarped Ramsgate cliffs to the minutely truthful family group in balmorals and linsey-woolsey petticoats, gathering shells by the sea margin, shells of which a conchologist might discriminate the species, though the basket in which they lie is not half the size of a little-finger nail." "Exhibition of the Royal Academy," *Times* (London), 5 May 1860, 5.

161. Brande, *Dictionary of Science*, 217, 220.
162. Glanville, "Contemporary Colour Theory," 30.
163. Brewster, *Treatise on Optics*, 73, 305, 285, 290, 291.
164. Carlyle, *Artist's Assistant*, 12.
165. Field, *Chromatography*, 22, 27, 49.
166. Carlyle, "Contemporary Painting Manuals," 39; Gage, *George Field and His Circle*, 33.
167. Hunt to Mrs. Blackburn, 22 February 1858, William Holman Hunt Personal and Family Papers, ca. 1833–1930, acc. no. 860667, Getty Research Institute.
168. Quoted in Hackney, Townsend, and Ridge, "Background, Training, and Influences," 26.
169. Carlyle, "Contemporary Painting Manuals," 42–43.
170. Hackney, Townsend, and Ridge, "Pre-Raphaelite Methods and Materials," 66.
171. Ibid., 60–61; Ridge, Townsend, and Hackney, "Paintings," 114–15.
172. Hunt to Blackburn, 22 February 1858, as in n. 167; see also Carlyle, *Artist's Assistant*, 218–20.
173. Ruskin, "Realistic Schools of Painting: D. G. Rossetti and W. Holman Hunt (1883)," in *Works of John Ruskin*, 33:272–73. William Michael Rossetti praised the work when it was exhibited, "in virtue of its showing sentiment of the highest order belonging to its subject, absolute truth, and beauty and finish of execution." *Fine Art, Chiefly Contemporary*, 237. Stephens offered a more specific reading, arguing that *Our English Coasts* was "at the surface a satire on the reported defenceless state of the country against foreign invasion," and "had a deeper meaning . . . of men and not sheep" as a riposte to human behavior, in that some of the sheep follow one another blindly into danger at the edge of the cliff and tangles of brambles, while others sleep or frolic. [Stephens], "Two Pre-Raphaelitisms . . . Concluded," 354. Such readings, as Jonathan Ribner, Alison Smith, and Allen Staley have argued, were sustained by the social and historical context—specifically, fears of a potential French invasion and the factionalization of the Church of England. Smith, entry for *Our English Coasts*, 34–35; Ribner, "'Our English Coasts, 1852'"; Staley, *Pre-Raphaelite Landscape*, 79–80. See also Bronkhurst, *William Holman Hunt*, 1:156–58.
174. Stephens, *William Holman Hunt*, 20. Stephens made this claim in the context of his discussion of Hunt's *Hireling Shepherd* (1851).
175. Gage, *George Field and His Circle*, 65.
176. Hay, *Laws of Harmonious Colouring*, 8, 37, 10, 12.
177. Gage, *George Field and His Circle*, 65.
178. Hunt, *Pre-Raphaelitism*, 2:470–71.
179. Talbot to Hooker, 26 March 1839, and Hooker to Talbot, 21 June 1839, in Schaaf, *Correspondence of William Henry Fox Talbot*, doc. nos. 3845 and 3895, respectively. Anne Secord argues persuasively that for Talbot, "botany . . . provided a model for both observation and the exercise of judgment," and that this early exposure to observational science informed his development of photography. "Talbot's First Lens," 42. See also Secord, "Pressed into Service."
180. Hooker verified Fitch's images by explaining in the introductory notice, "we think we have a warrant for the fidelity of our figures in the name of our artist, Mr. Fitch." Hooker, *British Ferns*, unpaginated (introductory notice).
181. See "Ferns and Their Portraits"; Bradbury, *Nature-Printing*; and Bradbury, *Printing*. See also Wakeman, "Henry Bradbury's Nature Printed Books"; Cave and Wakeman, *Typographia Naturalis*; Harris, "Experimental Graphic Processes"; and Hume, "Nature Print and Photography."
182. Armstrong, *Scenes in a Library*, 158–60.
183. Bradbury, *Nature-Printing*, 9.
184. Lindley and Moore, "Preface," unpaginated (1–2).
185. Hamerton, *Painter's Camp in the Highlands*, 2:211. For more on this painting, see Bronkhurst, *William Holman Hunt*, 1:159–60.
186. "Fine Arts: The Winter Exhibition in Pall Mall," 558.
187. Hamerton, *Painter's Camp in the Highlands*, 2:210–11.
188. Talbot to Herschel, 18 March 1841, doc. no. 04218, in Schaaf, *Correspondence of William Henry Fox Talbot*. Anne McCauley has noted Talbot's awareness of the picturesque; see her essay "Talbot's Rouen Window." See also Batchen, *Burning with Desire*, 71–72. Gillen D'Arcy Wood has argued that Talbot's Scottish landscapes are unpicturesque because they are overly clear, differ greatly from Turner's dynamic representations of the same scenes, and do not include tourists. The difficulty with this argument lies in the ways in which Turner's work is made to stand in for the picturesque, whereas Turner's project was arguably very different from that of Gilpin or Uvedale Price, who were key figures in conceptualizing the picturesque and distinguishing it from the sublime (with which Turner's views of Scotland are more convincingly linked). Wood, *Shock of the Real*, 185–94.
189. Gilpin, *Three Essays* (1792), ii, 26, 27.
190. Batchen, *Burning with Desire*, 69, 71, 75, 77.
191. Newton, "Upon Photography in an Artistic View," 6–7. R. W. Buss, in a short essay in the *Journal of the Photographic Society of London*, offered a brief outline of photography's gradual departure from its scientific origins. When the Photographic Society began in 1853, he explained, members had "two distinct paths . . . to choose from—*one* in which the perfecting of the process of photography by subtle chemical investigation, is proposed, the correcting of lenses and various improvements in the form of the camera, affording great scope for the observation of the scientific members,—while the *other* is that in which the processors and amateurs in art would

recreate themselves." To stimulate the latter, he recommended that "the artist with his camera plunge into Pre-Raffaelism, or Rembrandtism, or Reynoldsism," and stressed Sir Joshua Reynolds as a model above all. Buss, "On the Use of Photography," 75.

192. Brewster, "Photographic Society of Scotland," 48.
193. Smith, entry for *The Brent at Hendon*, 40; and Staley, *Pre-Raphaelite Landscape*, 43. See also Bennett, *Ford Madox Brown*, 179.
194. For more on analogies between Pre-Raphaelite painting and contemporary developments in natural history, see Rosenfeld, "New Languages of Nature."
195. Brown, *Diary of Ford Madox Brown*, 88.
196. See [Brewster], "Sight and How to See."
197. Brown, *Diary of Ford Madox Brown*, 93.
198. Ibid., 102, 96, 141. For more on the painting of the turnip field, see Bennett, *Ford Madox Brown*, 180. Brown's text has been reproduced here as in the original, without spelling corrections.
199. Evans, *Forging of the Modern State*, 278.
200. Brown, *Diary of Ford Madox Brown*, 145. For more on this painting, see Bennett, *Ford Madox Brown*, 181–82.
201. [Brewster], "Sight and How to See," 147.
202. "Fine Arts: Royal Academy," 22 May 1852, 582.
203. "Fine Arts: Royal Academy," 12 May 1860, 653.
204. For a reproduction of this work, see Bronkhurst, *William Holman Hunt*, 2:136, catalogue entry D260.
205. [Stephens], "Society of Painters," 586. For more on *Sunset at Chimalditi* and *Festa at Fiesole*, see Bronkhurst, *William Holman Hunt*, 2:132–33, 135–36.
206. Helmholtz, "Sensation of Sight," in *Popular Lectures on Scientific Subjects*, 230.
207. "Society of Painters in Water Colours," 174. The British painter Charles Eastlake, in his notes to his English translation of Goethe's treatise, reported on just such an experiment undertaken by William Herschel: "If, after gazing at the sun, the eyes are closed so as to exclude the light, the image of the sun appears at first as a luminous or white spectrum upon a dark ground, but it gradually passes through the series of colours to black, that is to say, until it can no longer be distinguished from the dark field of vision; and the colours which it assumes are successively those intermediate between white and black in the order of their illuminating power or brightness, namely, yellow, orange, red, violet, and blue." Goethe, *Goethe's Theory of Colours*, 388.
208. Ruskin to Charles Eliot Norton, August 1870, in *Works of John Ruskin*, 37:20.
209. Hunt, *Pre-Raphaelitism*, 1:150.

### Chapter 2

1. Tennyson, *Alfred Lord Tennyson,* 1:380, 381.
2. Brett, "Realism in Painting," 823–25.
3. Roos, "Matthew Arnold and Thomas Henry Huxley"; DeYoung, *Vision of Modern Science*, 94–95.
4. "Banquet at the Royal Academy," *Times* (London), 6 May 1867, 6.
5. Ibid.
6. Lightman, "Science and Culture," 20; Lightman, *Victorian Popularizers of Science*, 361.
7. "Banquet at the Royal Academy," *Times* (London), 2 May 1870, 10.
8. For more on Huxley, see Jensen, *Thomas Henry Huxley*; White, *Thomas Huxley.*
9. Huxley described the audience as "at once the best and the worst in London. The best because you have all the first scientific men there—the worst, because you have a great number of fashionable ladies—The only plan is to take a profound subject—and play at battle and shuttle-cock with it so as to suit both." Jensen, *Thomas Henry Huxley*, 54. See also Secord, "How Scientific Conversation Became Shop Talk," 45.
10. White, *Thomas Huxley,* 77.
11. Turner, *Between Science and Religion*, 9.
12. Quoted in Jensen, *Thomas Henry Huxley*, 71–73.
13. For more, see Lightman's chapter "Practitioners Enter the Field: Huxley and Ball as Popularizers," in *Victorian Popularizers of Science*, 353–421; and Turner's chapter "Victorian Scientific Naturalism," in *Between Science and Religion*, 8–37.
14. Lightman, "Introduction," in *Evolutionary Naturalism in Victorian Britain*, xii.
15. Lightman, *Victorian Popularizers of Science*, 368.
16. Lightman, "Science and Culture," 14.
17. "Banquet at the Royal Academy," *Times* (London), 1 May 1871, 6. The following year, Tyndall echoed such sentiments when he proclaimed that there was "no reason why Art and Science should not dwell together in amity . . . they are both suitors of the same mistress, Nature. You love her for her beauty, we for her order and her truth," points that Sir James Paget made again in his toast of 1876. "Banquet at the Royal Academy," ibid., 6 May 1872, 9; "Banquet of the Royal Academy," ibid., 1 May 1876, 12.
18. "Banquet at the Royal Academy," ibid., 6 May 1872, 10.
19. DeYoung, *Vision of Modern Science,* 39.
20. Ibid., 29; Ruskin, "Letter 34, October 1873," in *Works of John Ruskin*, 27:445. In 1874, Ruskin delivered the Oxford Lectures on Glaciers, which criticized Tyndall and supported Forbes. See E. T. Cook, "Introduction to Vol. XXVI," ibid., 26:xxxix, xl; and Trowbridge, "'Speakers Concerning the Earth.'"

21. Smith, *Fact and Feeling*, 21, 35.

22. Tyndall, "Scope and Limit of Scientific Materialism: An Address Delivered in the Mathematical and Physical Section of the British Association in Norwich, 19 August, 1868," in *Fragments of Science*, 110, 111.

23. Ibid., 130. In 1869, Huxley had similarly drawn an anti-materialist conclusion from Goethe; as David Roos explains in reference to an article on Goethe's aphorisms on nature that Huxley wrote for the first issue of *Nature*, "Huxley was, in effect, reminding his readers of the Kantian lesson that science can never fully explain 'the wonder and the mystery of Nature,' and that, therefore, some intuitive and affective response to nature must always supplement any method of explanation." Roos, "'Aims and Intentions' of Nature," 169.

24. He wrote, for example, "All investigations of internal anatomy, whether in plants, rocks, or animals, are hurtful to the finest sensibilities and instincts of form. But very few of us have any such sensibilities to be injured; and that we may distinguish the excellent art which they have produced, we must, by duller processes, become cognizant of the facts. The Torso of the Vatican was not wrought by help from dissection; yet all its supreme qualities could only be explained by an anatomical master. And these drawings of the Alps by Turner are in landscape, what the Elgin Marbles or the Torso are in sculpture. There is nothing else approaching them, or of their order. Turner made them before geology existed; but it is only by help of geology that I can prove their power." Ruskin, *Deucalion and Other Studies in Rocks and Stones*, in *Works of John Ruskin*, 26:102.

25. Smith, *Fact and Feeling*, 169. For a fuller exploration of Ruskin's reaction to Tyndall, see Smith's chapter "Ruskin's 'Analysis of Natural and Pictorial Forms,'" ibid., 152–79. Ruskin claimed in *Deucalion and Other Studies in Rocks and Stones* that "Professor Tyndall has confused himself to the extreme point of not distinguishing these two conditions of deductive and impulsive flux. His incapacity of drawing, and ignorance of perspective, prevented him from constructing diagrams either clearly enough to show him his own mistakes, or prettily enough to direct the attention of his friends to them" (*Works of John Ruskin*, 26:189). "That Professor Tyndall is unable to draw anything as seen from anywhere, I observe to be a matter of much self-congratulation to him; such inability serving farther to establish the sense of his proud position as a man of science, above us poor artists, who labour under the disadvantage of being able with some accuracy to see, and with some fidelity to represent, what we wish to talk about" (26:161).

26. Ruskin, *The Eagle's Nest*, in *Works of John Ruskin*, 22:221.

27. Smith, *Fact and Feeling*, 172.

28. Quoted in DeYoung, *Vision of Modern Science*, 114.

29. Ruskin, "The Three Aeras, Part of a Lecture Given at the London Institution in March, 1875, with Added Pieces from Lectures in Oxford," in *Works of John Ruskin*, 26:116.

30. "The Royal Academy," *Times* (London), 7 May 1877, 10.

31. Willis, *Vision, Science, and Literature*, 2, 5.

32. "The Royal Academy Banquet," *Times* (London), 7 May 1883, 6. As David Roos has explained, Huxley's speech was part of a longer debate with Matthew Arnold, who had defended classical humanism against propositions that university education be expanded to include the sciences. Roos, "Matthew Arnold and Thomas Henry Huxley," 317. In 1887, Huxley again addressed the Royal Academy and used the opportunity to expand on his arguments regarding education. He insisted that art, literature, and science (and, with regard to the last, "the methods of inductive logic and scientific inquiry") were the three essential components of a sound education. "Royal Academy Banquet," *Times* (London), 2 May 1887, 7.

33. "The Academy Banquet," *Times* (London), 7 May 1888, 12.

34. Quoted in Ruskin, "Notes on Some of the Principal Pictures Exhibited in the Rooms of the Royal Academy (1855)," in *Works of John Ruskin*, 14:22n2.

35. Warner, "John Everett Millais's 'Autumn Leaves,'" 131, 142, 129–31.

36. "Of all the pictures Millais has painted, unquestionably the most impressive was that which we shall notice as the last he has yet produced,—'The Burning of Autumn Leaves.' By the margin of a valley-wood stand four fate-like children, who are burning fallen leaves with fire; they have gathered a fresh heap, from which the smoke creeps upward, while one continues to add more and more. The sun has sunk and dark night cometh, the whole valley is full of a luminous mist, out of which stark, denuded poplars rise at intervals, standing sharp against the sky, which has been golden, but now fadeth to a dun brassiness, while in the zenith is the black-purple of night: 'For the night cometh in which no man can work.' The children's faces are turned from the glowing west, and are in the shadow;—there is a strange impassivity upon them, as if they knew not what they did, senseless instruments of fate, not foolish, but awfully still and composed;—they gather the leaves and cast them upon the pile, half unconscious of the awful threat:

"'For wickedness burneth as the fire: it shall devour the briers and thorns, and shall kindle in the thickets of the forest; and they shall mount up like the lifting up of smoke.'

"'Through the wrath of the Lord of hosts is the land darkened, and the people shall be as the fuel of the fire: no man shall spare his brother.'

"Nothing more awful than this picture can be conceived, or out of fewer materials have we seen so much expressed. You might take it either way;—as a beautiful study of a peculiar effect of nature, such is rarely painted, and a triumph that way; or might conceive it as we have attempted to describe." [Stephens], "Two Pre-Raphaelitisms . . . Modern Pre-Raphaelites," 324.

37. Quoted in Warner, *Victorians,* 73.
38. Quoted in Warner, "John Everett Millais's 'Autumn Leaves,'" 137.
39. Corbett, *World in Paint*, 58.
40. Ruskin, "Notes on Some of the Principal Pictures Exhibited in the Rooms of the Royal Academy, and the Society of Painters in Water-Colours, No. II-1856," in *Works of John Ruskin*, 14:64–67.
41. Lightman, "Science and Culture," 33.
42. Taylor and Shuttleworth, "Unconscious Mind," 67. For more on the eighteenth-century lineage, see Hatfield, "Remaking the Science of Mind."
43. Turner, *Between Science and Religion*, 212.
44. Lightman, "Science and Culture," 33–34.
45. Mill, *Dissertations and Discussions*, 2:89. Mill reconfirmed Berkeley's finding that "the information obtained through the eye consists of two things—sensations, and inferences from those sensations: that the sensations are merely colours variously arranged, and changes of colour; that all else is inference, the work of the intellect, not of the eye."
46. Mill, *System of Logic*, 2:498–501. "Thus, if we have once seen or touched an object, we can afterwards think of the object although it be absent from our sight or from our touch" (501).
47. Bain, *Senses and the Intellect*, 1–5.
48. "Feeling is exemplified by pleasures and pains; Volition is action prompted by Feelings; Thought, or Intellect, contains the processes known as Memory, Reason, Imagination, &c. All our emotions are included under Feeling; our sensations are partly Feelings and partly Intellectual States." Bain, *Logic*, 2:275.
49. Bain, *Senses and the Intellect*, 246–47.
50. Ibid., 571.
51. In *Logic*, 1:133–227, Bain provided an extensive treatment of deduction.
52. Bain, *Senses and the Intellect*, 606–7.
53. Ibid., 606–9.
54. In *Logic*, Bain offered a definition of realism in the context of philosophy: "It has been supposed that the points of community of agreeing things exist apart from the things. This view is called *Realism*. . . . Realism is still exemplified, however, in the doctrine of an Independent External World, and also in the doctrine of the separate existence of Mind or Soul. In strictness, the External World is known only as perceived by our senses; Mind is known only as conjoined with body" (1:5–6).
55. Dale, *In Pursuit of a Scientific Culture*, 59–84. Before he wrote *The Physiology of Common Life*, Lewes completed *Comte's Philosophy of the Sciences* (1853), on the French philosopher Auguste Comte, who allowed for hypothesis and deduction within a positivist system that privileged experiential, fact-based knowledge. Smith, *Fact and Feeling*, 29–32.
56. Lewes to Millais, undated (Tuesday), Mary Millais album, accession no. 7601; Morgan Library, New York. The return address of the letter is The Priory, 21 North Bank, Regents Park, which suggests that it was written after 1864, when George Eliot and George Lewes took up residence there. Collins, *George Eliot*, 111.
57. Dale, *In Pursuit of a Scientific Culture*, 76.
58. "The correspondence, therefore, between the external world and the Perception of Sight rests, either in whole or in part, upon the same foundation as all our knowledge of the actual world—on *experience*, and on constant *verification* of its accuracy by experiments which we perform with every movement of our body," what Helmholtz referred to as the empirical theory of vision (as opposed to nativism or the innate argument). Helmholtz, "The Eye as an Optical Instrument," in *Popular Lectures on Scientific Subjects*, 315.
59. Lenoir, "Eye as Mathematician," 110.
60. Bain, *Senses and the Intellect*, 246.
61. Lewes, *Physiology of Common Life*, 2:275, 278, 292–343. "For centuries metaphysicians have been misled by the notion that the idea of an object is a *copy* of that object—an image. The source of this fallacy is their having taken vision as their standard. If it seemed plausible to say that the image seen of the fire is an accurate copy of the fire, the plausibility vanished when the pain of a burn was the sensation spoken of. When I see a rose, my eye is affected; when I smell it, my nose is affected; when I touch it, my fingers are affected; these three sensations are entirely different; yet how can I be entitled to assume that the eye gives me a more accurate copy of the rose, than the hand or nose gives me?" (329).
62. In a letter of 26 August 1858 from Turin, Ruskin wrote, "Mr. Brett was with me at La Tour. He has been here a week to-day. I sent for him at Villeneuve, Val d'Aosta, because I didn't like what he said in his letter about his present work, and thought he wanted some lecturing like Inchbold: besides that, he could give me some useful hints. He is much tougher and stronger than Inchbold, and takes more hammering; but I think he looks more miserable every day, and have good hope of making him completely wretched in a day or two more—and then I shall send him back to his castle. He is living *in*

that castle which I sketched so long ago in Val d'Aosta—Château St. Pierre." Quoted in E. T. Cook, "Introduction to Vol. XIV," in *Works of John Ruskin,* 14:xxiii–xxiv.

63. Brett, Hickox, and Payne, *John Brett*, 39.
64. John Brett Sketchbooks, 8 July 1867, no. 23, reference no. PR 86–28, PAF 9235. For Brett's notes on how best to capture moving water in paint, see ibid., September 1863, no. 13, reference no. PR 86–19, PAF 8966–70.
65. Taylor was also a popular playwright and the secretary of the Board of Health from 1854 to 1871. While simultaneous posts were not unusual for critics at the time, in that criticism was not yet a profession, Taylor was atypical of critics of his day because of his university degree, having studied mathematics and classics at Trinity College, Cambridge.
66. "Exhibition of the Royal Academy, Fourth Notice," *Times* (London), 18 May 1864, 10.
67. *Cornhill Magazine* had been started in 1859 by George Murray Smith and was edited first by William Thackeray (1860–63) and then by George Henry Lewes. In addition to articles of general interest, it featured serialized fiction for which Millais, among other artists, contributed illustrations.
68. "Present Position of Landscape Painting in England," 283, 284. This article has been attributed to John Addington Symonds.
69. Ibid., 285, 286, 287, 288.
70. Hamerton, *Autobiography*, 128, 115.
71. Hamerton, "Can Science Help Art," 285–88.
72. Hamerton, *Painter's Camp in the Highlands*, 2:418–20, 425, 431–38, 440, 458. He described the transposition as taking place over three stages: "once from the passing effect to the one selected, which usually involves a complete change of colour, and very frequently even of form; and again from the natural scale of light to the pictorial subdivided scale." The final stage was the application of imagination to the scene.
73. Taylor and Shuttleworth, "Unconscious Mind," 69. The authors explain that Carpenter was interested in questions of individual autonomy and will and understood that some human actions are habitual or "spontaneous," although "it may be ultimately possible to control them."
74. Carpenter, *Principles of Human Physiology*, 617. Carpenter established the laws of association, which include the law of contiguity ("two or more states of consciousness, habitually existing together or in immediate succession, tend to cohere, so that the future occurrence of any one of them is sufficient to restore or revive the other") and the law of similarity ("which consists in the general fact that any present state of consciousness tends to revive previous states that are similar to it") (593–94).
75. Carpenter, *Principles of Mental Physiology*, 221–22. Carpenter, upon his assumption of the presidency of the British Association for the Advancement of Science in 1872, delivered a lecture on man as an interpreter of nature. To explain the role of the scientist, he turned to the artist, whom he described as serving as "the interpreter of Nature, not when he works as the mere copyist, delineating that which he sees with his bodily eyes, and which we would see as well for ourselves, but when he endeavours to awaken within us the perception of those beauties and harmonies which his own trained sense has recognized, and thus impart to us the pleasure he has himself derived from their contemplation." This analogy between art and science led Carpenter to conclude that both disciplines sought "a representation" of nature "framed by the mind itself out of the materials supplied by the impressions which external objects make upon the senses, so that, to each man of science, nature is what he individually believes her to be." Quoted in "The British Association," *Times* (London), 15 August 1872, 4.
76. Carpenter, *Principles of Mental Physiology*, 222.
77. Colvin, "English Painters of the Present Day," 2. On the previous page, Colvin asserted, "But nature is a word of many meanings, and the idea of her entertained by these reformers [the Pre-Raphaelites] and their prophets implied too rigid a conception of things mutable as Proteus. It implied the conception of an outside world concerning which absolute agreement was possible—of something quite constant and identical lying there for sight to explore and the hand to imitate, and ready to render its very and true self wholly up to the grasp of the artist, provided only he brought to his work sufficient power and faithfulness of eye and touch" (1).
78. Ibid., 3, 4.
79. Colvin, *Memories and Notes*, 90.
80. Colvin, "English Painters of the Present Day," 6.
81. For more on this painting, see Helmreich, "Poetry in Nature."
82. Millais, *Life and Letters*, 2:29. This account is from a note, reportedly pasted to the back of the stretcher for *Chill October*, signed by Millais and dated 18 May 1882.
83. "Exhibition of the Royal Academy," *Times* (London), 29 April 1871, 12.
84. "Fine Arts: The Royal Academy (First Notice)," 29 April 1871, 531.
85. Millais, *Life and Letters*, 2:285.
86. The reviewer for the *Art Journal* proposed that the fisherman was the miller. "Royal Academy: The One Hundred and Fourth Exhibition," 151. The *Athenaeum* suggested that he was a lad from the mill covered in flour dust. "Fine Arts: The Royal Academy (First Notice)," 4 May 1872, 565. For a

reproduction of the painting, see Rosenfeld and Smith, *Millais*, 223, catalogue entry 226.

87. Fleming-Williams and Parris, *Discovery of Constable*, 72.
88. "Royal Academy: The One Hundred and Fourth Exhibition," 151; "Fine Arts: The Royal Academy (First Notice)," 4 May 1872, 565; "Art, the Royal Academy," 562; "Exhibition of the Royal Academy—First Notice," *Times* (London), 4 May 1872, 4. The critic for the *Times* wrote, "His 'Flowing to the River' (56) cannot be ranked higher than a clever sketch. It is confused and flat, and misses the greatest charm of nature, its breadth and consequent dignity. Nature, it needs no reminder, is full of details; but these are not what first take the eye. They gradually make themselves felt, stealing one by one on the attention, and perpetually refreshing the mind with new matter. It is the great masses that first arrest attention and excite the imagination. A great landscape-painter acts accordingly, and it needs only a short visit to the Liber Studiorum at South Kensington Museum, or to the rare collection of that series now at the Burlington Fine Arts Club, to see how firmly Turner held to this principle, and what exquisite truthful-ness he thus attained. But Mr. Millais seems rather to disdain study, or at least not to devote much thought to any of his pictures. What he sees at a glance (and that is not a little) he sets down with surprising and enviable ability; and it is rare indeed for him to paint a picture of any sort that does not possess some merit of high order."
89. For a reproduction of this painting, see Rosenfeld and Smith, *Millais*, 227, catalogue entry 128.
90. "Exhibition of the Royal Academy," *Times* (London), 2 May 1874, 12.
91. Sully, *Sensation and Intuition*, 340, 361, 363, 365, 366–68.
92. Ruskin, "Notes on the Principal Pictures Exhibited in the Rooms of the Royal Academy, 1875," in *Works of John Ruskin*, 14:195, 199.
93. Ibid., 14:293, 294. Ruskin also attacked another of Millais's submissions of 1875, *The Deserted Garden*, for its mechanistic painting technique and lack of recognition of labor's moral value. See Barlow, *Time Present and Time Past*, 132; Helmreich, "Poetry in Nature," 170. For a reproduction of *The Fringe of the Moor*, see Rosenfeld and Smith, *Millais*, 229, catalogue entry 129.
94. "The Royal Academy," *Times* (London), 1 May 1875, 12.
95. *"The Moon is Up, and Yet it is not Night"* was shown at the Royal Academy's annual spring exhibition, and *Dew-Drenched Furze* was exhibited at the New Gallery.
96. [Stephens], "Fine-Art Gossip," 313.
97. [Stephens], "New Gallery (First Notice)," 575.
98. Quoted in "Royal Academy Banquet," *Times* (London), 1 May 1893, 10.
99. Tennyson, *In Memoriam*, 16, 80–81.
100. See, for example, John Brett's letter to the astronomer Arthur Cowper Ranyard on his observations of the satellites of Mars, 13 January 1870, Royal Astronomical Society (RAS) Papers 34.10. Brett also borrowed telescopes from the Royal Astronomical Society. See, for example, Brett's letters to the RAS, 12 January 1875, 19 December 1876, 7 January 1878, 21 December 1878, in RAS Letters, 1875, 1876, 1878.
101. Brett to Lockyer, 18 November 1870, RAS Papers 5.1; letter book relating to the total eclipse expedition, 1870. See also Brett's letter to Ranyard, cited in the previous note.
102. Brett, letter book relating to the total eclipse expedition, 18 November 1870, RAS Papers, 51.1.
103. See Brett, "Extract from a Report of the Eclipse Expedition of 1870, at Augusta, Sicily," 163–66 (quotation on 164).
104. See, for example, all by Brett, "On Certain Phenomena Surrounding the Sun's Limb"; "Memorandum of Observations of Jupiter"; "On Certain Phenomena Seen During Eclipses"; "On the Proper Motion of Bright Spots"; "Great Patch on Jupiter"; "Specular Reflexion Hypothesis"; and "Physical Condition of Mars." Olson and Pasachoff note that Brett contributed at least eleven essays to the *Monthly Notices of the Royal Astronomical Society*. See *Fire in the Sky*, 284.
105. T., "John Brett," 239–40.
106. Ruskin to Dante Gabriel Rossetti, Denmark Hill [July? 1865], in *Works of John Ruskin*, 36:494.
107. Payne, *John Brett*, 112–13.
108. Lightman, *Victorian Popularizers of Science*, 299.
109. In response to Brett's conclusions regarding Venus, Proctor wrote, "what is coming to the Royal Astronomical Society when such a paper as Mr. Brett's, on the bright ring seen round *Venus* during transit, is allowed by the Council to be read before a meeting of the Fellows? . . . All that the spectroscope has taught us, all that we have learned of meteoric and cometic systems and of the laws of cosmical evolution, the laws of optics, and in fine the teachings of every department of modern astronomy, seem of no weight whatever with the excellent artist whose paintings of Scilly [*sic*] scenes delighted us so much a year or so since. Not content with seeing the corona in the full blaze of sunlight, and setting globes as large as the Earth bobbing in the atmosphere of *Jupiter*, he would now set astronomers to look during the transit of *Venus* in 1882 for an image of the Earth reflected in *Venus's* looking glass! What next? What next?" Quoted in T., "John Brett," 239–40.
110. Payne, *John Brett*, 107, 188, 133, 136.
111. Lightman, *Victorian Popularizers of Science*, 329.
112. Quoted in Lightman, "Knowledge Confronts Nature," 203.
113. Kjaergaard, "'Within the Bounds of Science,'" 212–13.
114. Roos, "'Aims and Intentions' of Nature," 172.

115. Lightman, "Knowledge Confronts Nature," 203, 205.

116. Christiana Payne proposes that Lockyer and Brett met in 1868 through the New Club. *John Brett*, 107.

117. Brett, "Natural Science at the Royal Academy." Brett seems to anticipate Ruskin's arguments in *The Eagle's Nest* (1872) and recalls those of *The Queen of the Air* (1868). Indeed, Brett's example was the scientific analysis of the sky—which he argued would benefit from the painter's practice "of comparing the intervals between different colours, and estimating the intensities as well as the extent of several elements that go to make up any given scenes in the out-of-doors world" (158). In *The Queen of the Air*, Ruskin had drawn on Tyndall's recent lecture on the behavior of light to shore up the Greek mythical explanation of why the sky was blue. Ruskin, *The Queen of the Air: Being a Study of the Greek Myths of Cloud and Storm*, in *Works of John Ruskin*, 19:292. Brett continued into the 1870s to call for uniting the two disciplines, still believing that they could learn from each other: "if either would go to school with the other, they would, in my opinion, each stimulate and aid the labours of the other and divide between them a fair larger share of the spoils of the world" (158).

118. John Brett Sketchbooks, no. 32 (Ireland 1869), 1–3 and inside front cover, reference no. PR 86–38.

119. Daston and Galison, "Image of Objectivity," 82, 83.

120. "Exhibition of the Royal Academy," 236.

121. P., "Royal Academy Exhibition," 79.

122. White, "Royal Academy Exhibition of 1873," 295. John Forbes White was an important Scottish art collector who favored the work of George Reid, Corot, and the Hague school, and who also knew Alexander Bain and was associated with the British Association for the Advancement of Science. He had studied at the University of Aberdeen and recalled his natural history professor as a great influence, particularly in teaching skills in "observation, comparison, classification, and generalization." His income derived from his father's flour-milling business, which he eventually took over. Harrower, *John Forbes White*, 57, 69, 27, 28.

123. Quilter, *Opinions on Men, Women, and Things*, 110 (my thanks to Pamela Fletcher for drawing my attention to this work). Quilter contributed to general interest periodicals such as the *Contemporary Review* and *Cornhill Magazine*, was the critic for the *Times* for a year upon Tom Taylor's death, and served as the art critic for the *Spectator* from 1876 to 1887.

124. Ruskin, "Notes on the Principal Pictures Exhibited in the Rooms of the Royal Academy, 1875," in *Works of John Ruskin*, 14:293–94.

125. Hackney, "Cleaning John Brett's *The British Channel*."

126. 1. paint reflections (correctly near coast) in soft streaks downwards (diluted into grey towards foreground—feeble general reflections). Flat & dry.
2. Paint blue element all over- darker & greyer towards foreground.
3. While wet knife off parts where the calm is complete, in flecks and wisps over broad spaces especially near the coasts. Flat & dry.
4. Knife out reflections
4 [*sic*] Touch blue element with darker blue patches?
The blue element sometimes bordered with dark green
In glassy seas the blue element has soft edges
Where cloud shadows are the blue elements becomes purple
Oct. 14/63.

John Brett Sketchbooks, no. 13, reference no. PR 86–19, between PAF 8985 and PAF 8986, PR 86–19.

127. Hackney, "Cleaning John Brett's *The British Channel*."

128. Brett conscientiously attempted to explain optical phenomena and their equivalents in art; for example, he explained the distinction between "dispersion of light" and "scattering of light" and how these methods translated into articulating texture or a smooth polished surface in a painting or sculpture. Brett, "Functions of Texture in the Arts," 118.

129. Lockyer, "Physical Science for Artists," 30.

130. Fenn, "Fact in Landscape," 395.

131. Hodgson, *Fifty Years of British Art*, 93. The Chantrey Bequest denotes funds left by the sculptor Francis Chantrey for the purchase of works of art for the nation; it was overseen at this time by the Royal Academy.

132. Cook, *Popular Handbook to the Tate Gallery*, 231–32.

133. Hunt, "Modern English Landscape Painting," 570–71, 574, 579. Hunt drew attention, for example, to "realistic treatment—of that wish to look closely and keenly at nature in the fields as well as in the laboratory or herbarium, which has wrought confusion and change in practice and criticism and everything connected with landscape painting" (570–71).

134. "State of Art in England," 618.

135. For example, there was no visual evidence for plate tectonics, which therefore did not occur: "the sooner you disabuse your mind of the childish conception of 'upheaval' the better. You cannot open any book of geology without encountering this figment, yet any landscape-painter should know not only that there is no sufficient evidence of any thing so outrageous, but that it is wholly unnecessary as an explanation of the visible facts." Brett, *Three Months on the Scottish Coast*, 12.

136. Ibid., 4, 7, 8. A sketch, he explained, is "the painting of a single observation unadulterated, comprising whatever happens to occur at the time, and as much of it as circumstances

allow to be recorded at one sitting; and it is usually a hurried performance." In contrast, Brett noted, a finished work "is deliberately done under favourable conditions of light and shelter, and is generally an abstract of several observations, supplemented by whatever is deemed favourable to the presentment of the subject" (8).

137. Ibid., 10, 9, 12. As Brett explained, "when you have to register an observation in definite symbols, such as words or numbers, you are impelled to accuracy, and your memory is cultivated" (9).
138. Ibid., 11, 12.
139. Robins, *Fragile Modernism*, 9, 27, 23.
140. Quoted in ibid., 25.
141. Quoted in ibid., 31.
142. Whistler to Helen Euphrosyne Whistler, January 1884, system no. 08164. MacDonald, Montfort, and Thorp, *Correspondence of James McNeill Whistler*.
143. Whistler, "Ten O'Clock," 84.
144. For the original series of essays, see *Portfolio* 17 (1886): 1–7, 26–32, 53–58, 70–76, 85–91, 117–23, 125–29, and 152–59. I quote from the published collection of 1887 in Hamerton, *Imagination in Landscape Painting*.
145. Hamerton, *Imagination in Landscape Painting*, 1, 4, 20–21, 47, 66, 70.
146. Ibid., 6.
147. Galton, *Inquiries into Human Faculty*, 57–79, 66, 73. Hamerton countered that "the memory of an artist is distinguished from that of other people chiefly by this, that as he alone has really the knowledge of appearances, he is the only person who is able to call up images before the mind's eye which are distinct enough, and permanent enough, to be painted." Hamerton, *Imagination in Landscape Painting*, 19. Boisbaudran's treatise on drawing was translated into English in 1911 as *The Training of the Memory in Art and the Education of the Artist*. The French painter Henri Fantin-Latour studied with Boisbaudran, and Whistler became familiar with his work through his good friend Fantin-Latour, with whom he formed the Societé de Trois in 1858, along with Alphonse Legros.
148. Hamerton, *Imagination in Landscape Painting*, 72, 73.
149. Merrill, *Pot of Paint*, 148.
150. As stated by a critic for the *Pall Mall Gazette* in 1885, quoted in Robins, *Fragile Modernism*, 3.
151. "Art and Science in a New Light," 251.
152. Payne, *John Brett*, 161–63, 173.
153. Brett, "Relation of the Pictorial," 47.
154. Brett, "Relation of Photography," 83, 84.
155. For more, see Brett, Hickox, and Payne, *John Brett*, 54, 132.
156. Brett, "Realism in Painting," 825.
157. Ibid., 829–30.

## Chapter 3

1. Handy, "Art and Science," 181.
2. Marien, "Peter Henry Emerson," 102.
3. "[He] took up a volume of Comte and was soon lost in the 'Classification of the Sciences,' and wondering whether he was not something of a Positivist, for long since he had arrived at a state when he ceased to take everything for granted, but rather tried in his own mind to gauge and weigh it all, often finding many fallacies." Emerson, *Paul Ray at the Hospital*, 17.
4. Ibid., 21. The missionary is reminiscent of Philip Gosse.
5. Ibid., 17, 69–70, 79, 124, 140.
6. For more on photographic societies, see Harker, *Linked Ring*, 43–45.
7. Emerson, "Photography: A Lecture Delivered at the Priory Hall, Great Yarmouth," *Amateur Photographer* (2 April 1886), in Emerson, "Articles and Notices," 19.
8. Lenman, "British Photographers and Tourism," 92.
9. Taylor, *Old Order and the New*, 45.
10. For example, Nathaniel Cook published a new edition in 1853 edited by the naturalist William Jardine, who also edited the *Naturalist's Library*.
11. Thornton, *Fifty Years of the New English Art Club*, 3.
12. Quoted in McConkey, "Rustic Naturalism at the Grosvenor," 134.
13. "Art: The New English Art Club," *Spectator*, 16 April 1887, in New English Art Club, press clipping albums, 1:8–9. As the *Sunday Times* put it, "it was time to assert their artistic independence, to defy cheap popularity, and by producing work in accordance with their individual feelings and perception of truth, make an earnest effort to free English art from sentimentalism, technical conventionality, and commonplace." *Times* (London), 15 April 1888, ibid., 2:43.
14. *Glasgow Herald*, 19 April 1889, ibid., 3:112.
15. *Bohemian*, 23 April 1887, ibid., 1:3.
16. In 1886, the *Art Journal* highlighted the "unusual excellence of the wintry landscape by Mr. T. F. Goodall, shown in the New English Art Club annual exhibition," and described the painting *The Last Load* as "imbued with the spirit of Nature, and characteristically true in its rendering of the atmospheric effect dealt with." "New Art Club," 252.
17. Goodall, "Landscape," in Emerson and Goodall, *Life and Landscape*, 73, 74. Goodall critiqued much contemporary British art, arguing that it was produced only to achieve a visual spectacle and, furthermore, was "belittled by an infinity of imitative detail which passes for finish" (75).
18. Ibid., 78.

19. Emerson, "Photography: A Pictorial Art," *Amateur Photographer* (10 March 1886), in Emerson, "Articles and Notices," 18.
20. Helmholtz, "On the Relation of Optics to Painting," in *Science and Culture*, 279. Helmholtz described the aim of the artist as "to produce a kind of optical illusion," that is, an "artistic representation [that] produces in us a conception of their objects as vivid and as powerful as if we had them actually before us" (280). In the section devoted to shade, he elaborated: "It might at first appear that of the requisite truth to nature of a picture, so much would remain that, seen from the proper point of view, it would at least produce the same distribution of light, colour, and shadow in its field of view, and would produce in the interior of the eye exactly the same image on the retina as the object represented would do if we had it actually before us, and looked at it from a definite, fixed point of view. It might seem to be an object of pictorial skill to aim at producing, under the given limitations, the *same* effect as is produced by the object itself" (288).
21. Emerson, "Photography: A Pictorial Art," 17–18 (quoting Helmholtz, "On the Relation of Optics to Painting," 288, 290, 291). Helmholtz asked, "How now is it possible that, under such circumstances, we can imagine there is any similarity between the picture and reality?" (290).
22. Ibid., 18 (quoting Helmholtz, "On the Relation of Optics to Painting," 293; Helmholtz was drawing here on the work of Gustav Fechner).
23. Ibid.
24. A Member, "Photography and Art," 285.
25. John Tyndall became Helmholtz's major interlocutor in Great Britain, helping to translate and popularize his work; Helmholtz in turn translated Tyndall's research into German. Beer, *Open Fields*, 246. Emerson is referring here to what is known as the Tyndall effect—the scattering of light by particles suspended in a medium, such as water or air, through which light can pass. Tyndall referred to this effect in his lecture "Scientific Use of the Imagination": "Thus, while the reflected light gives us at noon the deep azure of the Alpine skies, the transmitted light gives us at sunset the warm crimson of the Alpine snows. The phenomena certainly occur *as if* our atmosphere were a medium rendered slightly turbid by the mechanical suspension of exceedingly small foreign particles." Tyndall, *Fragments of Science*, 142.
26. Emerson, "Photography: A Pictorial Art," 18.
27. See Emerson, "The Ideal Photographic Exhibition," *Amateur Photographer* (23 October 1885), in Emerson, "Articles and Notices," 15, and "Photography: A Pictorial Art," 17. Works by members of the Barbizon school were frequently on display in London in the last quarter of the nineteenth century, largely through the efforts of the Goupil Gallery. See Helmreich, "Goupil Gallery at the Intersection."
28. Emerson, "Photography: A Lecture Delivered at the Priory Hall," *Amateur Photographer* (2 April 1886), in Emerson, "Articles and Notices," 19. Works by these artists were in the collections of the National Gallery of Art by this date. Meindert Hobbema's *Stream by a Wood* (ca. 1663) had been acquired in 1871; *A Woody Landscape* (ca. 1665) was acquired in 1862; *A Woody Landscape with a Cottage* (ca. 1665) was part of the Wynn Ellis Bequest (1876); *The Avenue at Middelharnis* (1689) was purchased in 1871; *The Ruins of Brederode Castle* (1671) was acquired in 1871; and *The Watermills at Singraven near Denekamp* (ca. 1665–70) was purchased in 1871. Jacob van Ruisdael's *Waterfall at the Foot of a Hill, Near a Village* (ca. 1665–70) was purchased in 1871; *Ruins in a Dune Landscape* (ca. 1650–55) was purchased in 1875; and *A Landscape with Ruined Castle and a Church* (ca. 1665–70), *Two Watermills and an Open Sluice at Singraven* (ca. 1650–52), *A Torrent in a Mountainous Landscape* (ca. 1665–70), *A Landscape with a Ruined Building at the Foot of a Hill by a River* (ca. 1655), *A Road Winding Between Trees Towards a Distant Cottage* (ca. 1645–50), and *Three Watermills with Washerwomen* (ca. 1670) were part of the Wynn Ellis Bequest (1876); *A Waterfall in a Rocky Landscape* (ca. 1660–70) was purchased in 1859; *A Bleaching Ground in a Hollow by a Cottage* (ca. 1645–50) was bequeathed by Sir John May in 1854; *A Pool Surrounded by Trees, and Two Sportsmen Coursing a Hare* (ca. 1665) was acquired in 1871; and *A Landscape with a Waterfall and a Castle on a Hill* was bequeathed by Johann Moritz Oppenheim in 1864.
29. Goodall, "Landscape," in Emerson and Goodall, *Life and Landscape*, 79. This was not an original approach. Sir William Newton, at a meeting of the Photographic Society of London in 1853, offered similar advice about putting the subject slightly out of focus, "thereby giving a greater breadth of effect, and consequently, more *suggestive* of the true character of nature." Newton, "Upon Photography in an Artistic View," 6.
30. "The rule of focussing, therefore, should be, focus for the principal object of the picture, but all else must not be sharp; and even that principal object must not be as perfectly sharp as the optical lens will make it." Emerson, *Naturalistic Photography,* 119.
31. Goodall, "Landscape," in Emerson and Goodall, *Life and Landscape*, 81.
32. Ibid., 75.
33. Emerson, *Pictures from Life*, 10.
34. Green, "'Right Thing in the Right Place,'" 91. See also Lukacher, "Powers of Sight," 31.
35. "It is an old canon of art that every scene worth painting must have something of the sublime, the beautiful, or the picturesque. By its nature, photography can make no pretensions

to represent the first, but beauty can be represented by its means, and picturesqueness has never had so perfect an interpreter." Robinson, *Pictorial Effect in Photography*, 15.

36. Goodall, "Landscape," in Emerson and Goodall, *Life and Landscape*, 77.
37. Green, "'Right Thing in the Right Place,'" 95.
38. As John Taylor notes with respect to *A Dame's School* and other photographs by Emerson in which the figures are obviously posed, "setting-up was necessary" in order to avoid manipulating the negative as Robinson did, but "it was not a solution to the perfect naturalistic photograph," and to achieve the latter Emerson would have to renounce "storytelling" photography. Taylor, *Old Order and the New,* 28.
39. Emerson, *Idyls of the Norfolk Broads*, vii, x.
40. Ibid., vi. This stanza is the third of the third section of the poem. The poem had first been published in 1880 in Swinburne's collection *Studies in Song.*
41. McSweeney, "Swinburne's 'By the North Sea,'" 227, 222.
42. Emerson added, "In short, I trust the complete work may form a humble contribution to a Natural History of the English Peasantry and Fisherfolk." *Pictures of East Anglian Life,* iii.
43. Ibid.
44. Palermo, "World in the Ground Glass," 144.
45. Emerson, *Pictures of East Anglian Life,* 69, 133–34.
46. McConkey, *George Clausen*, 19–41.
47. See Lübbren, *Rural Artists' Colonies in Europe.*
48. Robins, "Living the Simple Life," 1–27, esp. 3, 9.
49. McConkey, *George Clausen*, 54, 61.
50. Ibid., 80.
51. *Pall Mall Gazette*, 9 April 1887, in New English Art Club, press clippings albums, 1:12.
52. Stevenson also contributed notices and essays to the *Pall Mall Gazette*, *Saturday Review*, *Magazine of Art*, *New Review*, and *Portfolio.*
53. Clausen, "Bastien-Lepage and Modern Realism," 114. Clausen was not alone in regarding Bastien-Lepage as a truthful painter; this was the dominant discourse in English criticism of the painter. See Wright, "Bastien-Lepage and English Critical Taste," 102.
54. Clausen to David Croal Thomson, 30 October 1889, George Clausen Letters, 1888–1909, reference no. MSL/1991/4, pressmark 86.YY.19.
55. Quilter, "Tendencies of French Art," 867, 868.
56. Frith, "Crazes in Art," 191; Frith, "'Realism' Versus 'Sloppiness,'" 7.
57. Phillips acknowledged that Clausen's work had been traced "ad nauseam" to these two painters. Nonetheless, Phillips wanted to reiterate Clausen's indebtedness to them, as Bastien-Lepage was recognized as the "pathetic realist *par excellence* of the present generation," a note of high praise. Phillips, "Summer Exhibitions at Home and Abroad," 171, 170.
58. See Helmreich, "John Singer Sargent."
59. McConkey, *George Clausen*, 92.
60. Handy, "Pictorial Beauties, Natural Truths," 3.
61. Emerson, "Science and Art," *Photographic News* (3 May 1889): 295, in Emerson, "Articles and Notices," 32–34. This aligns with Sully's declared aim for art.
62. Scientific photography was a distinct genre by the end of the nineteenth century; scientific photographic images, as Jennifer Tucker has established, were regarded as empirical proof and as a "neutral mirror of reality." This characterization, Tucker reminds us, was based on complicated arguments about "selectivity, judgment, skill and aesthetics," arguments akin to those in which Emerson was engaged. Tucker, *Nature Exposed,* 235.
63. Emerson, *Naturalistic Photography,* 10–11, 97.
64. Ibid., 106–7; Helmholtz, "On the Relation of Optics to Painting," 307.
65. For example, in his lectures on light, Tyndall asserted, "The eye is by no means a perfect optical instrument. It suffers from spherical aberration; a scattered luminosity, more or less strong, always surrounding the defined images of luminous objects upon the retina." Tyndall, *Notes of a Course of Nine Lectures*, 26. Quoting from Helmholtz, Emerson explained the difference between direct and indirect vision, which helped to account for flaws in human vision: "To *look* at anything means to place the eye in such a position that the image of the object falls on the small region of perfectly clear vision. This we may call *direct* vision, applying the term *indirect* to that exercised with the lateral parts of the retina, indeed with all except the central spot." He continued, quoting again from Helmholtz, "Whatever we want to see we look at and see it accurately; what we do not look at, we do not as a rule care for at the moment, and so do not notice how imperfectly we see it." Emerson, *Naturalistic Photography,* 102.
66. Emerson, *Naturalistic Photography*, 114.
67. Whewell, *Philosophy of the Inductive Sciences* (1847), 1:42.
68. Helmholtz, *Popular Lectures on Scientific Subjects*, 274.
69. Emerson, *Naturalistic Photography*, 114, 116, 118, 119.
70. Ibid., 22. The use of the term *impression* recalls the language Helmholtz used to describe the operation of the eye: "all that we apprehend of the external world is brought to our consciousness by means of certain changes which are produced in our organs of sense by external impressions, and transmitted to the brain by the nerves." Helmholtz, *Popular Lectures on Scientific Subjects*, 230.

71. "The method of rendering a picture as it appears to the eye has been called interpreting nature." Emerson, *Naturalistic Photography*, 22, 24–25.
72. Ibid., 22, 23.
73. Quoted in Newhall, *P. H. Emerson*, 65, 66.
74. See, for example, Emerson, "Correspondence," 380–81, in which Emerson responds to Robinson's critique.
75. Charles to Emerson, 27 March 1889, Papers of Peter Henry Emerson, 1886–1920, reference no. MC 2235, 1/3/1.
76. Crucially, Brett did not rehearse the tired argument that because photography was better than painting at achieving truth to nature, painting should eschew naturalism. Brett maintained that painting and photography should pursue the same aims, but photography forced him to acknowledge the degree to which he might idealize "natural images" —what he called "develop[ing] their utmost capacity"—by eliminating flaws. Drawing on memory and the artist's "experiences of the range of natural variations," the painter could derive the best version of the thing he sought to depict. Brett, "Relation of Photography," 84.
77. Emerson, "Science and Art," 87, 88.
78. "Discussion," 90.
79. Clausen to Emerson, n.d., Papers of Peter Henry Emerson, 1886–1920, reference no. MC 2235, 1/4/1. In a subsequent letter, Clausen praised Emerson's essay "Science and Art," generally agreeing with his distinction between science and art, while nonetheless qualifying some of Emerson's assumptions. He noted, for example, that artists had long been careful observers of external nature, pointing out that Rembrandt's images show that Rembrandt understood the behavior of light rays. Clausen to Emerson, 19 May 1889, ibid., 1/4/2. In another letter, Clausen criticized Emerson's photographs—he was probably referring to *Pictures of East Anglian Life*—for failing to achieve full truthfulness. One laborer's shirt, for example, was too bright for the environmental conditions, and the figures of *In the Barley Harvest* were too obviously posed. Clausen to Emerson, n.d., ibid., 1/4/4.
80. Balfour, "Artistic Focus," 71; Little, "Art, Nature, and Photography," 91.
81. Robinson, "Use of Nature and Idealism," 106–7. George Davison, whose photography practice owed a great deal to Emerson, challenged Robinson and, in particular, his nomenclature. Davison, "Limits and Possibilities," 213–16. For more on Davison, see Brian Coe, "George Davison."
82. Taylor, "Phase of Naturalistic Focussing," 281; Murchison, "Memory and Imagination," 195–96.
83. Emerson, "Renunciation," 55.
84. Clausen attended the 1889 Paris Exhibition, as documented in a letter to the art dealer David Croal Thomson; he described the exhibition as a "very wonderful show—and a valuable lesson." Clausen to Thomson, 30 October 1889, George Clausen Letters, 1888–1909, reference no. MSL/1991/4, pressmark 86.YY.19. See also Stone, "Peter Henry Emerson," 13.
85. Emerson to Thomas, 26 January 1890, "Letters to James Harvard Thomas."
86. Callender, "'Dear Mr. Driffield,'" 319.
87. Emerson, "Renunciation," 54.
88. Ibid., 54–55.
89. The critical response to Emerson's text was swift. Davison set about undermining Emerson's polemic, which he attributed to an "outcome of violent egotism and offended vanity." Most significantly, he insisted that "the limitations of photography are not so great as he [Emerson] now . . . wishes to make out, for the individuality of the worker is freer than he asserts, and he has allowed himself to be misled by an absurd confusion about Messrs. Hurter & Driffield's experimental results." Davison, "Reply to Dr. Emerson," 70. The *Amateur Photographer* weighed in with a response probably written by Robinson, who was clearly delighted that Emerson's aesthetic had been discredited. Newhall, *P. H. Emerson*, 95.
90. Emerson to Thomas, n.d., "Letters to James Harvard Thomas."
91. Emerson and Goodall, *Perspective Drawing and Vision*, 1.
92. Ibid., 3. They explained in conclusion, "Having shown how we see forms, it only remains to say that a mathematical perspective drawing or the drawing of an aplanatic photographic lens does not give forms as we see them. They are altogether false to the visual impression of the proportions of things, and therefore give a wrong idea of the original scene. On the other hand, a perspective drawing or correct photography gives the *actual facts* scientifically, *i.e.,* the pillars of the temple as leaning, the paper in experiment as *square*. All such drawings are, therefore, purely scientific diagrams, and artists, who wish to render what they see, must not rely upon them" (7).
93. In the 1892 article, Emerson divided the process of taking a photograph into two phases—the first was inherently artistic in that it required the aesthetic discernment and knowledge to select the scene, "arrange it on the screen," and choose the focus; the second, which began at the moment the photograph was taken, was inherently mechanical in that "A MACHINE DOES IT ALL FOR HIM." This second phase meant that the artist's individuality left only a trace presence in the image and, above all, that "machine work can never be art." Emerson, "Photography Not Art," 149, 150, 154.
94. This essay was first delivered as a paper to the Photographic Society of Great Britain and was subsequently published in

the *Photographic Journal* and in the third edition of *Naturalistic Photography for Students* (1899).

95. Emerson, "Naturalistic Photography," 156.
96. Ibid., 157.
97. "Now, gentlemen, I will quote you a paragraph from a text-book on Psychology, published only last year; a passage which shows how this view is now so far accepted that it has entered psychological text-books for students. The quotation is:—'Almost all the VISIBLE shapes of things are what we call perspective *distortions.* Square table tops constantly present two acute and two obtuse angles, circles drawn on our wall-papers show like ellipses, &c., and the transition from one to another of these altering forms are infinite and continual.'" Ibid. See James, *Psychology*, 344.
98. James, *Psychology*, 1, 7, 312, 316, 319, 329.
99. Emerson, "Naturalistic Photography," 161–62.
100. By contrast, Brian Lukacher notes, "For Emerson, turbidity and luminescence were heightened conditions of vision that verged wondrously on the eradication of sight itself." The resulting images were "temperamental, iconoclastic gestures against his own powers of sight." Lukacher, "Powers of Sight," 50.
101. McConkey, *George Clausen*, 79.
102. Clausen to Emerson, 19 May 1889, Papers of Peter Henry Emerson, 1886–1920, reference no. MC 2235, 1/4/2; Clausen to Emerson, n.d., ibid., 1/4/1. In the latter letter, Clausen explained, for example, that perception of distance, mass, and volume was shaped by binocular vision—that is, an object examined with one eye appears differently than when examined with both eyes—but in a painting, rather than attempt to replicate the viewpoint of the single eye (because a picture represented a single, fixed point of view), Clausen advocated "the subtle mingling of 2 different points of view." Emerson had argued that the lens of the human eye was differentiated from top to bottom, and to prove his point he used the example of looking at a tree upside down. Clausen countered that the tree, when examined upside down, might appear differently not because of flaws in the lens of the eye but simply because "we see things for a moment *freshly.* The impression of the sensuous image is not modified in any way by consciousness of what the object *is.*" Such commentary reveals that Clausen's conception of optical truth allowed for the role of mental perception.
103. Looking at painting exhibitions prompted Clausen to wonder "in regard to pictures (& especially 'realistic' ones) whether they were worth doing & [to] what extent? I don't think every subject requires a large canvas—or finished minuteness of work." Clausen to Emerson, n.d., ibid., 1/4/4.
104. Clausen to Emerson, n.d., ibid., 1/4/5.
105. "Impressionism and advertising seem to go together—too much so for me—I don't see why I shd ~~show~~ continue to share the responsibility for proceedings of wh. I don't approve. I've protested against it to no effect." Clausen to David Croal Thomson, 8 June 1890, George Clausen Letters, 1888–1909, reference no. MSL/1991/4, pressmark 86 YY 19.
106. *Pall Mall Gazette*, 9 April 1887, in New English Art Club, press clippings albums, 1:12. For more on Steer's career, see Holt, *Philip Wilson Steer.*
107. For more on Whistler's engagement with the next generation of painters in Britain, see Robins, *Fragile Modernism*, particularly chaps. 1 and 2.
108. See, for example, Sickert's reviews of the 1889 NEAC exhibition in his *Complete Writings on Art*, 33–44, 53–56.
109. Ibid., 57. For more on Sickert's admiration of Degas, see Robins, "Greatest Artist."
110. Robins, *Fragile Modernism*, 150–59.
111. *Pall Mall Gazette*, 16 April 1889, in New English Art Club, press clippings albums, 3:110.
112. Robins, *Fragile Modernism,* 112.
113. "From our Lady Correspondent," *Belfast News Letter,* 9 April 1888, in New English Art Club, press clippings albums, 2:33.
114. "Once more the New English Art Club is to the fore, and with a new exhibition the rumour of secessions is again in the air. . . . If the party of Messrs. Sickert, &c., continue to sky [i.e., place very high on the exhibition wall], reject, and otherwise maltreat the works of the naturalistic school, and, in fact, of every other school but that of the London impressionists, then we fear that the club will fail for want of funds, for it is an open secret that it has been the naturalistic painters rather than the present committee who have 'run the show.'" "Art Gossip," *Pictorial World*, 21 November 1891, ibid., 5:289.
115. Robins, "London Impressionists."
116. Sickert, *London Impressionists*, 5–7. Sickert added this clarification: "It has no wish to record anything merely because it exists. It is not occupied in a struggle to make intensely real and solid the sordid or superficial details of the subjects it selects" (7).
117. Robins, *Fragile Modernism*, 177. For more on the ambiguity of the term *impressionism* in British art circles, see ibid., 6.
118. Ibid., 162.
119. Becker, "George Moore," 51. See also, Flint, "'Philistine' and the New Art Critic."
120. Steer, "Impressionism in Art," unpaginated.
121. Holt, *British Artists and the Modernist Landscape*, 34.
122. Robins, *Fragile Modernism*, 166, 167, 168–72; McConkey, *Sir George Clausen*, 55, 76.
123. Stevenson, "George Clausen," 289, 291.

124. The *Saturday Review* commented, "In Mr. Clausen's picture the light wraps the figures and swamps detail; whereas even in Bastien's *Potato Gatherer* [*sic*] the evidence of light is weakened by many over-nice differences of tone." Quoted in McConkey, *Sir George Clausen,* 63.
125. Quoted in Jenkins, *Painters and Peasants*, 126.
126. McConkey, *Memory and Desire*, 150.
127. [Moore], "Royal Academy," 832–33. Moore compared Clausen's painting favorably to Stanhope Forbes's *Forging the Anchor*, in which the artist "copies the trousers seam by seam, patch by patch; and the ugliness of the garment bores you in the picture, exactly as it would in nature" (836).
128. Clausen, "Jules Bastien-Lepage as Artist," 111, 115, 119, 120, 124, 125, 126.
129. Ibid., 126.
130. Ibid.
131. Sickert, "Modern Realism in Painting," 141.
132. [Moore], "Book About Bastien-Lepage," 227.
133. "The New English Art Club," *Speaker*, 15 April 1893, in New English Art Club, press clippings albums, 8:658.
134. Clausen to Thomson, 11 November 1894, George Clausen Letters, 1888–1909, reference no. MSL/1991/4, pressmark 86.YY.19.
135. [Moore], "Exteriority," 685. Steer's work underwent a similar change, as described by Roger Fry. In 1902, when reviewing Philip Wilson Steer's exhibition at the Carfax Gallery, Fry recognized that the painter, while still working from the motif, could no longer be classified as a "convinced naturalist." Instead, Steer now wrestled with "the conflicting claims of nature and style," and he resolved the debate in favor of "a more formal mode of composition, a more deliberate design." Like others of his generation, Steer had come to realize that his "atmospheric colour" studies had only a "slight . . . leverage upon the emotions." He chose to chart a new path, studying nature to extract an overriding design or composition—what Fry referred to as "the shorthand in which Mr. Steer describes his vision." [Fry], "Fine Arts," 761. See also Laughton, *Philip Wilson Steer,* 91.

### Chapter 4

1. For more on the Camden Town Group, see Holt, "The Camden Town Group Then and Now," in Bonett, Holt, and Mundy, *Camden Town Group in Context.*
2. Ginner, "Neo-Realism," in *Catalogue of an Exhibition*, 7.
3. B., "Art."
4. For a full account of these exhibitions, see Robins, *Modern Art in Britain.*
5. For more on the galleries related to the Camden Town Group, see Shaw, "Carfax Gallery"; and Helmreich, "Socio-Geography of Art Dealers."
6. Tickner, *Modern Life and Modern Subjects*, 52–54.
7. John, "Autobiographical Fragments," 319.
8. "Tis," "Augustus John," 148.
9. John, "J. Dickson Innes," 3–4.
10. For example, John fostered a very close relationship with Lady Ottoline Violet Anne Morrell. See their correspondence in the Ottoline Morrell Collection, RLIN record no. TXRC98-A17, box 10, folders 7 and 8, box 11, folders 1 and 2, Harry Ransom Center, University of Texas at Austin.
11. "Tis," "Augustus John," 149.
12. See, for example, Bury, *Orientalisme Victorien.*
13. For example, Augustus John, in a series of letters written to his patron John Quinn, recounted his sexual encounters with a female servant in bravura language—at the same time that he was married and had a mistress (both women traveled with him in his gypsy caravan). John to Quinn, 11 January 1910 or 1911, 153 Church Street, Chelsea, John Quinn Papers. Innes, in a letter to his tutor John Fothergill, parodied a poem by Blake to delineate the influences on his life, including alcohol, tuberculosis, and his teachers Fothergill and Philip Wilson Steer, along with a childhood friend "who shewed me his cock." Innes to Fothergill, 28 January 1909, St. Ives, Cornwall, in Fothergill, "Correspondence and Drawings," reference no. TGA 7121.5.
14. John to Quinn, 2 February 1913, Alderney Manor, Dorset, John Quinn Papers.
15. Innes to Fothergill, January (?) 1909, The Retreat, Street-an-Pol, St. Ives, in Fothergill, "Correspondence and Drawings," reference no. TGA 7121.6.
16. John, *Chiaroscuro*, 203.
17. Holt, *British Artists and the Modernist Landscape*, 57, 59, 63.
18. John, *Chiaroscuro*, 104.
19. John to Quinn, 3 March 1910, Martigues, John Quinn Papers.
20. John to Quinn, 25 August 1910, Marseilles, ibid.
21. John, *Chiaroscuro*, 107.
22. Quinn to John, 19 May 1911, John Quinn Papers.
23. Robins, *Fragile Modernism*, 25. Menpes's statement is also quoted on this page.
24. John to Quinn, 15 June 1911, 181A King's Road, Chelsea, John Quinn Papers.
25. John referred here to the Coronation Exhibition of 1911, honoring George V and including scenes of the British Empire.
26. Quinn to John, 8 August 1911, John Quinn Papers.
27. Turner, *Between Science and Religion*, 228.
28. "Philosophy a Fashion: The Embarrassing Popularity of M. Bergson," *Times* (London), 7 February 1914, 8.

29. Antliff, *Inventing Bergson*, 6; Gillies, *Henry Bergson and British Modernism*, 29–30.
30. John to Quinn, 1910 (?), Villa Sainte Anne, Martigues, John Quinn Papers; John to Ottoline Morrell, May 1910, Villa Sainte Anne, Martigues, Ottoline Morrell Collection, box 11, folder 1.
31. Hulme, "New Philosophy," 198–99.
32. Bergson, *Time and Free Will*, 231.
33. Hulme, "New Philosophy."
34. Bergson, *Creative Evolution,* 196.
35. Hulme, "New Philosophy," 198.
36. Bergson, *Creative Evolution*, 176.
37. Ibid., 250.
38. Hulme, "New Philosophy," 198.
39. Bergson, *Matter and Memory*, 239, 241, 242.
40. Bergson, *Creative Evolution*, 177.
41. "Tis," "Augustus John," 148.
42. Bergson, *Introduction to Metaphysics*, 8, 9.
43. Bergson, *Creative Evolution*, 367, 368, 369.
44. Shiff, *Cézanne and the End of Impressionism*, 188.
45. Carter, "Independants and the New Intuition," 82.
46. Murry, "Bergson and Paris," 116.
47. Carter, "Independants and the New Intuition," 82–83.
48. Bergson, *Time and Free Will,* 12, 14–15, 16, 18.
49. Murry, "Art and Philosophy," 9, 10, 11.
50. Ibid., 12.
51. For more on the circle of artists around the journal, see Holt, *British Artists and the Modernist Landscape*, 81–94.
52. Antliff, *Inventing Bergson*, 13.
53. Ibid., 70.
54. Baker, "Mr. John and Others," 650.
55. Innes to Fothergill, January (?) 1909, St. Ives, in Fothergill, "Correspondence and Drawings," reference no. TGA 7121.8.
56. Fothergill, "James Dickson Innes," 6.
57. Innes to Fothergill, June 1908 or 1909, The Old Rd., Llanelly, S. Wales, in Fothergill, "Correspondence and Drawings," reference no. TGA 7121.1. Innes's original spelling has been retained.
58. John, "J. Dickson Innes," 5.
59. Binyon, "Mr. C. J. Holmes and Others," 265. Holt argues that Binyon was aware of Bergson's theories. *British Artists and the Modernist Landscape,* 86.
60. Quinn to John, 19 May 1911; John to Quinn, 15 June 1911, 181A King's Road, Chelsea, John Quinn Papers.
61. Drey, "New English Art Club," 375.
62. Fothergill, "James Dickson Innes," 11.
63. "Notes," *Colour* 15 (December 1921): viii.
64. Upstone, catalogue entry for Charles Ginner, *Clayhidon*, in Upstone, *Modern Painters*, 158. Note that Van Gogh used a perspective frame; see Vincent van Gogh to Theo van Gogh, 5 or 6 August 1882, in *Vincent van Gogh: The Letters*, http://vangoghletters.org/vg/letters.html, letter 254. For more on Ginner's biography, see Bonett, "Charles Ginner, 1878–1952."
65. Baron, *Perfect Moderns,* 45.
66. Holt, "Camden Town Group Then and Now."
67. Nicola Moorby, catalogue entry for Harold Gilman, *Canal Bridge, Flekkefjord*, in Upstone, *Modern Painters*, 61.
68. Richard Humphrys, catalogue entry for Harold Gilman, *An Eating House*, ibid., 92.
69. Corbett, *Modernity of English Art*, 244.
70. Upon Gilman's death, Ginner wrote a tribute in which he called attention to Gilman's late landscapes: "He concerned himself with the observation of each tone as a separate entity, opened his eyes to richness of colour and possessed a sound sense of draughtsmanship. One more thing he was to achieve which was the feeling for tones in larger and broader planes and more simplified masses, at the same time working up to a more rigid and firmer outline. It is especially in his beautiful wood scenes that he recently developed this sense of simplified design, without losing any of the other qualities that he had already absorbed." Ginner, "Harold Gilman: An Appreciation," 133.
71. Lewis and Fergusson, *Harold Gilman,* 30.
72. See Easton, "Camden Town Group into the London Group."
73. John supported his colleagues Eric Gill, Spencer Gore, and David Bomberg by recommending their work to Quinn for purchase. Michael Holroyd argues that John's absence from such enterprises as Fry's second postimpressionist exhibition stemmed from an argument between Fry and Rothenstein. Holroyd, *Augustus John,* 349.
74. This history is drawn largely from Wendy Baron's account in *Perfect Moderns*, 43–56. See also Upstone, "Painters of Modern Life: The Camden Town Group," in Upstone, *Modern Painters*, 9–25.
75. Ginner, "Neo-Realism," *New Age*, 272; Bergson, *Introduction to Metaphysics*, 7.
76. Ginner, "Neo-Realism," *New Age,* 272; Bergson, *Introduction to Metaphysics*, 10–11, 14, 56. Bergson explained (as translated by Hulme) that "there is one reality, at least, which we all seize from within, by intuition and not by simple analysis. It is our own personality in its flowing through time—our self which endures" (9).
77. Ginner, "Neo-Realism," *New Age,* 271, 272.
78. Ibid., 272.
79. Sickert, "Mr. Ginner's Preface," 819; Ginner, "Mr. Sickert and Mr. Poussin"; Sickert, "Thickest Painters in London," 155; Gilman, "Sickert and Neo-Realism," 190–91.
80. Maloney, "Humoreske."

81. Hulme, "Modern Art.—I," 341.
82. Beasley, "'Definite Meaning,'" 68–69.
83. Hulme, "Modern Art.—II," 467.
84. Worringer, "Selection from *Abstraction and Empathy*," 69.
85. Hulme, "Modern Art.—III," 661; Hulme, "Modern Art.—IV."
86. Hulme, "Modern Art.—II," 467, 468. Shiff's study *Cézanne and the End of Impressionism* reveals the multivalency of the critical reception of Cézanne's work at this time.
87. Hulme, "Modern Art.—II," 469.
88. Hulme, "Modern Art.—IV," 231.
89. Hulme, "Modern Art.—II," 469.
90. "Manifesto," 19, 21, 26.
91. As David Peters Corbett notes, however, Nevinson was complicit in "the fashionable assessment of modernism as an entertainment and novelty," which precluded a sustained critique of the conditions of modernity. *Modernity of English Art,* 25.
92. Nevinson, "Vital English Art," 160, 161.
93. Ginner, "Cézanne Stunt," 41, 42.
94. Ibid., 43.
95. Porter, "Fate of Scientific Naturalism."
96. Corbett, in *The Modernity of English Art*, argues that Nash's landscapes of the 1920s enact a form of displacement, denying modernity while drawing on its structuring influences (e.g., technology, the city, the abstract) (118).
97. For example, in its caption for Malcolm Milne's *Windermere*, which depicts the undulating hills and woods lining a waterway, the journal's editor wrote, "fifty years ago nobody would have attempted such a subject. It is only made possible in painting by grasping the structure, or anatomy, of the landscape—which enables the painter to reproduce it on canvas in decorative terms." *Colour* 12 (July 1920): 127.

**Conclusion**

1. Tickner, *Modern Life and Modern Subjects*, 193.

# { bibliography }


Alberti, Samuel. "The Status of Museums: Authority, Identity, and Material Culture." In *Geographies of Nineteenth-Century Science,* edited by David N. Livingstone and Charles W. J. Withers, 51–72. Chicago: University of Chicago Press, 2011.

Altick, Richard D. *The English Common Reader: A Social History of the Mass Reading Public.* 2nd ed. Columbus: Ohio University Press, 1957.

Antliff, Mark. *Inventing Bergson: Cultural Politics and the Parisian Avant-Garde.* Princeton: Princeton University Press, 1993.

Armstrong, Carol M. *Scenes in a Library: Reading the Photograph in the Book, 1843–1875.* Cambridge: MIT Press, 1998.

Arnold, H. P. *William Henry Fox Talbot: Pioneer of Photography and Man of Science.* London: Hutchinson Benham, 1977.

"Art, the Royal Academy." *Spectator*, 4 May 1872, 561–62.

"Art. I-1: First and Second Reports of the British Association for the Advancement of Science . . . Report of the Third Meeting . . . Fourth, Fifth, and Sixth Reports of Ditto." *British Critic and Quarterly Theological Review* 49 (January 1839): 1–48.

"Art. II.-1: The Ferns of Great Britain, &c. Nature-Printed by Henry Bradbury. . . ." *Quarterly Review* 101 (January 1857): 57–79.

"Art and Science in a New Light." *Nature,* 13 January 1887, 250–52.

Atkinson, J. Beavington. *English Painters of the Present Day.* London: Seeley, Jackson, and Halliday, 1871.

B., J. "Art: The New English Art Club." *Spectator*, 30 November 1912, 902.

Babbage, Charles. *Reflections on the Decline of Science in England, and on Some of Its Causes.* London: B. Fellowes and J. Booth, 1830.

Bain, Alexander. *Logic.* 2 vols. London: Longmans, Green, Reader & Dyer, 1870.

——. *The Senses and the Intellect.* London: J. W. Parker, 1855.

Baker, C. H. Collins. "Mr. John and Others." *Saturday Review*, 22 November 1913, 649–50.

Balfour, Graham. "Artistic Focus." *Photographic Quarterly* 1 (October 1889): 69–71.

Barber, Lynn. *The Heyday of Natural History, 1820–1870.* Garden City, N.Y.: Doubleday, 1980.

Barker, Elizabeth. "New Light on 'The Orrery': Joseph Wright and the Representation of Astronomy in Eighteenth-Century Britain." *British Art Journal* 1 (Spring 2000): 29–37.

Barlow, Paul. *Time Present and Time Past: The Art of John Everett Millais.* Aldershot: Ashgate, 2005.

Baron, Wendy. *Perfect Moderns: A History of the Camden Town Group.* Aldershot: Ashgate, 2000.

Barringer, Tim. *Men at Work: Art and Labour in Victorian Britain.* New Haven: Yale University Press, 2005.

——. *Reading the Pre-Raphaelites.* New Haven: Yale University Press, 1998.

——. "'Tracing the Finger of God': Landscape Painting in Victorian England." In *Pre-Raphaelite and Other Masters: The Andrew Lloyd Webber Collection,* edited by Mary Anne Stevens and Norman Rosenthal, 139–55. London: Royal Academy of Arts, 2003.

Barrow-Green, June. "'Merely a Speculation of the Mind'? William Henry Fox Talbot and Mathematics." In *William Henry Fox Talbot: Beyond Photography,* edited by Mirjam Brusius, Katrina Dean, and Chitra Ramalingam, 67–94. New Haven: Yale University Press, 2013.

Barthes, Roland. "The Reality Effect." In Barthes, *The Rustle of Language,* translated by Richard Howard, 141–68. London: Basil Blackwell, 1986.

Batchen, Geoffrey. *Burning with Desire: The Conception of Photography.* Cambridge: MIT Press, 1997.

Baudot, Laura. "An Air of History: Joseph Wright's and Robert Boyle's Air Pump Narratives." *Eighteenth-Century Studies* 46 (Fall 2012): 1–28.

Baxandall, Michael. *Patterns of Intention: On the Historical Explanation of Pictures.* New Haven: Yale University Press, 1985.

Beasley, Rebecca. "'A Definite Meaning': The Art Criticism of T. E. Hulme." In *T. E. Hulme and the Question of Modernism*, edited by Edward P. Comentale and Andrzej Gąsiorek, 57–71. Aldershot: Ashgate, 2006.

Becker, Robert Stephen. "George Moore: Artist and Art Critic." *Irish Arts Review* 2 (Autumn 1985): 49–55.

Beer, Gillian. *Darwin's Plots: Evolutionary Narrative in Darwin, George Eliot, and Nineteenth-Century Fiction.* Cambridge: Cambridge University Press, 2000.

———. *Open Fields: Science in Cultural Encounter.* Oxford: Clarendon Press, 1996.

Bennett, Mary. *Ford Madox Brown: A Catalogue Raisonné.* New Haven: Published for the Paul Mellon Centre for Studies in British Art by Yale University Press, 2010.

Bergson, Henri. *Creative Evolution.* Translated by Arthur Mitchell. New York: Henry Holt, 1911. Reprint, Mineola, N.Y.: Dover, 1998.

———. *An Introduction to Metaphysics.* Translated by T. E. Hulme. New York: G. P. Putnam's Sons, 1912.

———. *Matter and Memory.* Translated by Nancy Margaret Paul and W. Scott Palmer. London: George Allen, 1912. Reprint, Mineola, N.Y.: Dover, 2004.

———. *Time and Free Will: An Essay on the Immediate Data of Consciousness.* Translated by F. L. Pogson. London: George Allen, 1913. Reprint, Boston: Elibron Classics, 2005.

Bermingham, Ann. *Learning to Draw: Studies in the Cultural History of a Polite and Useful Art.* New Haven: Published for the Paul Mellon Centre for Studies in British Art by Yale University Press, 2000.

Binyon, Laurence. "Mr. C. J. Holmes and Others." *Saturday Review,* 4 March 1911, 265–66.

Bonett, Helena. "Charles Ginner, 1878–1952." In *The Camden Town Group in Context*, edited by Helena Bonett, Ysanne Holt, and Jennifer Mundy. http://www.tate.org.uk/art/research-publications/camden-town-group/charles-ginner-r1105346.

Bourdieu, Pierre. *The Field of Cultural Production: Essays on Art and Literature.* Edited by Randal Johnson. New York: Columbia University Press, 1993.

Bradbury, Henry. *Nature-Printing, Its Origin and Objects: A Lecture Delivered at the Royal Institution of Great Britain . . . May 11, 1855.* London: Bradbury and Evans, 1856.

———. *Printing: Its Dawn, Day, and Destiny. . . .* London: Bradbury and Evans, 1858.

Brake, Laurel. "Writing, Cultural Production, and the Periodical Press in the Nineteenth Century." In *Writing and Victorianism,* edited by J. B. Bullen, 54–72. London: Longman, 1997.

Brande, William Thomas, ed. *A Dictionary of Science, Literature, and Art, Comprising the History, Description, and Scientific Principles of Every Branch of Human Knowledge . . . .* London: Longman, Brown, Green, and Longmans, 1842.

Brett, Charles, Michael Hickox, and Christiana Payne. *John Brett: A Pre-Raphaelite in Cornwall.* Bristol, UK: Sansom, 2006. Exhibition catalogue.

Brett, John. "Extract from a Report of the Eclipse Expedition of 1870, at Augusta Sicily." *Monthly Notices of the Royal Astronomical Society* 31 (22 December 1870): 163–66.

———. "The Functions of Texture in the Arts." *Art Journal* 55 (April 1893): 117–20.

———. "The Great Patch on Jupiter." *Observatory: A Monthly Review of Astronomy* 3 (1879): 236–38.

———. "John Brett Diary, 1851–1861." Edited by Charles Brett. Transcription and typescript, August 2002. Private collection.

———. John Brett Sketchbooks, nos. 13 (reference no. PR 86–19, Scotia, Naples), 21 (reference no. PR 86–26, Towyn, 1866–67), 23 (reference no. PR 86–28, Victoria, 1867), 32 (reference no. PR 86–38, Ireland, 1869). National Maritime Museum, Royal Museums Greenwich.

———. "Memorandum of Observations of Jupiter Made During the Month of April, 1874." *Monthly Notices of the Royal Astronomical Society* 34 (8 May 1874): 359–60.

———. "Natural Science at the Royal Academy." *Nature* 2 (30 June 1870): 157–58.

———. "On Certain Phenomena Seen During Eclipses of the Sun, and Its Bearing on the Question of a Lunar Atmosphere." *Monthly Notices of the Royal Astronomical Society* 35 (13 November 1874): 14–16.

———. "On Certain Phenomena Surrounding the Sun's Limb as Seen in the Telescope." *Monthly Notices of the Royal Astronomical Society* 32 (8 June 1872): 297–300.

———. "On the Proper Motion of Bright Spots on Jupiter." *Monthly Notices of the Royal Astronomical Society* 36 (8 June 1876): 355–57.

———. "The Physical Condition of Mars." *Monthly Notices of the Royal Astronomical Society* 38 (2 December 1877): 58–61.

———. "Realism in Painting." *Contemporary Review* 75 (June 1899): 823–30.

———. "The Relation of Photography to the Pictorial Art." *Journal of the Camera Club* 2 (April 1889): 83–86.

———. "The Relation of the Pictorial to the Decorative Arts." In *Transactions of the National Association for the Advancement of Art and Its Application to Industry, 1888–1891*, 1:46–55. New York: Garland, 1979.

———. Royal Astronomical Society: RAS MSS 20 (a collection of miscellaneous astronomical drawings); RAS Papers 34.10 (correspondence); RAS Papers 51.1 (letter book relating to the total eclipse expedition, 1870); RAS Letters, 1875, 1876, 1878.

———. "The Specular Reflexion Hypothesis, and Its Bearing on the Transit of Venus." *Monthly Notices of the Royal Astronomical Society* 37 (12 January 1877): 126–27.

———. *Three Months on the Scottish Coast: A Series of Sketches and Pictures Painted During the Summer of the Present Year.* London: Fine Art Society, 1886.

Brewster, David. "The Photographic Society of Scotland, Address of Sir David Brewster." *Journal of the Photographic Society of London* 3 (21 May 1856): 48–50.

———. *A Treatise on Optics.* London: Printed for Longman, Rees, Orme, Brown & Green and John Taylor, 1831.

[Brewster, David]. "The Philosophy of the Inductive Sciences, Founded on Their History." *Edinburgh Review* 74 (January 1842): 265–306.

[———]. "The Sight and How to See." *North British Review* 26 (November 1856): 145–84.

Bronkhurst, Judith. *William Holman Hunt: A Catalogue Raisonné.* 2 vols. New Haven: Published for the Paul Mellon Centre for Studies in British Art by Yale University Press, 2006.

Brooks, Chris. *Signs for the Times: Symbolic Realism in the Mid-Victorian World.* London: Allen and Unwin, 1984.

Brown, Ford Madox. *The Diary of Ford Madox Brown.* Edited by Virginia Surtees. New Haven: Published for the Paul Mellon Centre for Studies in British Art by Yale University Press, 1981.

Brusius, Mirjam, and Chitra Ramalingam. "Beyond Photography." In *William Henry Fox Talbot: Beyond Photography,* edited by Mirjam Brusius, Katrina Dean, and Chitra Ramalingam, 1–24. New Haven: Yale University Press, 2013.

Burke, Edmund. *A Philosophical Inquiry into the Origin of Our Ideas of the Sublime and Beautiful, with an Introductory Discourse Concerning Taste and Several Other Additions.* London: R. and J. Dodsley, 1757.

Bury, Laurent. *L'Orientalisme Victorien dans les arts visuels et la littérature.* Grenoble: ELLUG Université Grenoble, 2010.

Buss, R. W. "On the Use of Photography to Artists." *Journal of the Photographic Society of London* 1 (21 June 1853): 75–76.

Callender, Ronald M. "'Dear Mr. Driffield . . .': Letters of Peter Henry Emerson and Hurter and Driffield." *History of Photography* 28 (Winter 2004): 315–22.

Carlyle, Leslie. *The Artist's Assistant: Oil Painting Instruction Manuals and Handbooks in Britain, 1800–1900, with Reference to Selected Eighteenth-Century Sources.* London: Archetype, 2001.

———. "Contemporary Painting Manuals." In Joyce H. Townsend, Jacqueline Ridge, and Stephen Hackney, *Pre-Raphaelite Painting Techniques, 1848–1856,* 39–49. London: Tate Publishing, 2004.

Carpenter, William. *Principles of Human Physiology.* 5th ed. London: John Churchill, 1855.

———. *Principles of Mental Physiology, with Their Applications to the Training and Discipline of the Mind, and the Study of Its Morbid Conditions.* 2nd ed. London: H. S. King, 1875.

Carter, Huntly. "The Independants and the New Intuition in Paris." *New Age,* 25 May 1911, 82–83.

Cave, Roderick, and Geoffrey Wakeman. *Typographia Naturalis.* Wymondham, UK: Brewhouse Press, 1967.

Chambers, Ephraim. *Cyclopaedia, Or, An Universal Dictionary of Arts and Sciences: Containing the Definitions of the Terms, and Accounts of the Things Signify'd Thereby* . . . . Vol. 1. London: J. and J. Knapton et al., 1728.

Christ, Carol T. *The Finer Optic: The Aesthetic of Particularity in Victorian Poetry.* New Haven: Yale University Press, 1975.

Christ, Carol T., and John O. Jordan, eds. *Victorian Literature and the Victorian Visual Imagination.* Berkeley: University of California Press, 1995.

Clausen, George. "Bastien-Lepage and Modern Realism." *Scottish Art Review* 1 (October 1888): 114–15.

———. "Jules Bastien-Lepage as Artist." In André Theuriet et al., *Jules Bastien-Lepage and His Art: A Memoir,* 107–27. London: T. Fisher Unwin, 1892.

———. Letters, 1888–1909, reference no. MSL/1991/4, pressmark 86.YY.19, National Art Library, Victoria and Albert Museum, London.

Codell, Julie F. "Artists' Professional Societies: Production, Consumption, and Aesthetics." In *Towards a Modern Art World,* edited by Brian Allen, 169–87. New Haven: Published for Studies in British Art, Yale Center for British Art, by Yale University Press, 1995.

———. *The Victorian Artist: Artists' Life Writings in Britain, ca. 1870–1910.* Cambridge: Cambridge University Press, 2003.

Coe, Brian. "George Davison: Impressionist and Anarchist." In *British Photography in the Nineteenth Century: The Fine Art Tradition,* edited by Mike Weaver, 215–41. Cambridge: Cambridge University Press, 1989.

Colvin, Sidney. "English Painters of the Present Day, XVIII—John Everett Millais." *Portfolio* 2, no. 1 (1871): 1–6.

———. *Memories and Notes of Persons and Places, 1852–1912.* New York: Charles Scribner's Sons, 1921.

Constable, John. *John Constable's Discourses.* Compiled by R. B. Beckett. Ipswich, England: Suffolk Record Office, 1970.

Cook, Edward T. *A Popular Handbook to the Tate Gallery, National Gallery of British Art.* London: Macmillan, 1898.

Corbett, David Peters. *The Modernity of English Art, 1914–1930.* Manchester: Manchester University Press, 1997.

———. *The World in Paint: Modern Art and Visuality in England, 1848–1914.* University Park: Pennsylvania State University Press, 2004.

Cormack, Malcolm. "Stubbs and Science." In *Fearful Symmetry: George Stubbs, Painter of the English Enlightenment,* edited by Nicholas H. J. Hall, 37–47. New York: Hall & Knight, 2000. Exhibition catalogue.

Crary, Jonathan. *Techniques of the Observer: On Vision and Modernity in the Nineteenth Century.* Cambridge: MIT Press, 1990.

Crosland, Maurice. "Chemistry and the Chemical Revolution." In *The Ferment of Knowledge: Studies in the Historiography of Eighteenth-Century Science,* edited by G. S. Rousseau and Roy Porter, 389–416. Cambridge: Cambridge University Press, 1980.

Dale, Peter Allan. *In Pursuit of a Scientific Culture: Science, Art, and Society in the Victorian Age.* Madison: University of Wisconsin Press, 1989.

Daston, Lorraine, and Peter Galison. "The Image of Objectivity." *Representations* 40 (Fall 1982): 81–128.

———. *Objectivity.* New York: Zone Books, 2007.

Daunton, Martin, ed. *The Organisation of Knowledge in Victorian Britain.* Oxford: Oxford University Press, 2005.

Davison, George. "The Limits and Possibilities of Art Photography." *Photographic Quarterly* 1 (April 1890): 207–16.

———. "A Reply to Dr. Emerson's 'Renunciation.'" *British Journal of Photography* 38 (30 January 1891): 70–71.

Dawson, Gowan, and Bernard Lightman. "Introduction." In *Victorian Scientific Naturalism: Community, Identity, Continuity*, edited by Gowan Dawson and Bernard Lightman, 1–24. Chicago: University of Chicago Press, 2014.

Delamotte, Philip Henry, ed. *The Sunbeam: A Book of Photographs from Nature.* London: Chapman and Hall, 1859.

DeYoung, Ursula. *A Vision of Modern Science: John Tyndall and the Role of the Scientist in Victorian Culture.* New York: Palgrave Macmillan, 2011.

"Discussion." *Journal of the Camera Club* 2 (April 1889): 90.

Doyle, Richard. *Bird's Eye Views of Society.* London: Smith & Elder, 1864.

Drey, O. Raymond. "The New English Art Club." *Rhythm* 2 (January 1913): 373–76.

Dulac, Anne. "William Hunter: A Brief Account of His Life as an Art Collector." In *"My Highest Pleasures": William Hunter's Art Collections,* edited by Peter Black, 19–62. Glasgow: The Hunterian, University of Glasgow in association with Paul Holberton, 2007. Exhibition catalogue.

Easton, Malcolm. "The Camden Town Group into the London Group: Some Intimate Glimpses of the Transition, 1911–14." In *The Camden Town Group in Context,* edited by Helena Bonett, Ysanne Holt, and Jennifer Mundy. http://www.tate.org.uk/art/research-publications/camden-town-group/malcolm-easton-the-camden-town-group-into-the-london-group-some-intimate-glimpses-of-the-r1104360.

Egerton, Judy. *Wright of Derby.* New York: Metropolitan Museum of Art, 1990. Exhibition catalogue.

Eliot, George. *George Eliot: Interviews and Recollections.* Edited by K. K. Collins. New York: Palgrave Macmillan, 2010.

Emerson, P. H. "Articles and Notices." Bound volume of articles from various periodicals and newspapers. Call no. C EME. Norfolk Heritage Center, Millennium Library, Norwich, UK.

———. "Correspondence: Dr. Emerson's Naturalistic Photography." *Photographic News* 33 (7 June 1889): 380–82.

———. *The Death of Naturalistic Photography.* London, 1890.

———. *Idyls [sic] of the Norfolk Broads.* London: Autotype, 1887.

———. "Letters to James Harvard Thomas." Transcribed by Fiona Pearson. Call number C EME. Norfolk Heritage Center, Millennium Library, Norwich, UK.

———. "Naturalistic Photography." *Journal and Transactions of the Photographic Society of Great Britain* 17 (March 1893): 156–63.

———. *Naturalistic Photography.* 2nd ed., rev. London: Sampson, Law, Marston, Searle & Rivington, 1890.

———. Papers of Peter Henry Emerson, 1886–1920. Reference no. MC 2235, Norfolk Record Office, Norwich, England.

———. *Paul Ray at the Hospital: A Picture of Student Life.* London: W. Hall & Lovitt, 1882.

———. "Photography Not Art." *Photographic Quarterly* 2 (January 1892): 148–55.

———. *Pictures from Life in Field and Fen.* London: George Bell and Sons, 1887.

———. *Pictures of East Anglian Life.* London: Sampson, Law, Marston, Searle & Rivington, 1888.

———. "A Renunciation." *British Journal of Photography* 38 (23 January 1891): 54–55.

———. "Science and Art." *Journal of the Camera Club* 2 (April 1889): 87–89.

Emerson, P. H., and T. F. Goodall. *Life and Landscape on the Norfolk Broads.* London: S. Low, Marston, Searle & Rivington, 1886.

———. *Notes on Perspective Drawing and Vision.* London: Warren, Hall and Lovitt, 1891.

Evans, Eric J. *The Forging of the Modern State: Early Industrial Britain, 1783–1870.* London: Longman, 1983.

"Exhibition of the Royal Academy." *Art Journal* 12 (August 1873): 236–41.

Fenn, William Wilthew. "Fact in Landscape." *Magazine of Art* 4 (June 1881): 393–95.

"Ferns and Their Portraits—Nature Printing." *Quarterly Review* 101 (January 1857): 57–79.

Field, George. *Chromatography, or A Treatise on Colours and Pigments: And of Their Powers in Painting, &c.* London: Charles Tilt, 1835.

"Fine Arts: Photogenic Drawing." *Literary Gazette; and Journal of the Belles' Lettres, Arts, Sciences, &c.,* no. 1217 (9 May 1840): 315–16.

"Fine Arts: Royal Academy." *Athenaeum*, 22 May 1852, 581–83.

"Fine Arts: Royal Academy." *Athenaeum*, 12 May 1860, 653–55.

"Fine Arts: The Royal Academy (First Notice)." *Athenaeum,* 29 April 1871, 531–33.

"Fine Arts: The Royal Academy (First Notice)." *Athenaeum,* 4 May 1872, 563–66.

"Fine Arts: The Winter Exhibition in Pall Mall." *Athenaeum,* 30 October 1858, 558–59.

Fleming-Williams, Ian, and Leslie Parris. *The Discovery of Constable.* London: Hamish Hamilton, 1984.

Fletcher, Pamela M., and Anne Helmreich, eds. *The Rise of the Modern Art Market in London, 1850–1939.* Manchester: Manchester University Press, 2011.

Flint, Kate. "The 'Philistine' and the New Art Critic: J. A. Spender and D. S. MacColl's Debate of 1893." *Victorian Periodicals Review* 21 (Spring 1988): 3–8.

———. *The Victorians and the Visual Imagination.* Cambridge: Cambridge University Press, 2000.

Forbes, James D. "Illustrations of the Viscous Theory of Glacier Motion—Part III." In *Philosophical Transactions of the Royal Society of London,* part 1, 177–210. London: Richard and John E. Taylor, 1846.

———. "Third Letter on Glaciers, Addressed to Professor Jameson (1842)." In Forbes, *Occasional Papers on the Theory of the Glaciers,* 17–35. Edinburgh: Adam and Charles Black, 1859.

Forrester, Gillian. *Turner's "Drawing Book": The "Liber Studiorum."* London: Tate Publishing, 1996.

Fothergill, John. "Correspondence and Drawings Between John Fothergill and Friends, 1905–1909." Reference no. TGA 7121, Tate Archive, Tate, London.

———. "James Dickson Innes." In *James Dickson Innes, Llanelly, 1887–Swanley, 1914,* edited by Lillian Browse, 5–13. London: Published for the Shenval Press by Faber and Faber, 1947.

———. "James Dickson Innes." In *Watercolour Drawings and Paintings by the Late James Dickson Innes, Memorial Exhibition, April–June 1923*, 5–7. Chelsea: Chenil Galleries, 1923. Exhibition catalogue.

Foucault, Michael. *The Archaeology of Knowledge.* Translated by A. M. Sheridan Smith. London: Tavistock, 1972.

Fraser, David. "Joseph Wright of Derby and the Lunar Society: An Essay on the Artist's Connections with Science and Industry." In Judy Egerton, *Wright of Derby,* 15–23. New York: Metropolitan Museum of Art, 1990. Exhibition catalogue.

Fredericksen, Andrea. *Vanishing Point: The Perspective Drawings of J. M. W. Turner.* London: Tate Publishing, 2004.

Frith, W. P. "Crazes in Art: 'Pre-Raphaelitism' and 'Impressionism.'" *Magazine of Art* 11 (April 1888): 187–91.

———. "'Realism' Versus 'Sloppiness.'" *Magazine of Art* 12 (November 1888): 6–8.

[Fry, Roger]. "Fine Arts, Drawings, and Sculpture at Messrs. Carfax's." *Athenaeum,* 14 June 1902, 761–62.

Funnell, Peter. *Millais: Portraits.* London: National Portrait Gallery, 1999. Exhibition catalogue.

Fyfe, Aileen, and Bernard Lightman. "Science in the Marketplace: An Introduction." In *Science in the Marketplace: Nineteenth-Century Sites and Experiences,* edited by Aileen Fyfe and Bernard Lightman, 1–19. Chicago: University of Chicago Press, 2007.

Gage, John. *George Field and His Circle: From Romanticism to the Pre-Raphaelite Brotherhood.* London: Christie's, 1989. Exhibition catalogue.

Galton, Francis. *Inquiries into Human Faculty and Its Development.* 2nd ed. London: J. M. Dent, 1907.

Gates, Barbara T. "Ordering Nature: Revisioning Victorian Science Culture." In *Victorian Science in Context,* edited by Bernard Lightman, 179–86. Chicago: University of Chicago Press, 1997.

Gibson-Wood, Carol. *Jonathan Richardson: Art Theorist of the English Enlightenment.* New Haven: Yale University Press, 2000.

Gillies, Mary Ann. *Henri Bergson and British Modernism.* Montreal: McGill-Queen's University Press, 1996.

Gilman, Harold. "Sickert and Neo-Realism." *New Age*, 30 April 1914, 190–91.

Gilpin, William. *Observations on the River Wye, and several parts of South Wales, &c., relative chiefly to picturesque beauty: made in the summer of 1770.* 3rd ed. London: Printed for R. Blamire, 1792.

———. *Three Essays: On Picturesque Beauty, On Picturesque Travel, and On Sketching Landscape, to Which Is Added a Poem, On Landscape Painting.* London: Printed for R. Blamire, 1792.

———. *Three Essays: On Picturesque Beauty, On Picturesque Travel, and On Sketching Landscape: with a Poem, on Landscape Painting. To These Are Now Added Two Essays Giving an Account of the Principles and Mode in Which the Author Executed His Own Drawings.* 3rd ed. London: T. Cadell and W. Davies, 1808.

Ginner, Charles. "The Cézanne Stunt." *Art and Letters, an Illustrated Quarterly* 2 (Winter 1919): 41–43.

———. "Harold Gilman: An Appreciation." *Art and Letters, an Illustrated Quarterly* 2 (Summer 1919): 129–34.

———. "Mr. Sickert and Mr. Poussin." *New Age,* 7 May 1914, 23.

———. "Neo-Realism." In *Catalogue of an Exhibition of Paintings by Harold Gilman and Charles Ginner,* 3–12. London: Goupil Gallery, 1914. Exhibition catalogue.

———. "Neo-Realism." *New Age,* 1 January 1914, 271–72.

Glanville, Helen. "Contemporary Colour Theory." In Joyce H. Townsend, Jacqueline Ridge, and Stephen Hackney, *Pre-Raphaelite Painting Techniques, 1848–1856,* 29–37. London: Tate Publishing, 2004.

Goethe, Johann Wolfgang von. *Goethe's Theory of Colours.* Translated by Charles Eastlake. London: J. Murray, 1840.

Goodall, T. F. "Landscape." In P. H. Emerson and T. F. Goodall, *Life and Landscape on the Norfolk Broads,* 71–81. London: S. Low, Marston, Searle & Rivington, 1886.

Green, Jennifer M. "'The Right Thing in the Right Place': P. H. Emerson and the Picturesque Photograph." In *Victorian Literature and the Victorian Visual Imagination,* edited by Carol T. Christ and John O. Jordan, 88–110. Berkeley: University of California Press, 1995.

Green-Lewis, Jennifer. *Framing the Victorians: Photography and the Culture of Realism.* Ithaca: Cornell University Press, 1996.

Hackney, Stephen. "Cleaning John Brett's *The British Channel Seen from the Dorsetshire Cliffs*, 1871." http://www.tate.org.uk/about/projects/cleaning-john-bretts-british-channel-seen-dorsetshire-cliffs-1871.

Hackney, Stephen, Joyce H. Townsend, and Jacqueline Ridge. "Background, Training, and Influences." In Joyce H. Townsend, Jacqueline Ridge, and Stephen Hackney, *Pre-Raphaelite Painting Techniques, 1848–1856,* 21–27. London: Tate Publishing, 2004.

———. "Pre-Raphaelite Methods and Manuals." In Joyce H. Townsend, Jacqueline Ridge, and Stephen Hackney, *Pre-Raphaelite Painting Techniques, 1848–1856,* 51–75. London: Tate Publishing, 2004.

Hall, Maria Boas. *All Scientists Now: The Royal Society in the Nineteenth Century.* Cambridge: Cambridge University Press, 1984.

Hamerton, Philip Gilbert. *An Autobiography, 1834–1858, and a Memoir by His Wife, 1858–1894.* Boston: Roberts Brothers, 1896.

———. "Can Science Help Art?" In Hamerton, *Portfolio Papers*, 285–89. London: Seeley, 1889.

———. *Imagination in Landscape Painting.* London: Seeley, 1887.

———. *A Painter's Camp in the Highlands, and Thoughts About Art.* 2 vols. Cambridge: Macmillan, 1862.

Hamilton, James. *Turner and the Scientists.* London: Tate Gallery Publications, 1998.

Handy, Ellen. "Art and Science in P. H. Emerson's Naturalistic Vision." In *British Photography in the Nineteenth Century: The Fine Art Tradition*, edited by Mike Weaver, 181–95. Cambridge: Cambridge University Press, 1989.

———. "Pictorial Beauties, Natural Truths, Photographic Practices." In Ellen Handy, with essays by Brian Lukacher and Shelley Rice, *Pictorial Effect, Naturalistic Vision: The Photographs and Theories of Henry Peach Robinson and Peter Henry Emerson*, 1–23. Norfolk, Va.: Chrysler Museum, 1994. Exhibition catalogue.

Hanson, Craig Ashley. *The English Virtuoso: Art, Medicine, and Antiquarianism in the Age of Empiricism.* Chicago: University of Chicago Press, 2009.

Harker, Margaret F. *The Linked Ring: The Secession Movement in Photography in Britain, 1892–1910.* London: Heinemann, 1979.

Harris, Elizabeth. "Experimental Graphic Processes in England, 1800–1859, Part IV." *Journal of the Printing Historical Society* 6 (1970): 53–89.

Harrison, Peter. *The Fall of Man and the Foundations of Science.* Cambridge: Cambridge University Press, 2007.

Harrower, Mary. *John Forbes White.* Edinburgh: T. N. Flouis, 1918.

Haskins, Katherine. *The Art Journal and Fine Art Publishing in Victorian England, 1850–1880.* Burlington, Vt.: Ashgate, 2012.

Hatfield, Gary. "Remaking the Science of Mind: Psychology as Natural Science." In *Inventing Human Science: Eighteenth-Century Domains,* edited by Christopher Fox, Roy Porter, and Robert Wokler, 184–231. Berkeley: University of California Press, 1995.

Hay, D. R. *The Laws of Harmonious Colouring, Adapted to Interior Decorations, Manufactures, and Other Useful Purposes.* 3rd ed. Edinburgh: William and Robert Chambers, 1836.

Heilbron, J. L. "Experimental Natural Philosophy." In *The Ferment of Knowledge: Studies in the Historiography of Eighteenth-Century Science,* edited by G. S. Rousseau and Roy Porter, 357–87. Cambridge: Cambridge University Press, 1980.

Helmholtz, Hermann von. *Popular Lectures on Scientific Subjects.* Translated by E. Atkinson. New York: D. Appleton, 1873.

———. *Science and Culture: Popular and Philosophical Essays*. Edited by David Cahan. Chicago: University of Chicago Press, 1995.

Helmreich, Anne. "The Death of the Victorian Art Periodical." *Visual Resources* 26 (September 2010): 242–53.

———. "The Goupil Gallery at the Intersection Between London, Continent, and Empire." In *The Rise of the Modern Art Market in London, 1850–1939,* edited by Pamela Fletcher and Anne Helmreich, 65–84. Manchester: Manchester University Press, 2011.

———. "John Singer Sargent, *Carnation, Lily, Lily, Rose*, and the Condition of Modernism in England, 1887." *Victorian Studies* 45, no. 3 (2003): 433–55.

———. "Poetry in Nature: Millais's Pure Landscapes." In *John Everett Millais: Beyond the Pre-Raphaelite Brotherhood,* edited by Debra N. Mancoff, 149–80. New Haven: Yale University Press, 2001.

———. "The Socio-Geography of Art Dealers and Commercial Galleries in Early-Twentieth-Century London." In *The Camden Town Group in Context,* edited by Helena Bonett, Ysanne Holt, and Jennifer Mundy. http://www.tate.org.uk/art/research-publications/camden-town-group/anne-helmreich-the-socio-geography-of-art-dealers-and-commercial-galleries-in-early-r1105658.

Herschel, J. F. W. *A Preliminary Discourse on the Study of Natural Philosophy.* London: Longman, Rees, Orme, Brown, and Green, 1830.

[Herschel, J. F. W.] "Art. VI-1: History of the Inductive Sciences from Earliest to the Present Time, by the Rev. William Whewell . . . 2. The Philosophy of the Inductive Sciences Founded upon Their History. . . ." *Quarterly Review* 68 (September 1841): 177–238.

Hodgson, J. E. *Fifty Years of British Art, as Illustrated by the Pictures and Drawings in the Manchester Royal Jubilee Exhibition, 1887.* London: John Heywood, 1887.

Holroyd, Michael. *Augustus John.* New York: Farrar, Straus and Giroux, 1996.

Holt, Ysanne. *British Artists and the Modernist Landscape.* Aldershot: Ashgate, 2003.

———. "The Camden Town Group Then and Now." In *The Camden Town Group in Context,* edited by Helena Bonett, Ysanne Holt, and Jennifer Mundy. http://www.tate.org.uk/art/research-publications/camden-town-group/ysanne-holt-the-camden-town- group-then-and-now-r1105679.

———. *Philip Wilson Steer.* Bridgend, Wales: Seren Books, 1992.

Hooker, William Jackson. *The British Ferns, or Coloured Figures and Descriptions, with the Needful Analyses of the Fructification and Venation, of the Ferns of Great Britain and Ireland, Systematically Arranged.* London: L. Reeve, 1861.

Hulme, T. E. "Modern Art.—I: The Grafton Group." *New Age,* 15 January 1914, 341–42.

———. "Modern Art.—II: A Preface Note and Neo-Realism." *New Age,* 12 February 1914, 467–69.

———. "Modern Art.—III: The London Group." *New Age,* 26 March 1914, 661–62.

———. "Modern Art.—IV: Mr. David Bomberg's Show." *New Age,* 9 July 1914, 230–32.

———. "The New Philosophy." *New Age,* 1 July 1909, 198–99.

Hume, Naomi. "The Nature Print and Photography in the 1850s." *History of Photography* 35 (January 2011): 44–58.

Hunt, A. W. "Modern English Landscape Painting." In *Victorian Painting, Essays, and Reviews,* edited by John Charles Olmsted, 3:569–85. New York: Garland, 1980.

Hunt, Robert. *A Popular Treatise on the Art of Photography, Including Daguerréotype, and All the New Methods of Producing Pictures by the Chemical Agency of Light.* Glasgow: Griffin, 1841.

Hunt, William Holman. *Pre-Raphaelitism and the Pre-Raphaelite Brotherhood.* 2 vols. London: Macmillan, 1905.

Jacobi, Carol. *William Holman Hunt: Painter, Painting, Paint.* Manchester: Manchester University Press, 2006.

James, William. *Psychology: The Briefer Course.* New York: Holt, 1892.

Jenkins, Adrian. *Painters and Peasants: Henry La Thangue and British Rural Naturalism, 1880–1905.* Bolton, UK: Bolton Museum and Art Gallery, 2000. Exhibition catalogue.

Jensen, J. Vernon. *Thomas Henry Huxley: Communicating for Science.* Newark: University of Delaware Press, 1991.

John, Augustus. "Autobiographical Fragments." Reference no. TAM 21E, Tate Archive, Tate, London.

———. *Chiaroscuro: Fragments of Autobiography.* London: Jonathan Cape, 1952.

———. "J. Dickson Innes." In *J. D. Innes, 1887–1914, Exhibition of Oil Paintings, Watercolours, and Drawings,* 3–7. Sheffield, UK: Graves Art Gallery, 1961. Exhibition catalogue.

Kase, Oliver. "The Anatomy of the Horse"; "'Make the Knife Go with the Pencil': Science and Art in George Stubbs's 'Anatomy of the Horse'"; and "No. 46 Bull Moose." In *George Stubbs, 1724–1806: Science into Art,* edited by Herbert Wilhelm Rott, 106–9, 42–59, and 182–83, respectively. Munich: Prestel, 2012. Exhibition catalogue.

Kember, Joe, John Plunkett, and Jill Sullivan, eds. *Popular Exhibitions, Science, and Showmanship, 1840–1910.* London: Pickering & Chatto, 2012.

Kjaergaard, Peter C. "'Within the Bounds of Science': Redirecting Controversies to Nature." In *Culture and Science in the Nineteenth-Century Media,* edited by Louise Henson, 211–21. Burlington, Vt.: Ashgate, 2004.

Klonk, Charlotte. *Science and the Perception of Nature: British Landscape Art in the Late Eighteenth and Early Nineteenth Century.* New Haven: Published for the Paul Mellon Centre for Studies in British Art by Yale University Press, 1996.

Kriz, Kay Dian. *The Idea of the English Landscape Painter: Genius as Alibi in the Early Nineteenth Century.* New Haven: Published for the Paul Mellon Centre for Studies in British Art by Yale University Press, 1997.

Lambart, Ray. *John Constable and the Theory of Landscape Painting.* Cambridge: Cambridge University Press, 2005.

Laughton, Bruce. *Philip Wilson Steer, 1860–1942.* Oxford: Clarendon Press, 1971.

Lecoq de Boisbaudran, Horace. *The Training of the Memory in Art and the Education of the Artist.* Translated by Lowes Dalbiac Luard. London: Macmillan, 1911.

L'Enfant, Julie. *William Rossetti's Art Criticism: The Search for Truth in Victorian Art.* Lanham, Md.: University Press of America, 1999.

Lenman, Robin. "British Photographers and Tourism in the Nineteenth Century: Three Case Studies." In *Visual Culture and Tourism,* edited by David Crouch and Nina Lübbren, 91–108. Oxford: Berg, 2003.

Lenoir, Timothy. "The Eye as Mathematician: Clinical Practice, Instrumentation, and Helmholtz's Construction of an Empiricist Theory of Vision." In *Hermann von Helmholtz and the Foundations of Nineteenth-Century Science,* edited by David Cahan, 109–53. Berkeley: University of California Press, 1993.

——. *Instituting Science: The Cultural Production of Scientific Disciplines.* Stanford: Stanford University Press, 1997.

Levine, George. *Darwin and the Novelists: Patterns of Science in Victorian Fiction.* Cambridge: Harvard University Press, 1988.

Lewes, George Henry. *The Physiology of Common Life.* Vol 2. Edinburgh: W. Blackwood, 1860.

Lewis, Wyndham, and Louis F. Fergusson. *Harold Gilman: An Appreciation.* London: Chatto & Windus, 1919.

Lightman, Bernard. *Evolutionary Naturalism in Victorian Britain: The "Darwinians" and Their Critics.* Burlington, Vt.: Ashgate/Variorum, 2009.

——. "Knowledge Confronts Nature: Richard Proctor and Popular Science Periodicals." In *Culture and Science in the Nineteenth-Century Media,* edited by Louise Henson, 199–210. Burlington, Vt.: Ashgate, 2004.

——. "Science and Culture." In *The Cambridge Companion to Victorian Culture,* edited by Francis O'Gorman, 12–42. Cambridge: Cambridge University Press, 2010.

——. *Victorian Popularizers of Science: Designing Nature for New Audiences.* Chicago: University of Chicago Press, 2007.

Lindley, John, and Thomas Moore. "Preface." In *The Ferns of Great Britain and Ireland, Nature Printed by Henry Bradbury,* ed. John Lindley and Thomas Moore. London: Bradbury and Evans, 1855.

Little, James Stanley. "Art, Nature, and Photography." *Photographic Quarterly* 1 (October 1889): 85–95.

Lockyer, J. Norman. "Physical Science for Artists." *Nature* 18 (9 May 1878): 30.

[Lockyer, J. Norman]. "Science and Art." *Nature* (24 May 1883): 73–76.

Lübbren, Nina. *Rural Artists' Colonies in Europe, 1870–1910.* Manchester: Manchester University Press, 2001.

Lukacher, Brian. "Powers of Sight: Robinson, Emerson, and the Polemics of Pictorial Photography." In Ellen Handy, with essays by Brian Lukacher, and Shelley Rice, *Pictorial Effect, Naturalistic Vision: The Photographs and Theories of Henry Peach Robinson and Peter Henry Emerson*, 29–53. Norfolk, Va.: Chrysler Museum, 1994. Exhibition catalogue.

Lyell, Charles. *Principles of Geology, Being an Attempt to Explain the Former Changes of the Earth's Surface, by Reference to Causes Now in Operation.* Vol. 3. London: J. Murray, 1833.

MacDonald, Margaret F., Patricia de Montfort, and Nigel Thorp, eds. *The Correspondence of James McNeill Whistler, 1855–1903.* Online edition. Glasgow: University of Glasgow. http://www.whistler.arts.gla.ac.uk/correspondence.

Macleod, Dianne Sachko. "F. G. Stephens, Pre-Raphaelite Critic and Art Historian." *Burlington Magazine* 128 (June 1986): 398–406.

Maimon, Vered. "Talbot's Art of Discovery." In *William Henry Fox Talbot: Beyond Photography,* edited by Mirjam Brusius, Katrina Dean, and Chitra Ramalingam, 143–60. New Haven: Yale University Press, 2013.

Maloney, J. "Humoreske." *New Age*, 16 July 1914, 263.

"Manifesto." *Blast* 1 (20 June 1914): 11–28.

Marien, Mary Warner. "Peter Henry Emerson: The Taxonomy of a Crow's Nest." *History of Photography* 21 (Summer 1997): 102–9.

McCauley, Anne. "Talbot's Rouen Window: Romanticism, Naturphilosophie, and the Invention of Photography." *History of Photography* 26 (Summer 2002): 124–31.

McConkey, Kenneth. *George Clausen and the Picture of English Rural Life.* Edinburgh: Atelier Books, 2012.

——. *Impressionism in Britain and Ireland.* Dublin: Hugh Lane Municipal Gallery of Modern Art, 1995. Exhibition catalogue.

——. *Memory and Desire: Painting in Britain and Ireland at the Turn of the Twentieth Century.* Aldershot: Ashgate, 2002.

——. "Rustic Naturalism at the Grosvenor Gallery." In *The Grosvenor Gallery: A Palace of Art in Victorian England,* edited by Susan P. Casteras and Colleen Denney, 129–45. New Haven: Yale University Press, 1996.

——. *Sir George Clausen, R.A., 1852–1944.* Bradford, UK: Bradford Art Galleries and Museums and Tyne and Wear County Council Museums, 1980. Exhibition catalogue.

McSweeney, Kerry. "Swinburne's 'By the North Sea.'" *Yearbook of English Studies* 3 (1973): 222–31.

A Member. "Photography and Art, to the Editor of the Photographic Journal." *Journal of the Photographic Society of London* 5 (7 May 1859): 285.

Merrill, Linda. *A Pot of Paint: Aesthetics on Trial in "Whistler vs. Ruskin."* Washington, D.C.: Smithsonian Institution Press, 1992.

Meyers, Amy R. W., ed. *Knowing Nature: Art and Science in Philadelphia, 1740–1840.* New Haven: Yale University Press, 2011.

Mill, John Stuart. *Dissertations and Discussions, Political, Philosophical, and Historical.* 2 vols. London: John W. Parker and Son, 1859.

——. *A System of Logic, Ratiocinative and Inductive, Being a Connected View of the Principles of Evidence, and the Methods of Scientific Investigation.* 2 vols. London: J. W. Parker, 1843.

Millais, John Guille. *The Life and Letters of Sir John Everett Millais, President of the Royal Academy.* 2 vols. London: Methuen, 1899.

[Moore, George]. "A Book About Bastien-Lepage." *Speaker* 5 (20 February 1892): 227.

[——]. "Exteriority." *Speaker* 15 (22 June 1895): 684–86.

[——]. "The Royal Academy." *Fortnightly Review* 51 (1 June 1892): 828–39.

Morrell, Jack, and Arnold Thackray. *Gentlemen of Science: Early Years of the British Association for the Advancement of Science.* Oxford: Clarendon Press, 1981.

Morrell, Ottoline. Correspondence with Augustus John, 1913–1924. Ottoline Morrell Collection, RLIN record no. TX-RC98-A17, box 10, folders 7 and 8, box 11, folders 1 and 2. Harry Ransom Center, University of Texas at Austin, Austin, Texas.

Murchison, H. Ernest. "Memory and Imagination." *Photographic Quarterly* 2 (April 1891): 187–96.

Murry, John Middleton. "Art and Philosophy." *Rhythm* 1 (Summer 1911): 9–12.

———. "Bergson and Paris." *New Age,* 1 June 1911, 115–16.

Myers, Frederic W. H. "Introduction." In *Phantasms of the Living,* by Edmund Gurney, Frederic W. H. Myers, and Frank Podmore, 1:xxxv–lxxi. London: Trübner, 1886.

National Gallery, Millbank. *Loan Exhibition of the Works of the Late J. D. Innes.* London: National Gallery, Millbank, 1921. Exhibition catalogue.

"Natural Science at the Royal Academy." *Nature* 2 (30 June 1870): 157.

Nevinson, C. R. W. "Vital English Art (a Lecture Delivered by Mr. C. R. W. Nevinson at the Dore Galleries)." *New Age,* 18 June 1914, 160–62.

"A New Art Club." *Art Journal,* new ser., 48 (July 1886): 252.

New English Art Club. Press clipping albums about the New English Art Club, 1887–1914. 28 vols. TGA 7310, Tate Archive, Tate, London.

Newhall, Nancy Wynne. *P. H. Emerson: The Fight for Photography as a Fine Art.* New York: Aperture, 1978.

Newton, Sir William. "Upon Photography in an Artistic View, and Its Relations to the Arts." *Journal of the Photographic Society of London* 1 (3 March 1853): 6–7.

"Notes." *Colour* 15 (December 1921): vii–viii.

Olson, Roberta J. M., and Jay M. Pasachoff. *Fire in the Sky: Comets and Meteors, the Decisive Centuries, in British Art and Science.* Cambridge: Cambridge University Press, 1998.

P. "Royal Academy Exhibition." *Fraser's Magazine,* new ser., 8 (July 1873): 74–85.

Palermo, Charles. "The World in the Ground Glass: Transformations in P. H. Emerson's Photography." *Art Bulletin* 89 (March 2007): 130–47.

Parris, Leslie, ed. *Pre-Raphaelite Papers.* London: Tate Gallery, 1984.

———, ed. *The Pre-Raphaelites.* London: Tate Gallery, 1984. Reprinted, with corrections, London: Tate Gallery, 1994. Exhibition catalogue.

Parsey, Arthur. *The Science of Vision, or Natural Perspective Containing the True Language of the Eye. . . .* 2nd ed. London: Longman, 1840.

Payne, Christiana. *John Brett: Pre-Raphaelite Landscape Painter.* New Haven: Yale University Press, 2010.

Phillips, Claude. "The Summer Exhibitions at Home and Abroad: II—The Royal Academy, the Grosvenor, and the New Gallery." *Art Journal* 52 (June 1890): 161–74.

"Photography and the Elder Fine Arts." *Journal of the Photographic Society of London* 3 (21 April 1856): 29–32.

Pointon, Marcia. "Geology and Landscape Painting in Nineteenth-Century England." In *Earth Science and Discipline Boundaries,* edited by L. J. Jordanova and Roy Porter, 99–104. Chalfont St. Giles, Buckinghamshire: Printed for The British Society for the History of Science by Halfpenny Furze, 1979.

Porter, Roy. *The Making of Geology: Earth Science in Britain, 1660–1815.* New York: Cambridge University Press, 1977.

———. "Medical Science and Human Science in the Enlightenment." In *Inventing Human Science: Eighteenth-Century Domains,* edited by Christopher Fox, Roy Porter, and Robert Wokler, 53–87. Berkeley: University of California Press, 1995.

———. "The Terraqueous Globe." In *The Ferment of Knowledge: Studies in the Historiography of Eighteenth-Century Science,* edited by G. S. Rousseau and Roy Porter, 285–324. Cambridge: Cambridge University Press, 1980.

Porter, Theodore M. "The Fate of Scientific Naturalism: From Public Sphere to Professional Exclusivity." In *Victorian Scientific Naturalism: Community, Identity, Continuity,* edited by Gowan Dawson and Bernard Lightman, 265–87. Chicago: University of Chicago Press, 2014.

Potter, Matthew. "Cambridge University and the Germanist Bridge: The Aesthetics and Politics of Internationalism at the *Fin de Siècle.*" In *Internationalism and the Arts,* edited by Grace Brockington, 157–76. Oxford: Peter Lang, 2009.

———. *The Inspirational Genius of Germany: British Art and Germanism, 1850–1939.* Manchester: Manchester University Press, 2012.

Powell, Baden. *Essays on the Spirit of the Inductive Philosophy, the Unity of Worlds, and the Philosophy of Creation.* London: Longman, Brown, Green, and Longmans, 1855.

Prettejohn, Elizabeth. "Aesthetic Value and the Professionalization of Victorian Art Criticism." *Journal of Victorian Culture* 2 (Spring 1997): 71–94.

———. *The Art of the Pre-Raphaelites.* Princeton: Princeton University Press, 2000.

Quilter, Harry. *Opinions on Men, Women, and Things.* London: Swan Sonnenschein, 1909.

———. "The Tendencies of French Art." *Contemporary Review* 51 (June 1887): 863–77.

Quinn, John. John Quinn Papers. Call no. *ZL-355, MSS Col. 2513, New York Public Library, New York, N.Y.

Reynolds, Joshua. *Discourses on Art.* Edited by Robert Wark. New Haven: Published for the Paul Mellon Centre for Studies in British Art by Yale University Press, 1975.

Ribner, Jonathan P. "'Our English Coasts, 1852': William Holman Hunt and the Invasion Fear at Midcentury." *Art Journal* 55 (Summer 1996): 45–54.

Ridge, Jacqueline, Joyce H. Townsend, and Stephen Hackney, with a contribution on frames by John Anderson. "The Paintings." In Joyce H. Townsend, Jacqueline Ridge, and Stephen Hackney, *Pre-Raphaelite Painting Techniques, 1848–1856*, 77–189. London: Tate Publishing, 2004.

Ritter, Heinrich. *The History of Ancient Philosophy.* Translated by Alexander J. W. Morrison. Vol. 1. Oxford: D. A. Talboys, 1838.

Roberts, Helene. "Exhibition and Review: The Periodical Press and the Victorian Art Exhibition System." In *The Victorian Periodical Press: Samplings and Soundings,* edited by Joanne Shattock and Michael Wolff, 79–107. Leicester: Leicester University Press, 1982.

Robins, Anna Gruetzner. *A Fragile Modernism: Whistler and His Impressionist Followers.* New Haven: Yale University Press, 2007.

———. "The Greatest Artist the World Has Ever Known." In Anna Gruetzner Robins and Richard Thomson, *Degas, Sickert, and Toulouse-Lautrec: London and Paris, 1870–1910,* 52–93. London: Tate Publishing, 2005. Exhibition catalogue.

———. "Living the Simple Life: George Clausen at Childwick Green, St. Albans." In *The Geographies of Englishness: Landscape and the National Past, 1880–1940,* edited by David Peters Corbett, Ysanne Holt, and Fiona Russell, 1–63. New Haven: Published for the Paul Mellon Centre for Studies in British Art by Yale University Press, 2002.

———. "The London Impressionists at the Goupil Gallery." In *Impressionism in Britain,* edited by Kenneth McConkey, 87–96. London: Yale University Press in association with the Barbican Gallery, 1995. Exhibition catalogue.

———. *Modern Art in Britain, 1910–1914.* London: M. Holberton in association with the Barbican Art Gallery, 1997. Exhibition catalogue.

Robinson, Henry Peach. *Pictorial Effect in Photography, Being Hints on Composition and Chiaro-scuro for Photographers.* London: Piper & Carter, 1869.

———. "The Use of Nature and Idealism in Art." *Photographic Quarterly* 1 (January 1890): 105–17.

Rogers, Catherine. "W. H. F. Talbot, *The Pencil of Nature*, and the Camera." *History of Photography* 26 (Summer 2002): 141–50.

Roos, David A. "The 'Aims and Intentions' of Nature." In *Victorian Science and Victorian Values: Literary Perspectives,* edited by James G. Paradis and Thomas Postlewait, 159–80. New York: New York Academy of Sciences, 1981.

———. "Matthew Arnold and Thomas Henry Huxley: Two Speeches at the Royal Academy, 1881 and 1883." *Modern Philology* 74 (February 1977): 316–24.

Rosenfeld, Jason. "New Languages of Nature in Victorian England: The Pre-Raphaelite Landscape, Natural History, and Modern Architecture in the 1850s." PhD diss., New York University, 1999.

Rosenfeld, Jason, and Alison Smith, with contributions by Heather Birchall. *Millais.* London: Tate Publishing, 2007. Exhibition catalogue.

Rosenthal, Michael. *Constable: The Painter and His Landscape.* New Haven: Yale University Press, 1983.

Rossetti, William Michael. *Fine Art, Chiefly Contemporary, Notices Re-Printed, with Revisions.* London: Macmillan, 1867.

Rott, Herbert Wilhelm, ed. *George Stubbs, 1724–1806: Science into Art.* Munich: Prestel, 2012. Exhibition catalogue.

"The Royal Academy." *Art Journal* (June 1874): 161–66.

"The Royal Academy: The One Hundred and Fourth Exhibition." *Art Journal* 11 (1 June 1872): 149–156.

Ruskin, John. *The Library Edition of the Works of John Ruskin.* Edited by E. T. Cook and Alexander Wedderburn. 39 vols. London: George Allen, 1903–12.

———. "The Pre-Raffaellites." *Times* (London), 13 May 1851, 8.

Schaaf, Larry J., ed. *The Correspondence of William Henry Fox Talbot.* http://foxtalbot.dmu.ac.uk/.

———. *Out of the Shadows: Herschel, Talbot, and the Invention of Photography.* New Haven: Yale University Press, 1992.

Schaffer, Simon. "Natural Philosophy." In *The Ferment of Knowledge: Studies in the Historiography of Eighteenth-Century Science,* edited by G. S. Rousseau and Roy Porter, 55–91. Cambridge: Cambridge University Press, 1980.

Scott, William Bell. *Autobiographical Notes of the Life of William Bell Scott, and Notices of His Artistic and Poetic Circle of Friends, 1830 to 1882.* New York: Harper & Brothers, 1892.

Secord, Anne. "Pressed into Service: Specimens, Space, and Seeing in Botanical Practice." In *Geographies of Nineteenth-Century Science,* edited by David N. Livingstone and Charles W. J. Withers, 283–310. Chicago: University of Chicago Press, 2011.

———. "Talbot's First Lens: Botanical Vision as an Exact Science." In *William Henry Fox Talbot: Beyond Photography,* edited by Mirjam Brusius, Katrina Dean, and Chitra Ramalingam, 41–66. New Haven: Yale University Press, 2013.

Secord, James. "How Scientific Conversation Became Shop Talk." In *Science in the Marketplace: Nineteenth-Century Sites and Experiences,* edited by Aileen Fyfe and Bernard Lightman, 23–59. Chicago: University of Chicago Press, 2007.

Shapin, Steven, and Simon Schaffer. *Leviathan and the Air-Pump: Hobbes, Boyle, and the Experimental Life.* Princeton: Princeton University Press, 2011.

Shaw, Samuel. "The Carfax Gallery and the Camden Town Group." In *The Camden Town Group in Context,* edited by Helena Bonett, Ysanne Holt, and Jennifer Mundy. http://www.tate.org.uk/art/research-publications/camden-town-group/samuel-shaw-the-carfax-gallery-and-the-camden-town-group-r1104371.

Shiff, Richard. *Cézanne and the End of Impressionism: A Study of the Theory, Technique, and Critical Evaluation of Modern Art.* Chicago: University of Chicago Press, 1984.

Sickert, Walter. *London Impressionists.* London: Goupil Gallery, 1889. Exhibition catalogue.

———. "Modern Realism in Painting." In André Theuriet et al., *Jules Bastien-Lepage and His Art: A Memoir*, 133–43. London: T. Fisher Unwin, 1892.

———. "Mr. Ginner's Preface." *New Age*, 30 April 1914, 819–20.

———. "The Thickest Painters in London." *New Age*, 18 June 1914, 155.

———. *Walter Sickert: The Complete Writings on Art.* Edited by Anna Gruetzner Robins. Oxford: Oxford University Press, 2000.

Sloan, Kim. *Alexander and John Robert Cozens.* New Haven: Published in association with the Art Gallery of Ontario by Yale University Press, 1986.

Smith, Alison. Entry for *The Glacier of Rosenlaui*, entry for *The Brent at Hendon*, and entry for *Our English Coasts 1852 (Strayed Sheep).* In *Pre-Raphaelite Vision*, edited by Allen Staley and Christopher Newall, 152–153, 40, and 34–35, respectively. London: Tate Publishing, 2004.

Smith, Jonathan. *Fact and Feeling: Baconian Science and the Nineteenth-Century Literary Imagination.* Madison: University of Wisconsin Press, 1994.

Smith, Lindsay. *Victorian Photography, Painting, and Poetry: The Enigma of Visibility in Ruskin, Morris, and the Pre-Raphaelites.* Cambridge: Cambridge University Press, 1995.

Smith, Pamela H. *The Body of the Artisan: Art and Experience in the Scientific Revolution.* Chicago: University of Chicago Press, 2004.

Snow, C. P. *The Two Cultures: And a Second Look; An Expanded Version of "The Two Cultures and the Scientific Revolution."* 2nd ed. London: Cambridge University Press, 1964.

Snyder, Joel, and Neil Walsh Allen. "Photography, Vision, and Representation." *Critical Inquiry* 2 (Autumn 1975): 143–69.

Snyder, Laura J. *Reforming Philosophy: A Victorian Debate on Science and Society.* Chicago: University of Chicago Press, 2006.

"Society of Painters in Water Colours, Sixty-Sixth Exhibition." *Art Journal* 32 (1870): 173–75.

Solkin, David H., ed. *Art on the Line: The Royal Academy Exhibitions at Somerset House, 1780–1836.* New Haven: Yale University Press, 2001. Exhibition catalogue.

———. *Painting for Money: The Visual Arts and the Public Sphere in Eighteenth-Century England.* New Haven: Published for the Paul Mellon Centre for Studies in British Art by Yale University Press, 1993.

Sprat, Thomas. *The History of the Royal-Society of London, for the Improving of Natural Knowledge.* London: Printed by T. R. for J. Martyn at the Bell without Temple-bar, and J. Allestry at the Rose and Crown in Duck-Lane, printers to the Royal Society, 1667.

Staley, Allen. *The Pre-Raphaelite Landscape.* New ed. New Haven: Published by the Paul Mellon Centre for Studies in British Art by Yale University Pres, 2001.

Staley, Allen, and Christopher Newall, eds. *Pre-Raphaelite Vision.* London: Tate Publishing, 2004. Exhibition catalogue.

"The State of Art in England." *Blackwood's Magazine* 131 (May 1882): 609–22.

Steer, Philip Wilson. "Impressionism in Art." Philip Wilson Steer Papers, Reference code GB 1518 CI/PWS 8/1, Courtauld Institute of Art, London.

Stephens, Frederic George. *William Holman Hunt and His Works: A Memoir of the Artist's Life with Description of His Pictures.* London: James Nisbet, 1860.

[Stephens, Frederic George]. "Fine-Art Gossip." *The Athenaeum,* 8 March 1890, 313–14.

[———]. [John Seward, pseud.] "The Purpose and Tendency of Early Italian Art." *Germ, the Literary Magazine of the Pre-Raphaelites,* no. 2 (1850): 58–64.

[———]. "The New Gallery (First Notice)." *Athenaeum,* 3 May 1890, 575–77.

[———]. "The Society of Painters of Water Colours (First Notice)." *Athenaeum,* 30 April 1870, 585–86.

[———]. "The Two Pre-Raphaelitisms: Third Article; Concluded." *Crayon* 3 (December 1856): 353–56.

[———]. "The Two Pre-Raphaelitisms: Third Article; The Modern Pre-Raphaelites." *Crayon* 3 (November 1856): 321–24.

Stevenson, R. A. M. "George Clausen." *Art Journal* 52 (October 1890): 289–94.

Stone, J. D. "Peter Henry Emerson and the Pictorial Representation of Perception." *PhotoHistorian* (September 2000): 8–15.

Sully, James. *Sensation and Intuition: Studies in Psychology and Aesthetics.* London: H. S. King, 1874.

Swinburne, Algernon Charles. *Studies in Song.* London: Chatto & Windus, 1880.

[Symonds, John Addington]. "The Present Position of Landscape Painting in England." *Cornhill Magazine* 11 (March 1865): 281–92.

T., H. H. "John Brett." *Monthly Notices of the Royal Astronomical Society* 62 (4 February 1902): 238–41.

Talbot, William Henry Fox. "Fine Arts: Calotype (Photogenic) Drawing." *Literary Gazette; and Journal of the Belles' Lettres, Arts, Sciences, &c.*, no. 1256 (13 February 1841): 108.

———. "Fine Arts: Calotype (Photogenic) Drawing." *Literary Gazette; and Journal of the Belles' Lettres, Arts, Sciences, &c.*, no. 1258 (27 February 1841): 139–40.

———. *The Pencil of Nature.* London: Longman, Brown, Green, and Longmans, 1844–46.

———. "Photogenic Drawing: Some Account of the Art of Photogenic Drawings, or The Process by Which Natural Objects May Be Made to Delineate Themselves Without the Aid of the Artist's Pencil." *Athenaeum*, 9 February 1839, 114–17.

———. *Sun Pictures in Scotland.* London, 1845.

[Talbot, William Henry Fox]. "Scientific and Literary, Royal Society." *Athenaeum*, 2 February 1839, 97.

Taylor, H. Dennis. "A Phase of Naturalistic Focussing." *Photographic Quarterly* 1 (April 1890): 267–82.

Taylor, Jenny Bourne, and Sally Shuttleworth. "The Unconscious Mind and the Workings of Memory: Introduction." In *Embodied Selves: An Anthology of Psychological Texts, 1830–1890*, edited by Jenny Bourne Taylor and Sally Shuttleworth, 67–72. Oxford: Clarendon Press, 1998.

Taylor, John. *A Dream of England: Landscape, Photography, and the Tourist's Imagination.* Manchester: Manchester University Press, 1994.

———. *The Old Order and the New: P. H. Emerson and Photography, 1885–1895.* Munich: Prestel, 2006.

Tennyson, Lord Alfred. *In Memoriam.* London: Edward Moxon, 1850.

Tennyson, Hallam. *Alfred Lord Tennyson: A Memoir by His Son.* 2 vols. London: Macmillan, 1897.

Theuriet, André, et al. *Jules Bastien-Lepage and His Art: A Memoir.* London: T. Fisher Unwin, 1892.

Thornes, John E. *John Constable's Skies: A Fusion of Art and Science.* Birmingham: University of Birmingham Press, 1999.

Thornton, Alfred Henry Robinson. *Fifty Years of the New English Art Club, 1886–1935.* London: Printed for the Club by the Curwen Press, 1935.

Tickner, Lisa. "English Modernism in the Cultural Field." In *English Art, 1860–1914: Modern Artists and Identity*, edited by David Peters Corbett and Lara Perry, 13–30. New Brunswick: Rutgers University Press, 2001.

———. *Modern Life and Modern Subjects: British Art in the Early Twentieth Century.* New Haven: Yale University Press, 2000.

"Tis." "Augustus John: His Place in Art." *Colour* 6 (July 1917): 148–50.

Todhunter, I. *William Whewell, D. D., Master of Trinity College, Cambridge, an Account of His Writings with Selections from His Literary and Scientific Correspondence.* 2 vols. London: Macmillan, 1876.

Townsend, Joyce H., Jacqueline Ridge, and Stephen Hackney. *Pre-Raphaelite Painting Techniques, 1848–1856.* London: Tate Publishing, 2004.

Trowbridge, Caroline. "'Speakers Concerning the Earth': Ruskin's Geology After 1860." In *Repositioning Victorian Science: Shifting Centres in Nineteenth-Century Scientific Thinking*, edited by James Clifford, Elisabeth Wadge, Alex Warwick, and Martin Willis, 17–30. London: Anthem Press, 2006.

Tucker, Jennifer. *Nature Exposed: Photography as Eyewitness in Victorian Science.* Baltimore: Johns Hopkins University Press, 2005.

Turner, Frank M. *Between Science and Religion: The Reaction to Scientific Naturalism in Late Victorian England.* New Haven: Yale University Press, 1974.

Tyndall, John. *Fragments of Science for Unscientific People: A Series of Detached Essays, Lectures, and Reviews.* New York: D. Appleton, 1871.

———. *Notes of a Course of Nine Lectures on Light Delivered at the Royal Institution of Great Britain, April 8–June 3, 1869.* London: Longmans, Green, 1870.

Upstone, Robert, ed. *Modern Painters: The Camden Town Group.* London: Tate Publishing, 2008. Exhibition catalogue.

Waggoner, Diane. *The Pre-Raphaelite Lens: British Photography and Painting, 1848–1875.* Washington, D.C.: National Gallery of Art, in association with Lund Humphries, Farnham, Surrey, 2009. Exhibition catalogue.

Wakeman, Geoffrey. "Henry Bradbury's Nature Printed Books." *Library* 21, no. 1 (1966): 63–67.

Warner, Malcolm. "John Everett Millais's 'Autumn Leaves,' 'A Picture Full of Beauty and Without Subject.'" In *Pre-Raphaelite Papers*, edited by Leslie Parris, 126–42. London: Tate Gallery, 1984.

———. *The Victorians: British Painting, 1837–1901.* Washington, D.C.: National Gallery of Art, 1997. Exhibition catalogue.

Weaver, Mike, ed. *British Photography in the Nineteenth Century: The Fine Art Tradition.* Cambridge: Cambridge University Press, 1989.

Whewell, William. *The Bridgewater Treatises on the Power, Wisdom, and Goodness of God, as Manifested in the Creation; Treatise III, On Astronomy and General Physics Considered with Reference to Natural Theology.* 7th ed. London: W. Pickering, 1839.

———. *The Philosophy of the Inductive Sciences.* 2 vols. London: John Parker, 1840.

———. *The Philosophy of the Inductive Sciences Founded upon Their History.* New ed. 2 vols. London: J. W. Parker, 1847.

[Whewell, William]. "On *The Connexion of the Physical Sciences*, by Mrs. Somerville." *Quarterly Review* 51, no. 101 (1834): 54–68.

Whistler, James McNeill. "Ten O'Clock." In *Whistler on Art: Selected Letters and Writings of James McNeill Whistler*, edited by Nigel Thorp, 79–95. Washington, D.C.: Smithsonian Press, 1994.

White, John Forbes. "The Royal Academy Exhibition of 1873." *Contemporary Review* 22 (July 1873): 266–96.

White, Paul. *Thomas Huxley: Making the "Man of Science."* Cambridge: Cambridge University Press, 2003.

Willis, Martin. *Vision, Science, and Literature, 1870–1920: Ocular Horizons.* London: Pickering & Chatto, 2011.

Willsdon, Clare. "Dyce 'in Camera': New Evidence of His Working Methods." *Burlington Magazine* 132 (November 1990): 760–65.

Wiseman, Nicholas. *Points of Contact Between Science and Art.* London: Hurst and Blackett, 1863.

Wolf, Herta. "Nature as Drawing Mistress." In *William Henry Fox Talbot: Beyond Photography,* edited by Mirjam Brusius, Katrina Dean, and Chitra Ramalingam, 119–42. New Haven: Yale University Press, 2013.

Wood, Gillen D'Arcy. *The Shock of the Real: Romanticism and Visual Culture, 1760–1860.* New York: Palgrave, 2001.

Wood, J. G. *The Common Objects of the Sea-Shore, Including Hints for an Aquarium.* London: George Routledge, 1857.

——. *The Natural History of Man.* 2 vols. London: G. Routledge and Sons, 1870.

Worringer, Wilhelm. "Selection from *Abstraction and Empathy*." In *Art in Theory, 1900–2000: An Anthology of Changing Ideas,* rev. ed., edited by Charles Harrison and Paul Wood, 66–69. Malden, Mass.: Blackwell Publishing, 2003.

Wright, Alastair Ian. "Bastien-Lepage and English Critical Taste." *Gazette des Beaux-Arts* 116 (September 1990): 94–104.

Yanni, Carla. *Nature's Museums: Victorian Science and the Architecture of Display.* Baltimore: Johns Hopkins University Press, 1999.

Yeo, Richard R. *Defining Science: William Whewell, Natural Knowledge, and Public Debate in Early Victorian Britain.* Cambridge: Cambridge University Press, 1993.

——. *Science in the Public Sphere: Natural Knowledge in British Culture, 1800–1860.* Aldershot: Ashgate, 2001.

# { index }

Page numbers in *italics* refer to figures.

*Typeset by*

REGINA STARACE

*Printed and bound by*

OCEANIC GRAPHIC INTERNATIONAL

*Composed in*

BASKERVILLE 10 PRO

*Printed on*

KINMARI FSC MATT

*Bound in*

JHT